PERSPECTIVES IN CONDENSED MATTER PHYSICS

A Critical Reprint Series

Condensed Matter Physics is certainly one of the scientific disciplines presently characterized by a high rate of growth, both qualitatively and quantitatively. As a matter of fact, being updated on several topics is getting harder and harder, especially for junior scientists. Thus, the requirement of providing the readers with a reliable guide into the forest of printed matter, while recovering in the original form some fundamental papers suggested us to edit critical selections on appealing subjects.

In particular, the present Series is conceived to fill a cultural and professional gap between University graduate studies and current research frontiers. To this end each volume provides the reader with a critical selection of reprinted papers on a specific topic, preceded by an introduction setting the historical view and the state of art. The choice of reprints and the perspective given in the introduction is left to the expert who edits the volume, under the full responsibility of the Editorial Board of the Series. Thus, even though an organic approach to each subject is pursued, some important papers may be omitted just because they lie outside the editor's goal.

The Editorial Board

OPTICALS EFFECTS IN LIQUID CRYSTALS

edited by

István Jánossy

KLUWER ACADEMIC PUBLISHERS

Jaca Book

prima edizione
maggio 1991

copertina e grafica
Ufficio grafico Jaca Book

ISBN 0-7923-1277-5 (Kluwer)
ISBN 88-16-96006-X (Jaca Book)

per informazioni sulle opere pubblicate e in programma
ci si può rivolgere a Editoriale Jaca Book spa - Servizio Lettori
via Gioberti, 7, 20123 Milano, telefono 4988927

Preface

In 1988 physicists and chemists commemorated the centenary of the discovery of the first liquid crystals. For a long period after this discovery, although many significant results were found, liquid crystal research remained a marginal topic of condensed matter physics. The situation changed in the sixties. At that time the remarkable electro-optical properties of liquid crystals were recognized and found soon widespread application in numeric displays. From a more fundamental point of view, the interest in disordered systems increased in general at the same time. Liquid crystals represented an important class of such systems. Among others, phase transitions, hydrodynamics and topological defects occurring in them attracted considerable attention. The connection between the liquid-crystalline state and the structure of biological membranes stimulated a lot of works also.

In the present volume we discuss a relatively new and rapidly developing branch of the field, namely *nonlinear optical effects* in liquid crystals. Optical studies have always played a significant role in liquid crystal science. Research of optical nonlinearities in liquid crystals began at the end of the sixties. Since then it became a powerful tool in the investigation of symmetry properties, interfacial phenomena or dynamic behaviour. Furthermore, several new aspects of nonlinear processes were demonstrated and studied extensively in liquid crystals. The subject covered in this book is therefore of importance both for liquid crystal research and for nonlinear optics itself.

The term "nonlinear optics" is used here in a broad sense. In addition to such a classic nonlinear process as harmonic generation we review optical reorientation, thermal effects and photoinduced changes in liquid crystals too. We devote also a part of the volume to linear optics, namely to the problem of light propagation in mesophases. The motivation to include this part was twofold. Firstly, the topic is important enough in itself from both theoretical and practical points of view. Secondly, the understanding of the basic laws of light propagation is necessary in order to interpret the nonlinear effects properly.

The papers reprinted in this volume represent of course only a small fraction of the literature on the subject. The selection of the articles is necessarily somewhat arbitrary and many other choices would have been possible. We think, however, that the selected papers cover the main results in the field regarding theory, experimental works and potential applications. Further papers are cited in this introductory review and many other references can be found in the reprinted articles from which the reader may learn more about the details.

Budapest, May 1990

István Jánossy

Table of Contents

I. Jánossy, Optical Effects in Liquid Crystals

Introduction

Liquid Crystals: Phases and Basic Properties

In certain organic substances, composed of anisotropic molecules, the transition from the crystalline to the liquid state takes place in two or more distinct steps. In these materials between the solid and liquid states additional phases are formed which exhibit both liquid-like behaviour (fluidity) and crystalline-like features (macroscopic anisotropy). The substances showing this phenomenon are called *liquid crystals*, the intermediate phases are termed *liquid-crystalline phases* or *mesophases*. By now thousands of liquid crystals are known and at least ten thermodynamically different mesophases are recognized.

In all mesophases there is a long-range orientational order of the molecules. The preferred direction of the molecular alignment can be described with the help of a unit vector, the *director*, which, roughly speaking, is parallel to the "long axis" of the elongated molecules. In reality the orientation of the individual molecules fluctuates significantly in space and time, therefore it is more correct to define the director as a symmetry axis of the orientational distribution of the molecules. In certain mesophases this distribution has a rotational symmetry around the director (uniaxial phases), in others the distribution function depends on the azimuthal angle too (biaxial phases).

The fluctuations around the director can be quantitatively described by an *orientational order parameter* which is 1 for a perfectly aligned system and 0 in the case of spherical symmetry, i.e. in the isotropic phase.

The mesophases differ from each other regarding the positional order of the molecules (Fig. 1). In the *nematic* phase there is no long range positional order at all just as in isotropic liquids. Nematics are normally uniaxial, however biaxial nematics were discovered very recently. In the *smectic* phases the centre of masses of the molecules are concentrated in layers forming a one-dimensional density wave. In the smectic A and C phases there is no long-range positional order within the layers. The smectic A phase is uniaxial, the director (**n**) is parallel with the layer normal, **l**. In the C phase the director is tilted with respect to the layer normal. This phase is biaxial although the deviation from uniaxiality is usually small. There are further smectic phases in which the molecules form two-dimensional lattices within the layers *(ordered smectic phases)*. The difference between ordered smectic phases and crystals lies in the fact that in the former systems the lattices are uncorrelated and can freely slide on each other.

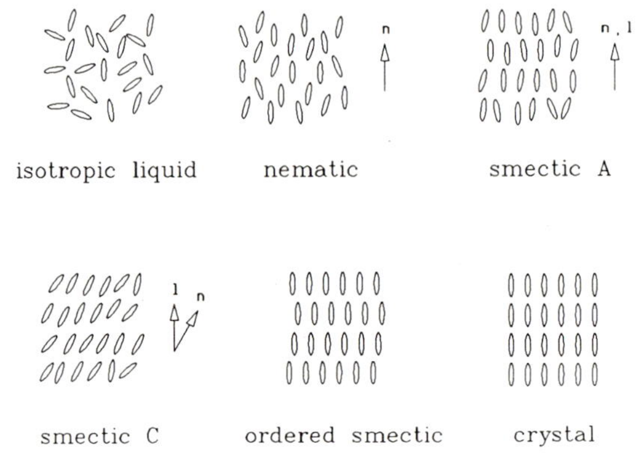

Fig.1. Schematic representation of mesophases.

Liquid crystalline phases formed by chiral molecules (i.e. molecules differing from their mirror image) show unique macroscopic properties. The best-known example is the *cholesteric* phase which is termodynamically equivalent to the nematic phase. In the later phase the free-energy of the system corresponds to a uniform director distribution in the whole sample. On the other hand in cholesterics the molecules tend to form a helical structure the helical axis being perpendicular to the director. A similar helical structure develops in the smectic C phase when the molecules are chiral. In this case the helical axis is parallel to the layer normal; the tilt angle is constant while the azimuthal angle is rotating in space. The pitch of the helix in these systems is typically in the order of a micron.

Liquid crystals can be composed both of polar and apolar molecules. An important fact in connection with polar substances is that in uniaxial phases there is no polar ordering of the molecules. In average the dipole moments aligned in a given direction are compensated by those aligned in the opposite direction. As a consequence no spontaneous macroscopic polarization develops. More generally one can state that rotation of the director by π does not affect the physical state of the liquid crystal. In biaxial phases built of chiral molecules, such as the chiral smectic C phase, the situation is different. In these systems the compensation of the dipole moments is not perfect, a macroscopic polarization appears in the direction perpendicular both to the layer normal and the director. These phases are therefore *ferroelectric*. Ferroelectric liquid crystals are currently perhaps the most intensively studied mesophases considering both fundamental research and potential applications.

The list of mesophases given in this section is far from being complete and we emphasized only a few important facts which are important in connection with the topic of the present book. For a detailed description of the physical and physicochemical properties of liquid crystals we refer to the monographies by de Gennes, Chandrashekar, Blinov and de Jeu.[1]

Propagation of Light in Liquid Crystals

Liquid crystals are anisotropic materials, hence their linear optical properties are determined by a symmetrical dielectric tensor, ε, rather than a scalar refractive index. In certain cases it is possible to prepare uniformly oriented liquid crystal films. In these films ε is a constant and light propagation can be described by the well-known laws of crystal optics.[2]

As an example let us consider nematics. We mentioned that in nematics the internal free-energy minimum corresponds to a constant director field within the entire sample. The actual director configuration in the layer is, however, influenced by the boundary conditions and, if present, by external fields. By now there are well-developed techniques to align nematics parallel or perpendicularly (or in any direction) relative to glass substrates. In this way it is possible to prepare "single-crystals" such as those shown in Fig. 2a and 2b, or films deformed in a controlled manner (Fig. 2c and 2d).

In nematics the dielectric tensor can be written in the form

$$\varepsilon_{ij} = \varepsilon_\perp \delta_{ij} + (\varepsilon_\parallel - \varepsilon_\perp) n_i n_j \tag{1}$$

where $\mathbf{n}$ is the director. In analogy to uniaxial solid crystals one can write

$$\varepsilon_\perp = n_o^2, \qquad \varepsilon_\parallel = n_e^2 \tag{2}$$

where n_o and n_e are the ordinary and extraordinary refractive indices respectively. In a nematic "single crystal" there is an ordinary wave propagating with a phase velocity c/n_o and an extraordinary one with a phase velocity c/n_{eff}. n_{eff} is given by the well-known relation

$$1/n_{eff}^2 = cos^2\theta/n_o^2 + sin^2\theta/n_e^2 \tag{3}$$

where θ is the angle between the director and the wave vector.

The problem of light propagation becomes much more complicated in spatially inhomogeneous liquid crystal layers. There is no general method to solve the Maxwell equations for an arbitrary director distribution or, more generally speaking, for an arbitrary spatial dependence of the dielectric tensor. On the other hand in some important special cases exact solutions were found and useful approximations were worked out for other conditions. In this section we survey these results.

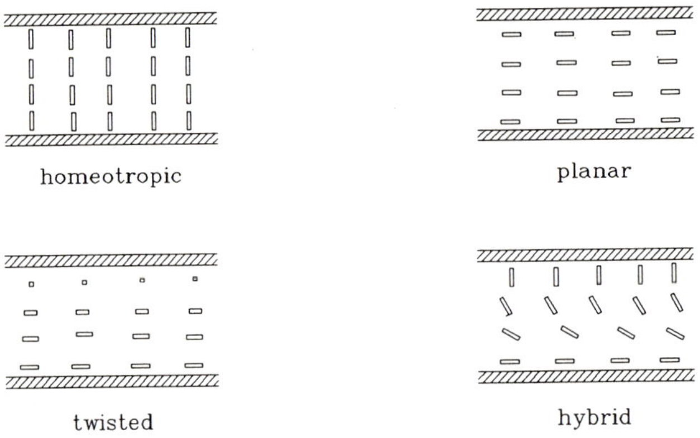

Fig. 2. Typical oriented nematic layers.

A classic problem in the optics of liquid crystals is the interpretation of the striking optical phenomena observed in cholesteric liquid crystals. As Sir William Bragg wrote in 1934 in connection with a cholesteric:[3]

> ...it reflects brilliant colours like those of a peacock's feather. ...But the most remarkable effect is that the reflected or more correctly speaking, scattered light, is circularly polarized. If the incident light is circularly polarized, it is reflected if the circulation is represented by a right-handed screw, and transmitted if the screw is left-handed. More remarkable still is the fact that the scattered light is right-handed, like the absorbed light to which it is due. In all other known cases of the reflection of circularly polarised light the sense of rotation is reversed.
>
> These substances when in their characteristic state, are optically active to an extraordinary degree, represented sometimes by as much as a whole turn in the hundredth of a millimetre.

He noted also that

> ...the causes and laws [of these effects] have never been fully explained.

The theory of wave propagation in cholesterics was first developed by Oseen[4] and later by de Vries[R1]. Their theories were based on the assumption that the cholesteric structure is simply a twisted version of the nematic. The director rotates in space around an axis forming thereby a helical structure with a pitch p.

For the case of a light beam propagating along the helical axis a simple analytical solution of the Maxwell equation was found. The important point in the derivation of the results was the transformation of the field equations into a frame following the rotation of the director. Relative to *this* frame the solutions are of the simple form

$$\mathbf{E} = \mathbf{E}_0 e^{i(kz-\omega t)}$$

For a given frequency there are four independent modes (normal modes) with different k values. Two of these normal modes describe forward propagating waves; the other two describe the corresponding backward propagating waves. The "wave vector", k, depends on the frequency and the pitch.

The analysis of the dispersion relation $k = k(\omega, p)$ and the polarization properties of the normal modes yielded a straightforward explanation of the optical phenomena described above. De Vries showed that as long as the ratio of the wavelength to the pitch is larger than the birefringence, $\lambda/p > n_e - n_o$, pure normal modes are excited by very nearly circularly polarized incident waves. *Outside* the region defined by the inequalities

$$n_o p < \lambda < n_e p$$

the normal waves are nearly circularly polarized within the cholesteric also, but the right and left circularly polarized waves propagate with different phase velocities. As well known in crystal optics, this fact leads to optical rotation. *Inside* this region one of the normal modes (for a right-handed helix the right circularly polarized one) is reflected at the surface of the film giving rise to the brilliant colours mentioned by Bragg. Of course this coloration occurs only in substances where the reflection band, $n_o p < \lambda < n_e p$, falls into the visible part of the spectrum.

The mathematical description of light propagation in an arbitrary direction relative to the helix is more laborious. Several authors dealt with this problem.[R2,5,6] It was found that the central wavelength of the reflection band can be given, in analogy to Bragg's law for X-ray diffraction, by the relation

$$\lambda = \bar{n} p \cos\theta$$

Here θ denotes the angle between the wave vector and the helical axis; $\bar{n} = (n_o + n_e)/2$. (Note that the reflection wavelength corresponds to a periodicity of $p/2$ rather than p; this is because a rotation of the director by π does not affect the optical behaviour.) It was shown furthermore that in the case of oblique incidence higher-order Bragg reflections should occur as well.

The predicted angular dependence of the reflection band was easily verified experimentally. On the other hand the observation of higher-order bands required careful experimental work. Such a study was carried out by Berreman and Sheffer[R2]. In contrast to previous experimental works which were carried out on films with a certain angular distribution of the helical axis, they prepared a single-domain cholesteric film. The use of a single domain made possible a detailed quantitative comparison of the measured reflection and transmission spectra with the theoretical calculations. The second-order reflection band was observed in agreement with the theory. From the data they could prove also that cholesterics can be regarded locally uniaxial.

Berreman and Scheffer applied a numerical method to solve the Maxwell equations for light propagation in cholesterics. The Maxwell equations were rewritten in the form

$$\partial\psi/\partial z = \omega/c\mathcal{D}\psi \tag{4}$$

where ψ is a four-component vector and $\mathcal{D}$ is a 4x4 matrix. The elements of ψ are obtained from the field vectors, those of $\mathcal{D}$ are derived from the components of the dielectric tensor. Eq. 4 can be solved by standard numerical integration.

Other authors used different mathematical procedures which provided practically analytical solutions in the specific case of light propagation in cholesterics. The significance of the Berreman-Scheffer method lies in the fact that it can be applied to any system in which the director changes only along one direction (stratified medium). This occurs for instance in liquid-crystalline electro-optic cells, hence the method is of importance in device applications also. Furthermore its extension to biaxial or absorbing media is straightforward thus it can be employed for smectic C or dyed films too.

Numerical methods on the other hand give little insight into the general features of the problem. Therefore efforts have been made to find approximate, simple solutions of the Maxwell equations in particular conditions. An important case occurs when the director changes only slightly over distances comparable to the wavelength. In the case of an isotropic medium with a slowly varying refractive index the geometrical-optics approximation can be applied. The question arises whether there is a similar approximation for liquid crystals?

In order to find such an approximation we return for a moment to the problem of light propagation in a helical structure. In the foregoing discussion the limit of long pitches, $\lambda/p \ll n_e - n_o$, was excluded. As a matter of fact this limit was considered as early as 1911 by Maugin[7] and was treated by de Vries also.[R1] They showed that in this regime the normal modes become linearly polarized; one is polarized along the director, the other perpendicularly to it. The corresponding phase velocities are c/n_e and c/n_o respectively. This sort of light propagation is often termed "adiabatic" referring to the fact that the polarization of the normal modes follow adiabatically the rotation of the director.

The concept of adiabatic propagation can be generalized for an arbitrary stratified film in the following way. Consider a medium in which the dielectric tensor varies along the z direction. At every z position there are four local normal modes; for a uniaxial system these are a forward propagating extraordinary and ordinary wave (e-mode and o-mode) plus the corresponding backward propagating modes. Let us consider an incident wave that excites only one of these modes on entering the liquid crystal. In the adiabatic approximation this wave is represented by the same single normal mode across the whole film. Quantitatively one can write for the field amplitude in the p-th mode

$$\mathbf{E}_p(z) = A_p \mathbf{e}_p e^{i(\phi(z)+\mathbf{k}_{\|}\mathbf{r}_{\|})} \tag{5}$$

with

$$\phi(z) = \int_0^z q_p(z')dz'.$$

Here $\mathbf{k}_{\|}$ and $\mathbf{r}_{\|}$ are the tangential components of the wave vector and the position vector respectively; q_p is the z component of the wave vector, $\mathbf{e}_p$ is a unit vector in the direction of the polarization. A_p is the "amplitude" of the p-th mode. We note that in the o-modes A_p is constant and $\mathbf{e}_p$ is perpendicular both to the director and the wave vector. In the e-modes the angle between $\mathbf{e}_p$ and $\mathbf{n}$ can be anything from 0 to $\pi/2$ and A_p varies in space. Apart from an arbitrary constant, the magnitude of A_p can be determined from the condition of constant energy-flow in the given mode.

The adiabatic approximation was used in a number of cases intuitively without an analysis of the criterion of its validity. A systematic investigation of this limit was presented, among others, by Allia *et al.*[R3] Below we try to emphasize the physics behind their procedure without going into mathematical details.

As a zeroth-order approximation light propagation is considered adiabatic in the sense described before. In a "single crystal" this is an exact solution of the problem. Spatial variations of the dielectric tensor introduce *couplings* between the normal waves. As a consequence of the coupling, energy is exchanged between different modes. The amount of energy transferred from a particular mode to another one is determined by two factors. The first one is the strength of the coupling which is of the order of $\partial\varepsilon/\partial z$. The second factor is a sort of "coherence length" i.e. the distance on which the phase difference between the two waves shifts by π. The energy exchange between the two modes can be neglected if the product of these two factors is much smaller than unity.

The coherence length and the coupling coefficient are different for each pair of normal modes. The coherence length for two counterpropagating modes is very short, it is of the order of $\lambda/4\bar{n}$. In nematics, where significant spatial variations occur typically over distances of several μm-s, the coupling between these beams can be almost always neglected. (An exception is the case when total internal reflection takes place inside the layer.) For two forward propagating modes (an e-mode and an o-mode) the coherence length is $\lambda/2(n_{eff} - n_o)$. When the director is perpendicular to the wave vector, $n_{eff} = n_e$ (see Eq. 3). The corresponding coherence length is typically 1.5μm, which is still rather small. However, if the director and the wave vector are parallel we have $n_{eff} = n_o$, hence the coherence length becomes infinite. In such situations the adiabatic approximation breaks down. A well-known example occurs in the twisted nematic electro-optic cells. Nonetheless, as pointed out by Allia et al., the coupling between forward and backward propagating beams can be neglected even in these cases. In this way the Berreman matrix (Eq. 4) can be reduced to a 2x2 matrix, which makes numerical calculations much simpler.

The approximation in which the couplings between counterpropagating beams are neglected can be regarded as an extension of the geometrical-optics approximation to liquid crystals. It is different from the adiabatic approximation in which no coupling is considered at all.

An elegant formulation of the geometrical-optics approximation for normal incidence and small birefringence was provided by Santamato and Shen.[R4] Following a general method developed by Kubo and Nayata[8] they transformed the Maxwell equations into a simple differential equation for the Stokes parameters. This equation is particularly useful for describing the orienting action of an optical field in a nematic; see later.

Zheleznyakov *et al.*[9] considered the problem of light propagation in an inhomogeneous cholesteric. They started from the normal modes corresponding to the local pitch of the helix. Spatial variations of the pitch were taken into account as couplings between the normal waves in a similar way as described above.

Ong[10] presented a rigorous treatment of the geometrical-optics approximation for a rather special case. He considered a stratified layer in which the director is oriented everywhere in the plane of incidence of the light beam. In this situation the e-ray and o-ray are fully decoupled and already the zeroth-order approximation gives an excellent agreement with the exact solution.

To our knowledge no generalization of the adiabatic or geometrical-optics approximation has been presented for the case of two or three-dimensional spatial variations of the director. This problem arises for instance in the interpretation of diffraction from liquid crystal gratings[11] or self-focusing. We believe that an extension of Ong's method to these more complex situations would be very useful.

Optical Harmonic Generation in Liquid Crystals

Optical harmonic generation is one of the earliest and most developed area in the field of nonlinear optics. Not surprisingly, investigations of nonlinear optical effects in liquid crystals started also with the study of harmonic generation. The aim of the first studies was to observe second harmonic generation (SHG) in cholesteric liquid crystals.

SHG is governed by the second-order susceptibility tensor connecting the nonlinear polarization with the field strengths:

$$P_i^{NL} = \sum_{j,k} \chi_{ijk}^{(2)} E_j E_k. \tag{6}$$

A number of the tensor elements $\chi_{ijk}^{(2)}$ vanish because of the symmetry properties of the nonlinear medium. For example in centrosymmetric substances SHG is forbidden -at least in the electric-dipole approximation on which Eq. 6 is based. Cholesterics differ from their mirror image, hence considering only this restriction SHG would appear possible. A closer look at the symmetry properties however leads to the opposite conclusion.

Cholesterics belong to the D_∞ symmetry group which allows for one independent non-vanishing element of $\chi^{(2)}$, namely $\chi_{zxy}^{(2)} = -\chi_{zyx}^{(2)}$. (The director is along the x axis). For non-absorbing materials Kleinman's rule holds also, according to which the elements of a nonlinear susceptibility tensor are invariant under any permutation of their indices.[12] Taking into account this latter restriction one finds that $\chi_{zxy}^{(2)}$ vanishes as well, so finally no SHG is allowed.

In contrast to the above expectation, in 1967 Freund and Rentzepis reported SHG in cholesteryl carbonate.[13] This observation, if correct, would question the accepted symmetry properties of cholesterics. However, subsequent investigations by Durand and Lee[14] and Goldberg and Schnur[15] did not confirm the previous observation. Durand and Lee demonstrated that SHG in cholesteryl carbonate can originate from small crystalline particles present in the liquid crystal. These crystals can be melted by heat treatment leading to a gradual disappearance of the SHG signal.

In the experiments mentioned above unoriented samples were used. In such samples the locally generated harmonic signals - if any - are incoherently superposed resulting in a weak total signal. Reliable studies require oriented layers in which spatially coherent generation - phase matching - of the harmonic signal can be achieved. The first study of harmonic generation under phase-matching conditions in cholesterics was carried out by Shelton and Shen.[R5]

Let us review briefly the condition of phase matching in homogeneous media. The generation of the m-th harmonic can be regarded as a result of the interaction of m waves, all vibrating at the fundamental frequency, ω , but not necessarily belonging to the same normal mode. For collinear waves phase matching occurs if

$$\sum_{l=1}^{m} k_l(\omega) = k_h(m\omega) \tag{7}$$

where k_l denotes the wave vector of the l-th fundamental component and k_h that of the harmonic wave. This relation can be interpreted as the conservation law for the linear momentum of the electromagnetic field.

In isotropic media Eq. 7 is satisfied only if the fundamental and harmonic waves propagate with the same phase velocity; this is normally not the case because of dispersion. Phase matching can be realized, however, in birefringent materials provided the fundamental and harmonic waves belong to different polarization modes. At a properly chosen angle of incidence or temperature the phase velocity difference due to dispersion is compensated by the birefringence. As liquid crystals have large inherent birefringence phase matching is obviously possible in these materials.

In cholesterics the situation is more complex. As shown in R5 the condition of phase matching is given again by Eq. 7 but k_l now denotes a wave vector relative to the rotating frame. As discussed in the previous section, this wave vector depends strongly on the pitch. Shelton and Shen realized that this circumstance can be used to achieve phase matching through an adjustment of the pitch. They pointed out furthermore that in cholesterics the momentum of the electromagnetic field is not necessarily conserved during harmonic generation; exchange of momentum between the field and the periodic structure is possible. The term "coherent optical umklapp process" was suggested for these cases in analogy to the electron-electron interaction in solid crystals. As a striking example of such a process let us mention that in cholesterics forward propagating fundamental waves can generate a backward propagating harmonic wave!

Shelton and Shen carried out careful experiments on harmonic generation in mixtures of cholesterol derivatives. From the linear optical data of the material it was possible to predict the pitch values at which the different phase matching conditions were satisfied. The pitch was adjusted by varying the composition and the temperature of the substance. No second harmonic signal was detected even in circumstances were phase matching was expected to occur. On the other hand third-harmonic generation - which is allowed for any material - *was* observed and phase matching took place close to the theoretically predicted pitch values. Both "normal" and "umklapp" processes were detected.

SHG can be induced in any medium by applying an external d.c. electric field. A number of such investigations were accomplished in various liquid crystalline phases. Saha[16] demonstrated field-induced SHG in cholesterics using the Shelton-Shen method to achieve phase matching. Barnik *et al.*[R6] detected SHG in oriented layers of nematic and smectic A liquid crystals in the presence of a d.c. field; phase matching was obtained by varying the angle of incidence of the pump beam.

A controversy arose in the case of the liquid crystal 4-methoxy-benzylidene-4-butylaniline (MBBA). As first reported by Arakelyan *et al.*[17] SHG takes place in oriented nematic MBBA layers even in the absence of an external field. From this fact they concluded that MBBA is composed of noncentrosymmetric "blocks".

Barnik *et al.*[R6] observed the same effect but deduced a much smaller nonlinear susceptibility from the measurements than Arakelyan *et al.* They provided a different interpretation as well: in their opinion SHG was due to electric- quadrupole effects. (Quadrupole effects originate from the coupling between the polarization **P** and the spatial derivatives of the electric field.) As second-order quadrupole effects are not forbidden by the existence of an inversion centre the interpretation of Barnik *et al.* is compatible with the generally accepted symmetry properties of nematics. A detailed theoretical analysis of the problem was given by Ou-Yang *et al.*[18] Their results indicate also that the quadrupole mechanism is responsible for SHG in MBBA rather than the existence of noncentrosymmetric blocks.

A further interesting system regarding SHG is the chiral smectic C phase.[R7,19] Symmetry in this mesophase (biaxiality and no centre of inversion) allows for SHG even in the electric-dipole approximation. Taking into account all constrains (including Kleinman's relations) the second-order susceptibility tensor has four independent non-vanishing elements.[19] Chiral smectic C liquid crystals are arranged in a similar helical structure as cholesterics, therefore phase matching could be achieved by utilizing the method of Shelton and Shen. To our knowledge no experiment of this type has been carried out up to now. Barnik *et al.*[R7] applied an electric field to unwind the helix and obtained phase matching in the unwound state by temperature tuning. Under such circumstances, however, the signal contains a field-induced as well as an inherent component. To separate the two contributions the authors investigated the time dependence of the intensity of the harmonic signal after the field was switched off. They found an initial step-like decrease which was attributed to the disappearance of the field-induced component. The residual signal disappeared in several seconds which corresponded to the formation of a helical structure from the unwound configuration. This latter part was therefore identified as the inherent contribution to SHG. Clearly, further experiments are desirable to get more information on this effect.

Finally we mention a work by Guyot-Sionnest *et al.*[20] in which SHG from liquid crystal - glass and liquid crystal - air interfaces were studied. This work demonstrated the value of such studies in the very important field of interfacial properties of liquid crystals.

Optical Reorientation in Nematics

A well-known nonlinear process taking place in the liquid state of anisotropic molecules is the optical-field induced birefringence (optical Kerr effect[21]). This nonlinearity results from the reorientation of the molecules in the electric field of a light beam. In the isotropic phase the optical field perturbs the orientational distribution of the molecules. In the perturbed state more molecules are aligned parallel to the electric field than perpendicularly to it and as a consequence the medium becomes birefringent. On the other hand in liquid crystals the orientational distribution of the molecules is inherently anisotropic. The optical field, just as a d.c. electric or magnetic field, induces a collective rotation of the molecules. This process can be described as a reorientation of the director.

The effect of correlated molecular motions can be observed even in the isotropic phase of nematic liquid crystals near the phase transition temperature. This phenomenon was first studied by Wong and Shen[22] and Prost and Lalanne.[23] They found both a pretransitional increase of the Kerr coefficient and a critical slowing-down in the relaxation process.

Systematic research on optical reorientation in the nematic phase itself started around 1980 in at least four groups simultaneously (Zolotko *et al.* in Moscow and Budapest; Zel'dovich *et al.* in Moscow and Yerevan; Durbin *et al.* in Berkeley; Khoo *et al.* in Pennsylvania). Since then it became one of the most intensively studied nonlinear optical effects in liquid crystals.

Before going into details of optical reorientation we survey the effects caused by static electric fields in nematic layers. (More precisely we consider "quasi-static" electric fields for which the complications arising from space-charges and flexoelectricity can be neglected.) The interaction between the field and the nematic can be described by including an interaction term in the free-energy density of the liquid crystal. In the electric-dipole approximation this term is

$$f_{int} = -\frac{1}{2}\mathbf{P}\mathbf{E} \tag{8}$$

where $\mathbf{P}$ is the polarization. For a nematic

$$\mathbf{P}/\varepsilon_0 = (\varepsilon_\perp^{(s)} - 1)\mathbf{E} + (\varepsilon_\parallel^{(s)} - \varepsilon_\perp^{(s)})(\mathbf{n}\mathbf{E})\mathbf{n}. \tag{9}$$

The superscript (s) indicates that the dielectric constants refer to static fields.

The part of the interaction free-energy that depends on the director is

$$f_{static} = -\frac{1}{2}\varepsilon_0\varepsilon_a^{(s)}(\mathbf{nE})^2 \tag{10}$$

with

$$\varepsilon_a^{(s)} = \varepsilon_{\parallel}^{(s)} - \varepsilon_{\perp}^{(s)}.$$

For materials with $\varepsilon_a^{(s)} > 0$ the interaction free-energy minimum corresponds to a parallel alignment of the director with respect to the field. In nematic cells -such as the ones shown in Fig. 2.- a competition takes place between the orienting action of the substrates and that of the external field (unless the initial alignment of the director coincides with the direction of the applied field). As a result the initial director pattern becomes distorted.

Associated with the distortion of the director field there is an "elastic" free-energy which was discussed by Ossen, Frank and others. (For details see the monographs.[1]) The equilibrium director configuration can be determined by minimizing the total free-energy, i.e. the sum of the interaction and the elastic contributions. An equivalent method is to determine the balance between the volume torques arising from the interaction with the field and from the distortion respectively. The former torque can be given as

$$\mathbf{\Gamma}_{static} = \varepsilon_0\varepsilon_a^{(s)}(\mathbf{nE})(\mathbf{nxE}). \tag{11}$$

The influence of an electromagnetic field on the molecular orientation can be described along similar lines.[R8,R11] For optical fields the interaction free-energy and the "optical" torque are

$$f_{opt} = -\frac{1}{2}\varepsilon_0\varepsilon_a(\mathbf{nE})^2, \tag{12}$$

$$\mathbf{\Gamma}_{opt} = \varepsilon_0\varepsilon_a(\mathbf{nE})(\mathbf{nxE}) \tag{13}$$

with

$$\varepsilon_a = \varepsilon_{\parallel} - \varepsilon_{\perp} = n_e^2 - n_o^2.$$

Note that as these expressions are quadratic functions of the field strength their time averages are normally not zero. For linearly polarized beams the time average of f_{opt} can be given as

$$f_{opt} = -g(\mathbf{e}_p\mathbf{n})^2 I$$

where g is a factor depending on the refractive indices; $\mathbf{e}_p$ is a unit vector in the direction of the polarization, I is the intensity.

As a rule for nematics ε_a is positive. Thus the interaction free-energy is minimum when the molecules are oriented along the polarization direction of the light beam. Should this condition not correspond to the initial alignment in the cell, "optical reorientation" occurs under the influence of the electromagnetic wave.

The expressions for the optical free-energy and torque are very similar to the corresponding terms describing the action of an applied static or low-frequency electric field. The only difference is that in the present case the dielectric constants refer to optical frequencies. Because of this similarity one might have the impression that optical reorientation is a trivial analogue of the well-known low-frequency field effects. Certainly, in some situations the analogy between optical and low-frequency fields works quite well. In other cases, however, this analogy breaks down completely. We come back to this problem later. Here we only note that the distinction between "optical" and "low-frequency" fields is based on the ratio of the wavelength of the electromagnetic wave to the sample thickness, L. For the former fields $\lambda/L \ll 1$, while for the latter ones $\lambda/L \gg 1$.

A simple case of optical reorientation was investigated theoretically and experimentally by Pilipetskii *et al.*[R8] They studied the influence of a laser beam on a planar nematic layer. Reorientation was found at oblique incidence and extraordinary polarization of the light wave as expected. They compared the self-focusing effect in the nematic with that in isotropic liquids, such as CS_2. It was found that the corresponding nonlinear coefficient is nine orders of magnitude larger in the former case than in the latter one. Hence they proposed the term "giant optical nonlinearity" for optical reorientation in nematics. The huge enhancement of the nonlinearity is of course a result of the collective nature of the molecular motions. It should be mentioned, however, that in nematics optical reorientation is a non-local process thus the nonlinear coefficient depends on the sample thickness and the beam diameter also.

Zolotko *et al.*[R9] and Durbin *et al.*[R10] carried out corresponding experiments on homeotropic films. For oblique incidence and *e*-polarization the situation is very similar to the case of planar alignment. On the other hand an important difference occurs at normal incidence. The torque exerted by a normally incident light beam is zero both in a planar film (at *e* polarization) and in a homeotropic film. In the first case the electric field vector is parallel while in the second case it is perpendicular to the director. $\mathbf{\Gamma}_{opt}$ is zero in both cases, see Eq. 13. The latter case is however, unstable. Above a certain threshold power reorientation of the homeotropic layer takes place. This threshold phenomenon is analogous to the Freedericksz transition induced by a low-frequency electric or magnetic field.[24]

Durbin *et al.* derived a simple expression for the threshold intensity which applies for very broad light beams (plane-wave limit). As pointed out by Ong[25] they did not treat quite correctly the optical part of the free-energy and overestimated the threshold by the factor n_e/n_o. Yet an excellent agreement was found between theory and experiment; probably the omission of transverse effects more or less compensated the numerical error mentioned above.

Detailed theories of the "optical" Freedericksz transition were provided by Zel'dovich *et al.*[R11] and Csillag *et al.*[R12] In these works the influence of the finite laser-spot size was taken into account. It was shown that the transverse variation of the laser intensity leads to an increase of the threshold power. This increase becomes especially significant when the spot size is comparable to the layer thickness or smaller than it. Csillag *et al.* measured the spot size dependence of the Freedericksz threshold in a cyano-biphenyl nematic and a good fit between theory and experiment was found.

In the theories presented in R11 and R12 the Gaussian intensity distribution of the laser beam was replaced by a rectangular shape. Khoo *et al.*[R13] presented a theory for a Gaussian beam and measured the director distribution within the illuminated area.

Transverse effects play an important role in the self-focusing associated with optical reorientation also. Self-focusing was discussed by several authors.[26,27,28] For strong deformations the far-field diffraction pattern consists of a series of concentric bright rings. The number of rings is simply the light-induced phase shift at the centre of the beam divided by 2π. In an optical Freedericksz transition for a typical layer thickness ($L \approx 100\mu m$) this number can be as high as 50 producing a most impressive diffraction pattern (see e.g. in R9).

Superposition of an external field on the optical field provides the interesting possibility to control the beam divergence electrically.[R14] Such a superposition is interesting in another aspect too. Ong[25,R15] investigated theoretically the deformation of a nematic layer just beyond the Freedericksz threshold. He found that at certain values of the material parameters the orientation changes discontinuously. He pointed out in addition that the Freedericksz transition becomes discontinuous for any nematic if a sufficiently strong stabilizing external field is superposed on the light field. This latter prediction was verified experimentally by Karn *et al.*[29] with the help of an external magnetic field and by Wu *et al.*[R16] who applied a low-frequency electric field.

The optical reorientation processes discussed up to now were qualitatively similar to the corresponding low-frequency field effects. As mentioned earlier this is not always the case. A breakdown of the analogy with static fields was first reported by Zolotko *et al.*[R9] who observed in a homeotropic layer a drastic increase of the Freedericksz threshold power for an *o*-ray as the angle of incidence was increased. Durbin *et al.* mentioned that in a planar cell Freedericksz transition cannot be induced by a light beam polarized perpendicularly to the director. From a simple analogy one would expect for these cases a threshold not deviating significantly from the "homeotropic" Freedericksz threshold.

The explanation of these apparent anomalies is simple and tricky at the same time; it is connected with the "adiabatic" propagation of the light. A systematic description of the problem was first given by Csillag *et al.*[30] They pointed out that in the adiabatic approximation of light propagation reorientation cannot be caused by an *o*-polarized input. In the *o*-ray the electric field is everywhere perpendicular to the director, hence the optical torque is zero (Eq. 13). In the case of an applied external field the Freedericksz transition is initiated by a small fluctuation in the orientation; this fluctuation is then reinforced by the external field. However, there is no such reinforcement in the case of an adiabatically propagating *o*-ray as the polarization follows the rotation of the director, remaining perpendicular to it at every point.

Deviations from adiabatic propagation can lead to the deformation of the layer even with an ordinary input. These deformations can be, however, very different from the ones generated by static fields. The interaction of an *o*-ray with a homeotropic layer at small angles of incidence may serve as an illustration of such situations. As mentioned previously, a strong increase of the threshold compared to the case of normal incidence was found. Zolotko *et al.*[R9,R17] observed in addition that above the threshold the director field had no stationary configuration, self-oscillation occurred.

A strongly related effect was described by Santamato *et al.*[R18] for normal incidence and circularly polarized input. In this case again continuous rotation of the molecules was observed above the threshold. It was demonstrated that this rotation corresponded to a precession of the director around the normal axis. The direction of the precession was reversed when the input was switched from right to left circular polarization.

In order to interpret these remarkable effects one has to consider the optical torque exerted by an elliptically polarized beam. Note that because $\mathbf{\Gamma}_{opt}$ is a quadratic function of the field strength this torque is not the sum of the torques exerted separately by the two orthogonal components of the electric field vector. As shown in R18 and R22 the optical torque of an elliptically polarized wave has a component along the wave vector. It can be given as

$$\Gamma^z_{opt} = \frac{I}{\omega} \partial s_3 / \partial z. \tag{14}$$

Here s_3 is a normalized Stokes parameter:

$$s_3 = i < E_x E_y^* - E_x^* E_y > / < E_x E_x^* + E_y E_y^* > .$$

The quantity $I/\omega s_3$ gives the internal angular momentum ("spin") of the electromagnetic wave; Γ^z_{opt} is equal to the angular momentum transferred from the light beam to the liquid crystal per unit time. Hence Eq. 14 can be interpreted as a conservation law for the angular momentum.

Santamato *et al.* assumed that with homeotropic boundary conditions the normal component of the optical torque cannot be balanced by an elastic torque, hence the molecules are set into rotation. Γ^z_{opt} is balanced by a "viscous" torque arising from the precession of the director. They established a simple relation between the angular velocity of the director rotation and the total change of the Stokes parameter, s_3. This relation was experimentally verified.

In connection with the experiments described above there are still many open questions. E.g. it is not clear what are the precise conditions in which there is no steady-state configuration for the director. Transverse effects may have a significant influence on the oscillatory behaviour also. We believe that further work, both theoretical and experimental, would be worthwhile in this direction.

Another interesting orientational effect for which the adiabatic approximation cannot be applied occurs in hybrid-aligned nematic cells. In these cells the molecules are oriented homeotropically at one of the surfaces and parallel to the substrate at the other one (Fig. 2d). Such layers were first considered in the context of optical reorientation by Barbero and Simoni.[R19] They investigated the orienting action of an *e*-ray and called attention to the fact that reorientation occurs without threshold even at normal incidence. On the other hand for an *o*-polarized input beam the reorientation has a threshold character. Zel'dovich *et al.*[R20] calculated this threshold considering the deviations from the adiabatic propagation. The strange result was found that the threshold is different for a light beam entering the cell at the homeotropic surface and for that entering the cell at the planar side. This asymmetry was observed experimentally.[R21]

The problem of deformation of a planar layer by a normally incident light beam was discussed theoretically by Santamato *et al.*[R22] They calculated the threshold intensity for reorientation of the director by an *o*-ray. The predicted threshold is so high for a typical nematic layer that it would be difficult to observe it experimentally, at least with c.w. lasers. It is interesting to note that if the input beam is circularly polarized or polarized in 45^0 with respect to the director a twist deformation occurs without threshold. In this case, due to the birefringence, the ellipticity of the beam varies spatially within the sample. As discussed previously such a variation is associated with an optical torque parallel to the wave vector (see Eq. 14). This torque causes a distortion in the planar orientation which could be detected perhaps more easily than the threshold phenomenon.

In this section we concentrated on nematic liquid crystals. Much less work has been carried out in connection with optical reorientation in other mesophases. Interesting effects were predicted for cholesterics[31,R20] and smectic C layers.[R23] It seems that corresponding experimental studies have not yet been accomplished.

Dynamics of Optical Reorientation

The dynamics of optical reorientation in nematics has been studied much less extensively than the steady-state effects. The theoretical description of transient phenomena can be given in the framework of the non-equilibrium version of the continuum theory (Ericksen-Leslie hydrodynamic theory).

The dynamics of reorientation in static fields was studied by Pieranski *et al.*[32] Durbin *et al.*[R10] investigated the dynamic behaviour of the optical Freedericksz transition and found a complete analogy with the corresponding d.c. case. They measured in a 250μm thick layer a relaxation time of the order of a minute. This very slow relaxation -just as the "giant" nonlinear optical coefficient- is of course a consequence of the collective motion of the molecules. It should be noted that because of the non-local nature of optical reorientation in nematics, the relaxation time depends on the geometrical factors. The relaxation process can be speeded up significantly if smaller characteristic dimensions are used than in the Durbin experiment. It is difficult, however, to achieve a relaxation time below a second. This circumstance is a serious obstacle in any practical application of optical reorientation.

The slow relaxation of the nematic layers does not exclude the possibility to induce an observable reorientation in them by short laser pulses. Such a study was first carried out by Hsiung *et al.*[R24] who detected the response of a nematic film to 6 nanosec long pulses from a Q-switched laser.

Hsiung *et al.* pointed out that the application of laser pulses offers a unique possibility to study transient phenomena in liquid crystals. During the laser pulse energy, linear momentum and angular momentum is transferred from the light wave to the liquid crystal resulting in temperature rise, flow and rotation of the molecules respectively. After the laser pulse is over the three processes relax with three different time constants. The relaxation can be followed by a weak probe beam and the different processes - if the relaxation times differ sufficiently - can be separated. In the experiments of Hsiung *et al.* the thermal and the "slow" orientational modes were identified. They did not observe the "fast" mode connected with the induced flow; probably the refractive index change associated with this mode was too weak to detect it in the given geometry.

Eichler and Macdonald[33] carried out experiments with 80 psec pulses in a nematic. The energy was a few mJ per pulse. They produced an intensity grating and detected the self-diffraction of the laser beam. The diffraction efficiency was measured both in the nematic and in the isotropic phases. In the nematic phase the diffracted intensity depended on the angle between the grating and the director in a characteristic way. This dependence corresponded to a collective reorientation of the molecules indicating the validity of the Ericksen-Leslie continuum theory even for this very short time scale. In the isotropic phase naturally no angular dependence was observed, yet interestingly the diffraction efficiency was of the same order of magnitude as in the nematic phase.

On extremely short time scales the moment of inertia of the liquid crystal molecules should play a role also. The estimated relaxation time of the associated "librational" mode is of the order of a picosecond.[R24] Hence sub-picosecond exciting laser pulses would be necessary to study such effects.

Thermo-Optic and Related Effects

In the optical reorientation experiments described earlier thermally induced changes were only of secondary importance. Nevertheless different authors noted the presence of thermal effects even in transparent nematics such as the cyanobiphenyl compounds. At this point it should be noted that the distinction between orientational and thermal effects is straightforward. In a typical experiment the relaxation time for thermal effects is around milliseconds while it is a few seconds for reorientation. The polarization dependence and the influence of external fields are very different for the two processes also.

A nice example of the influence of laser heating on reorientation was found by Cheung *et al.*[34] They observed in a Fabry-Perot resonator filled with a nematic liquid crystal self-oscillation. It was shown that the oscillation resulted from the competition between the fast thermal and slow orientational mechanisms. Thermal processes are "fast" of course only in comparison with reorientation. In semiconductors similar self-oscillations occur in which heating is the "slow" process and optical nonlinearities of electronic origin play the role of the fast mechanism.[35]

Thermo-optic effects in liquid crystals were investigated extensively from the point of view of device applications. Nematics are particularly suitable for such purposes because in this phase the refractive indices have an unusually large temperature dependence, especially near the nematic-isotropic phase transition.[R25,R26] This strong temperature dependence is connected to the corresponding variation of the order parameter, S. The order parameter is a measure of the degree to which the molecules are oriented along the director. The birefringence, $n_e - n_o$ is proportional to the order parameter while the "average" index, $(n_e + 2n_o)/3$, is only weakly temperature dependent. Considering these relations one finds

$$\partial n_e/\partial T \approx \frac{2}{3}\Delta n' \partial S/\partial T, \qquad \partial n_o/\partial T \approx -\frac{1}{3}\Delta n' \partial S/\partial T \tag{15}$$

where $\Delta n'$ is a constant. Near the nematic-isotropic phase transition $\partial S/\partial T$ almost diverges and this fact is reflected in the strong temperature dependence of the refractive indices. One degree below the phase transition temperature $\partial n_e/\partial T$ and $\partial n_o/\partial T$ are of the order of $10^{-2}K^{-1}$ in contrast to the values $10^{-4}K^{-1}$ for isotropic liquids and semiconductors.

In R25, R26 and R27 the reader finds detailed descriptions of phase-conjugation, self-diffraction and optical bistability realized with the help of thermal effects in nematics. The absorption of the cells were controlled by adding dyes to the nematic[R25,R26] or coating metallic layers onto the substrates.[R27] Due to the large thermo-optic coefficients very low operating power-levels could be used. As an example, Lloyd and Wherrett observed optical bistability at an input power as low as $20\mu W$.[R27]

The study of thermo-optic effects -besides of potential applications- can give some new insight into the physics of liquid crystals. For instance, Armitage and Delwart[R25] analyzed the dynamics of the refractive index changes in an absorbing nematic. They showed that related to the large increase of the thermo-optic coefficient there is a critical slowing-down phenomenon. The optical changes follow the onset of light absorption with a certain delay; this delay is due to the finite relaxation time of the order parameter. The delay time increases sharply as the clearing temperature is approached, however it is only around 100 nsec even at half a degree below the phase transition.

Another interesting, somewhat related effect was found by Odulov *et al.* in MBBA.[R28] They observed self-diffraction of a laser beam under circumstances in which reorientation could not occur and heating was negligible. This nonlinear behavior was interpreted in terms of molecular conformation changes of the MBBA molecules induced by the laser radiation. The conformation changes influence the refractive indices in two ways. Firstly, there is a direct contribution from the metastable molecules as their (linear) susceptibilities are obviously different from that of the ground state molecules. Secondly and more importantly, the transformed molecules act as a sort of "impurities" which suppress the clearing point of the material. The shift in the phase transition temperature has a very similar effect on the refractive indices as the change of the actual temperature. When the experiment is carried out near the clearing point the induced refractive index change can be large, just as in the case of the thermo-optic effects. Indeed, Odulov *et al.* observed a strong enhancement of the nonlinearity in the vicinity of the phase transition.

A further effect connected to the absorption of light was reported very recently by Jánossy *et al.*[36] They found in a commercial nematic guest-host mixture optical reorientation at anomalously low power levels. E.g. in a homeotropic sample at normal incidence the optical Freedericksz threshold was around 1.5 mW in contrast to the 100mW expected for a corresponding transparent nematic layer. The underlying mechanism of the anomaly is not known yet.

Thermo-optic and other absorption induced effects can take place of course in any liquid crystalline phase. Thermally-induced texture changes in the smectic A and cholesteric phases are well-known and are used for display purposes.[37] Recently Becker *et al.*[38] studied thermo-optic effects in cholesterics induced by nanosecond laser pulses. They found a relaxation process on μsec time scale which is probably due to thermal diffusion within single turns of the helix. At the other end of the time scale a work by Zolotko *et al.*[39] should be mentioned. They observed in a planar cholesterics that prolonged laser illumination led to the formation of a periodic grid resembling those induced by electric field or dilation of the layer. This instability was seen only if the absorption of the light in the layer was sufficiently strong. No interpretation of the effect has been given; maybe it is due to a sort of photochemical reaction which changes the equilibrium value of the pitch in the material.

Concluding Remarks

As demonstrated in this volume, the study of optical effects in liquid crystals is motivated by three main purposes. Firstly, it is used as a tool in the basic research of different liquid crystalline states. Secondly, in mesophases new types of optical nonlinearities occur or new aspects of nonlinear processes become apparent. The study of these effects contributes to the progress of nonlinear optics. Thirdly, liquid crystals are investigated from the point of view of applications in certain nonlinear optical devices.

There are of course many open questions and further possibilities in the field. Some specific points were emphasized in the text. It should be remarked that up to now most researchers concentrated on the nematic phase. Although there are still many important aspects to be investigated even in this phase, the study of other mesophases looks very promising as well. We called attention already to the problem of optical reorientation in the cholesteric and smectic C phases. Regarding thermal effects we remind that the interesting point about nematics is the nearly critical behaviour near the nematic - isotropic phase transition. Similar phenomena can be expected to take place at other second-order phase transitions such as the smectic A - smectic C or some of the nematic- smectic A transitions.

It seems that no investigation has been carried out on nonlinear optical effects in more ordered smectic phases, polymeric liquid crystals or in blue phases. Some preliminary work has been performed on polymer dispersed nematic droplets.[40] The study of these systems may develop in the future.

Concerning potential applications we considered only all-optical nonlinear effects. It should be mentioned that liquid crystals can be utilized in the so-called self-electrooptic-devices (SEED)[41] too. In these devices there is an electrical feedback of the transmitted light to the layer. Such systems can be constructed by combining a photoconducting layer with a liquid crystal film. Ferroelectric liquid crystals are especially suitable for this purpose and may represent an alternative to the multiple quantum-well structures on which SEED-s are normally based. Whether all-optical effects will find a widespread application is still an open question.

References

Reprinted Articles:

R1. Hl. de Vries, Acta Cryst. **4**, 219 (1951).

R2. D.W. Berreman and T.J. Scheffer, Mol. Cryst. Liq. Cryst. **11**, 395 (1970).

R3. P. Allia, C. Oldano and L. Trossi, Mol. Cryst. Liq. Cryst. **143**, 17 (1987).

R4. E. Santamato and Y.R. Shen, J. Opt. Soc. Am. **A4**, 356 (1987).

R5. J.W. Shelton and Y.R. Shen, Phys. Rev. **A5**, 1867 (1972).

R6. M.I. Barnik, L.M. Blinov, A.M. Dorozhkin and N.M. Shtykov, Mol. Cryst. Liq. Cryst. **98**, 1 (1983).

R7. M.I. Barnik, L.M. Blinov and N.M. Shtykov, Sov. Phys. JETP **59**, 980 (1984).

R8. N.F. Pilipetskii, A.V.Sukhov, N.V. Tabiryan and B.Y. Zel'dovich, Optics Comm. **37**, 280 (1981).

R9. A.S. Zolotko, V.F. Kitaeva, N. Kroo, N.N. Sobolev and L. Csillag, Sov. Phys. JETP Letters **32**, 158 (1980).

R10. S.D. Durbin, S.M. Arakelian and Y.R. Shen, Phys. Rev. Letters **47**, 1411 (1981).

R11. B.Ya. Zel'dovich, N.V. Tabiryan and Yu.S. Chilingarian, Sov. Phys. JETP **54**, 32 (1981).

R12. L. Csillag, I. Jánossy, V.F. Kitaeva, N. Kroó and N.N. Sobolev, Mol. Cryst. Liq. Cryst. **84**, 125 (1982).

R13. I.C. Khoo, T.H. Liu and P.Y. Yan, J. Opt. Soc. Am. **B4**, 115 (1987).

R14. L. Csillag, N. Éber, I. Jánossy, N. Kroó, V.F. Kitaeva and N.N. Sobolev, Mol. Cryst. Liq. Cryst. **89**, 287 (1982).

R15. H.L. Ong, Applied Phys. Letters **46**, 822 (1985).

R16. J.J. Wu, G.S. Ong and S.H. Chen, Applied Phys. Letters **53**, 1999 (1988).

R17. A.S. Zolotko, V.F. Kitaeva, N. Kroo, N.N. Sobolev, A.P. Sukhorukov, V.A. Troshkin and L. Csillag, Sov. Phys. JETP **60**, 488 (1984).

R18. E. Santamato, B. Danio, M. Romagnoli, M. Settembre and Y.R. Shen, Phys. Rev. Letters **57**, 2423 (1986).

R19. G. Barbero and F. Simoni, Applied Phys. Letters **41**, 504 (1982).

R20. B.Ya. Zel'dovich and N.V. Tabiryan, Sov. Phys. JETP **63**, 80 (1986).

R21. B.Ya. Zel'dovich, N.F. Pilipetskii and A.V. Sukhov, Sov. J. Quantum Electron. **17**, 120 (1987).

R22. E. Santamato, G. Abbate, P. Maddalane and Y.R. Shen, Phys. Rev. **A36**, 2389 (1987).

R23. H.L. Ong and C.Y. Young, Phys. Rev. **A29**, 297 (1984).

R24. H. Hsiung, L.P. Shi and Y.R. Shen, Phys. Rev. **A30**, 1453 (1984).

R25. D. Armitage and S.M. Delwart, Mol. Cryst. Liq. Cryst. **122**, 59 (1985).

R26. I.C. Khoo, IEEE J. Quantum Electron. **QE-22**, 1268 (1986).

R27. A.D. Lloyd and B. S. Wherrett, Applied Phys. Letters **53**, 460 (1988).

R28. S.G. Odulov, Yu. A. Reznikov, M.S. Soskin and A.I. Khizhnyak, Sov. Phys. JETP **58**, 1154 (1983).

Other References:

1. P.G. deGennes, *The Physics of Liquid Crystals* (Clarendon, Oxford 1974); S. Chandrasekhar, *Liquid Crystals* (Cambridge University Press, Cambridge 1977); L.M. Blinov, *Electro-Optical and Magneto-Optical Properties of Liquid Crystals* (Wiley, New York 1983); W.H. deJeu, *Physical Properties of Liquid Crystals* (Gordon and Breach, London 1980)

2. M. Born and E. Wolf, *Principles of Optics* (Pergamon, Oxford 1975), Chapter 14.

3. W.H. Bragg, Nature **133**, 445 (1934).

4. C.W. Oseen, Trans. Faraday Soc. **29**, 883 (1933).

5. R. Dreher and G. Meier, Phys. Rev. **A8**, 1616 (1973).

6. C. Oldano, E. Miraldi and T. Taverna Valaberga, Phys. Rev. **A27**, 3291 (1983).

7. C. Maugin, Bull. Soc. franç. Miner. **34**, 71 (1911).

8. H. Kubo and R. Nagata, J. Opt. Soc. Am. **73**, 1719 (1983).

9. V.V. Zheleznyakov, V.V. Kocharovski and V.Vl. Kocharovski, Zh. Eksp. Teor. Fiz. **79**, 1735 (1980). (Translation in Sov. Phys. JETP (USA)).

10. H. L. Ong, Mol. Cryst. Liq. Cryst. **143**, 83 (1987).

11. R.A. Kashnow and J.E. Bigelow, Appl. Optics **12**, 2302(1973); E. Guyon, I. Janossy, P. Pieranski and J. Jonathan, J. Optics (France) **8**, 357 (1977).

12. *Nonlinear Optics*, edited by P.G. Harper and B.S. Wherrett, (Academic Press, London 1977).

13. I. Freund and P.M. Rentzepis, Phys.Rev.Letters **18**, 393 (1967).

14. G. Durand and C.H. Lee, Mol.Cryst. **5**, 171 (1968).

15. L.S. Goldberg and J.M. Schnur, Radio Electr. Eng. (GB) **39**, 279 (1969).

16. S.K. Saha, Optics Comm. **37**, 373 (1981).

17. S.M. Arakelyan, Yu.S. Chilingaryan, G.L. Grigoryan, G.A. Lyakhov, S.Ts. Nersisyan and Yu.P. Svirko, Mol.Cryst.Liq.Cryst. **71**, 137 (1981).

18. Ou-Yang, Zhong-can and Xie Yu-zhang, Phys.Rev. **A32**, 1189 (1985).

19. A. Taguchi, K. Kajikawa, Y. Ouchi, H. Takezoe and A. Fukuda, in *Nonlinear Optics of Organics and Semiconductors*, Springer Proceedings of Physics (Springer-Verlag, Berlin-Heidelberg 1989), page 250.

20. P. Guyot-Sionnest, H. Hsiung and Y.R. Shen, Phys.Rev.Letters **57**, 2963 (1986).

21. Y.R. Shen, *The Principles of Nonlinear Optics* (Wiley, New York 1984).

22. G.K.L. Wong and Y.R. Shen, Phys. Rev. Lett. **30**, 895 (1973).

23. J. Prost and J. R. Lalanne, Phys. Rev. **A8**, 2090 (1973).

24. V. Freedericksz and V. Zolina, Z. Cryst. **79**, 255 (1931); H. Deuling, Mol. Cryst. Liq. Cryst. **19**, 123 (1972).

25. H.L. Ong, Phys. Rev. **A28**, 2393 (1983).

26. A.S. Zolotko, V.F. Kitaeva, N. Kroo, N. N. Sobolev, A.P Sukhorukov and L. Csillag, Sov. Phys. JETP **56**, 786 (1982).

27. E. Santamato and Y.R. Shen, Optics Lett. **9**, 564 (1984).

28. F. Bloisi, L. Vicari, F. Simoni, G. Cipparrone and C. Umeton, J. Opt. Soc. Am. **B5**, (1988).

29. A.J. Karn, S. M. Arakelian, Y. R. Shen and H. L. Ong, Phys. Rev. Lett. **57**, 448 (1986).

30. L. Csillag, I. Jánossy, V. F. Kitaeva, N. Kroó, N. N. Sobolev and A. S. Zolotko, Mol. Cryst. Liq. Cryst. **78**, 173 (1981).

31. H.G. Winful, Phys. Rev. Letters **49**, 1179 (1982).

32. P. Pieranski, F. Brochard and E. Guyon, J. Physique **34**, 35 (1973).

33. H.J. Eichler and R. Macdonald, *Proc. Int. Conf. Lasers 88*, p.511 (1989).

34. M.M. Cheung, S.D. Durbin and Y.R. Shen, Optics Lett. **8**, 39 (1983).

35. H. M. Gibbs, *Optical Bistability: Controlling Light with Light* (Academic Press Inc., London 1985), Chapter 6.

36. I. Jánossy, A. D. Lloyd and B. S. Wherrett, Mol. Cryst. Liq. Cryst. **179**, 1 (1990).

37. E. Kaneko, *Liquid Crystal TV Displays* (KTK Scientific Publishers, Tokyo 1987), Chapter 8.

38. R.S. Becker, S. Chakravoti and S. Das, J. Chem. Phys. **90**, 2802 (1989).

39. A. S. Zolotko, V. F. Kitaeva, N. N. Sobolev, V. Yu. Fedorovich and N. M. Shtykov, Sov. Phys. JETP Lett. **43**, 614 (1986).

40. F. Simoni, G. Cipparrone, C. Umeton, G. Arabia and G. Chidichino, Applied Phys. Letters **54**, 896 (1989); P. Palffy-Muhoray, B.J.Frisken, J. Kelly and H.J. Yuan, *SPIE* **1105**, 33 (1989).

41. D.A.B. Miller and T. H. Wood, Opt. News **10**, 19 (1984).

REPRINTED ARTICLES

Acta Cryst. (1951). **4**, 219

Rotatory Power and Other Optical Properties of Certain Liquid Crystals

By Hl. de Vries

Physical Laboratory of the University at Groningen, Netherlands

(*Received* 7 *July* 1950)

A group of liquid crystals, mainly derivates of cholesterol, shows remarkable optical properties, including strong rotatory power and selective reflexion of circularly polarized light in a narrow region of wave-lengths. In this paper it is shown how these properties can be explained by the assumption that the molecules are arranged in a special way, so that the electrical axes rotate screw-like. It is inessential whether this occurs in small steps or continuously. When the axes make one revolution over a thickness p, then light in a region around $\lambda = pn$ will be reflected (n = refractive index). The second important parameter is the value of the double refraction $\alpha = (n_2 - n_1)/n$. From p and α all optical properties can be calculated. No accurate data for testing the theory are available but qualitatively the agreement is complete.

1. Description of the phenomena

From a monograph by Friedel (1922), we may summarize the phenomena occurring for a group of liquid crystals, mainly cholesterol derivates, as follows:

(1) Strong rotatory power (see Fig. 1). It amounts to more than ten and even to hundreds of revolutions per mm., whereas quartz gives only 24°/mm.

(2) Whereas in normal substances wave-length regions of opposite sign of the rotatory power are separated by a region of *absorption*, the liquid crystals have a region of *reflexion* of circularly polarized light. One circularly polarized component of the incident beam is completely unaffected; for the substance characterized by Fig. 1 (type 'dextro') it is only the

circularly polarized beam with anticlockwise rotating electrical vector which is reflected.* For substances of the type 'laevo' all signs are reversed.

(3) The electrical vectors of the incident and the reflected light rotate in opposite directions,* whereas the direction of rotation is unaffected by reflexion at normal substances.

(4) The mean wave-length, λ_0, of the reflexion band depends on the angle of incidence ϕ (the angle between the surface and the incident beam). Approximately the relation is represented by $\lambda = 2d \sin \phi$. Apparently a kind of Bragg reflexion occurs on internal planes with a distance d of several thousand Ångström units.

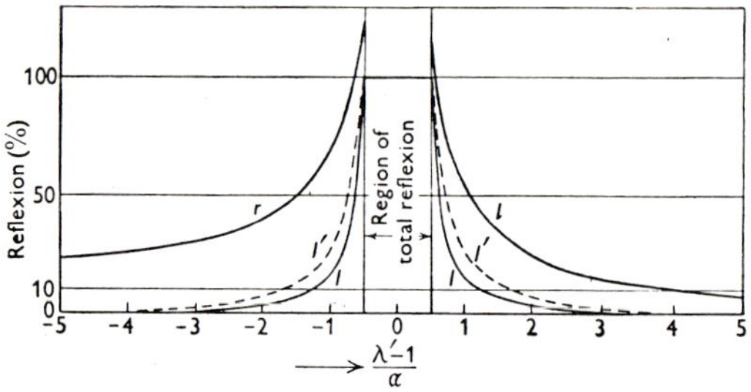

Fig. 1. Reflexion and rotatory power as a function of the wave-length $\lambda' = \lambda/\lambda_0$. A normal wave-length scale can be obtained by inserting for λ_0 and α some numerical value; say 550 mμ and 0·05. Curve r, right-handed rotation; l, left (arbitrary scale; sign of rotation according to *normal* usage). I, reflexion by a thick layer. I', intensity in the maxima (see also Fig. 5).

(5) Planes of Grandjean. When the substance fills a wedge-shaped space bright and dark lines appear, which follow the lines of equal thickness. Some authors interpreted these dark lines as intersections of *planes* in the crystal with the boundaries. It is remarkable that the distance d which can be found from these experiments agrees with d mentioned in the preceding paragraph.

2. Qualitative development of the model

The points (4) and (5) of the foregoing section suggest that these liquid crystals consist of a large number of thin layers, and that the light is reflected at the successive boundaries. Needless to say, however, reflexion at the boundary of two layers which make optical contact can only occur when there is a difference of the refractive indices of the layers. Suppose therefore that the crystal consists of thin *anisotropic* layers, and let the electrical axis in successive layers be turned through an angle ϕ. Similar piles of thin double-refracting sheets have often been discussed, but the reflexion was always neglected. Qualitatively this model easily explains several phenomena. The reflexion follows from the fact that both linear components of the wave in one layer will experience a change of refractive index when passing into the next layer. The fast component will be reflected with phase reversal; the other component is reflected without reversal. This involves that the sense of rotation of the electrical vector of circularly polarized light changes sign by the reflexion. The quantitative treatment in the next two sections will show that the angle ϕ between two successive layers is inessential for the phenomena, when at least the thickness p of the pile of layers in which the electrical axes make one complete revolution remains the same. Therefore the rigorous treatment in §§ 5–14 deals with a continuous rotation of the principal directions. Nevertheless, the elementary model is very useful since it gives a better insight into the mechanism.

* Contrary to ordinary usage we refer all signs of rotation to an observer who looks in the direction of the incident light, i.e. the positive z axis used in the theory (unless stated otherwise).

3. Quantitative treatment of the reflexion between two anisotropic layers

We confine ourselves to perpendicular incidence. The x and y axes are laid along the electrical axes of the first layer. (Dielectric constants ϵ_1 and ϵ_2 respectively; $\epsilon_1 < \epsilon_2$.) The ξ and η axes are along the corresponding directions of the second layer (Fig. 2), which is turned through an angle ϕ (see § 2). Suppose the incident wave to be linearly polarized (electrical vector E). The x component of the wave is given by

$$E_x \exp\left[2\pi i(t/T - n_1 z/\lambda) + i\delta_x\right]. \tag{1}$$

Here $E_x = E\cos\beta$ (Fig. 2); n_1 = refractive index = $\sqrt{\epsilon_1}$; λ = wave-length *in vacuo*; δ_x = the phase constant, which can be chosen zero for the incident wave by

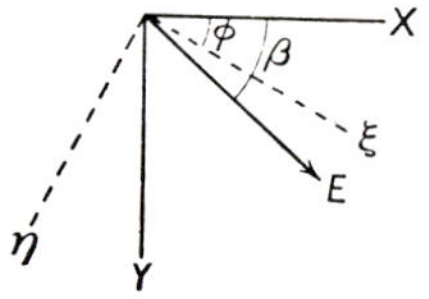

Fig. 2. Co-ordinate system used.

appropriate choice of t (so $\delta_x = \delta_y = 0$). *A priori* we do not know the phase constant δ for the reflected and transmitted waves. The formulae are very much simplified by combining these phase factors with the amplitude into a complex amplitude e. Then the formula (1) is converted into

$$e_x \exp\left[2\pi i(t/T - n_1 z/\lambda)\right]. \tag{1a}$$

Appropriate choice of t makes e_x and e_y real. For the y components n_2 should be substituted for n_1. For the reflected waves $-nz/\lambda$ is to be replaced by $+nz/\lambda$. Their amplitudes will be denoted by primes, e'_x, e'_y, etc. The reflexion can be calculated in the normal way from the boundary conditions at the plane between the two layers. The tangential components of E and H must be the same at both sides. H is calculated from

$$\frac{1}{c}\frac{\partial H_y}{\partial t} = -\frac{\partial E_x}{\partial z}, \quad \frac{1}{c}\frac{\partial H_x}{\partial t} = \frac{\partial E_y}{\partial z}. \tag{2}$$

Combining (1) and (2) we can express H in terms of E. In the plane $z=0$ all exponential factors in (1) or (1a) are the same, so that they can be omitted from the boundary conditions:

$$\left.\begin{array}{ll}\text{For } E_x: & e_x + e_x' = e_\xi \cos\phi - e_\eta \sin\phi,\\ \text{For } E_y: & e_y + e_y' = e_\xi \sin\phi + e_\eta \cos\phi,\\ \text{For } H_x: & n_2 e_y - n_2 e_y' = n_1 e_\xi \sin\phi + n_2 e_\eta \cos\phi,\\ \text{For } H_y: & n_1 e_x - n_1 e_x' = n_1 e_\xi \cos\phi - n_2 e_\eta \sin\phi.\end{array}\right\} \quad (3)$$

We give here only the amplitudes of the reflected wave

$$e_x' = (-e_x \sin\phi + e_y \cos\phi)\,\frac{n_2 - n_1}{2n}\sin\phi.$$

Here some simplification has been made, which is justified when $(n_2 - n_1)$ is small; this is generally the case. We have further put $n_1 + n_2 = 2n$. This formula, as well as the exact one, involves that there is no phase shift between the incident and the reflected wave; e_x' is real, since e_x and e_y could be chosen real. A phase reversal will be expressed by the sign of e_x'.

In the notation of §6 $(n_2 - n_1)/n = \alpha$. Inserting, furthermore, $e_x = E\cos\beta$ and $e_y = E\sin\beta$ (see Fig. 2), one obtains

$$e_x' = E\sin(\beta-\phi)\,.\,\tfrac{1}{2}\alpha\sin\phi,$$

and similarly

$$e_y' = E\cos(\beta-\phi)\,.\,\tfrac{1}{2}\alpha\sin\phi. \quad (4)$$

Obviously e_x' and e_y' are components of a vector e', which makes an angle $\beta' = 90 - \beta + \phi$ with the x axis. Its length is $\frac{1}{2}\alpha E \sin\phi$.

4. The interference of the reflected waves (discrete layers)

In order to find the combined effect of a pile of thin layers it should be noted that, approximately, the vector E has a constant length and direction throughout the system. This means that β decreases by ϕ for the successive layers, since we make use of a rotating co-ordinate system. Consequently β' increases by ϕ (see equation (3)), and relative to a fixed direction e' rotates 2ϕ at subsequent boundaries. When the waves arrive at the surface they have relative orientations as given in Fig. 3. (Multiple reflexions are neglected.) The phase difference of two successive waves is $2nb$, b being the thickness of one layer. It will be clear from Fig. 3 that there will be a maximum of intensity when E_1' and E_7', E_2' and E_8', etc., are in phase. This requires that $\frac{2\pi}{2\phi}\,2nb = \lambda$. The same formula is obtained when $\frac{2\pi}{2\phi}$ is not an integer. The thickness, p, of a pile in which there is a complete revolution of the principal directions equals $\frac{2\pi}{\phi}\,b$, so that the condition for a maximum becomes

$$pn = \lambda. \quad (5)$$

It should be noted that 2λ would not fit, since in this case E_1 and E_4, etc. would compensate each other. The next possibility is 7λ, 13λ, etc., but this leads to unreasonably large values of p.

According to (5) the wave-length λ of maximum reflexion depends on p only, whereas the value of b is inessential. The only value of b which would give special results is $b = \frac{1}{4}p$ and $\phi = \frac{1}{2}\pi$. In this case all partial waves E' have the same direction, and the resulting reflected wave will be linearly polarized. When $\phi \neq \frac{1}{2}\pi$ it can be proved that the resulting wave is circularly polarized. This can be understood qualitatively from Fig. 3. First E_1 reaches its maximum; then E_2, etc. So we see the maximum rotate in a *clockwise* direction, i.e. the sense in which the subsequent principal directions rotate. The length of the resulting vector is $\frac{1}{2}kE'$, k being the number of layers in the pile, and E' the amplitude of one partial wave. $E' = \frac{1}{2}\alpha E\sin\phi$. For small values of ϕ the amplitude of the circular wave reduces to

$$\tfrac{1}{4}\alpha E k\phi = \tfrac{1}{2}\alpha\pi d\lambda^{-1} nE,$$

where $d = kb$.

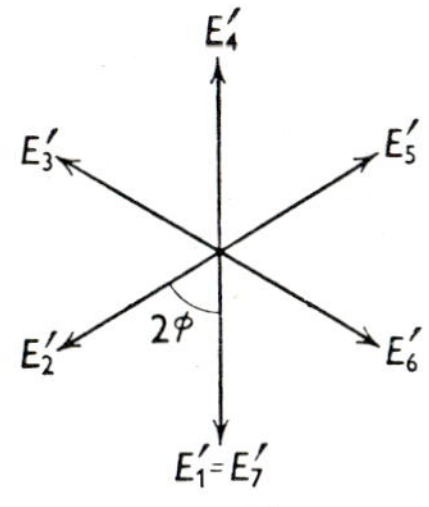

Fig. 3. The partial waves E_1', etc., from successive planes of reflexion, for $\phi = 30°$.

Here again the result does not depend on the thickness of one layer. For this reason we shall proceed in the following sections with a model in which the electrical axes turn continuously, though part of the results to be obtained there could perhaps also be derived from the present model.

Only one implication of the present theory has still to be mentioned. When the incident beam is circularly polarized with a clockwise rotating vector it is easily seen that the two reflected circular waves originating from its linear components just cancel out, whereas an incident beam with anticlockwise rotating vector gives waves reflected in phase with a clockwise rotating vector. This agrees with the properties summarized in §1. When the length of the rotating vector is E, the amplitude of the reflected wave is twice the value found above, i.e.

$$E' = \alpha\pi d\lambda^{-1} nE. \quad (6)$$

5. Rigorous treatment for continuously changing electrical axes

It was shown that the thickness b of one (hypothetical) layer was inessential for the phenomena. Moreover, the existence of a well-defined thickness would require a special explanation. It seems therefore more reasonable to assume continuous rotation of the principal directions. When the substance reflects visible light, p equals about $0{\cdot}5\,\mu$ and the rotation of the principal directions is about 1° for a monomolecular layer. This must be related with the form and the arrangement of the molecules in some way, but it is beyond the scope of the present paper to explain this. I have found that a similar continuous rotation was already supposed to exist in azoxyphenetol by Mauguin (1911). This substance showed remarkable optical properties when the cover-glass was rotated through an angle ϕ. These phenomena could be explained by supposing that the cover-glass dragged the principal directions of the adhering layer of the crystal with it, so that the crystal acquired a screw structure in which the electrical axes change their orientation continuously from the lower glass to the cover-glass. Our hypothesis for the cholesterol derivates differs from the model of Mauguin in two ways: (1) spontaneous rotation, (2) a large number of revolutions with a pitch $p \approx \lambda$, whereas the rotation in Mauguin's experiments was smaller than 90° and $p \gg \lambda$.

6. The normal waves in the screw structure

The well-known relations between D and E for a wave in the direction of the z axis are given by

$$\frac{1}{c^2}\frac{\partial^2 D_x}{\partial t^2}=\frac{\partial E_x}{\partial z^2}, \quad \frac{1}{c^2}\frac{\partial^2 D_y}{\partial t^2}=\frac{\partial^2 E_y}{\partial z^2}, \quad D_z=0. \tag{7}$$

The relations between D and E have a simple form in the ξ–η system (see Fig. 2), where the axes ξ and η are perpendicular to the z axis and in the direction of the electrical axes of the crystal. We suppose the angle ϕ between ξ and x to depend on z, $\phi=2\pi z/p$ (see also § 2). Here p stands for the pitch of the 'screw'. In this system we have $D_\xi=\epsilon_1 E_\xi$, $D_\eta=\epsilon_2 E_\eta$. Suppose $\epsilon_2>\epsilon_1$.

In order to use the relations (7) it is necessary to transform them into the ξ–η system. According to Fig. 2 one can write

$$\left.\begin{aligned} E_x&=E_\xi \cos 2\pi z/p - E_\eta \sin 2\pi z/p,\\ E_y&=E_\xi \sin 2\pi z/p + E_\eta \cos 2\pi z/p.\end{aligned}\right\} \tag{8}$$

Similar relations hold for D. With these values (7) takes the form

$$\left.\begin{aligned} \frac{\epsilon_1}{c^2}\frac{\partial^2 E_\xi}{\partial t^2}&=\frac{\partial^2 E_\xi}{\partial z^2}-\frac{4\pi}{p}\frac{\partial E_\eta}{\partial z}-\frac{4\pi^2}{p^2}E_\xi,\\ \frac{\epsilon_2}{c^2}\frac{\partial^2 E_\eta}{\partial t^2}&=\frac{\partial^2 E_\eta}{\partial z^2}+\frac{4\pi}{p}\frac{\partial E_\xi}{\partial z}-\frac{4\pi^2}{p^2}E_\eta.\end{aligned}\right\} \tag{9}$$

As a solution we try

$$\left.\begin{aligned} E_\xi&=A\\ E_\eta&=iB\end{aligned}\right\}\exp\left[2\pi i\left(\frac{t}{T}-\frac{mz}{\lambda}\right)\right]. \tag{10}$$

This is an elliptically polarized wave in the ξ–η system. It would be misleading to call m the refractive index, though it plays this role in (10). It has very uncommon properties, since (10) describes the wave motion in a 'screwed' system. By substituting (10) in (9) one obtains

$$\left.\begin{aligned} A\frac{\epsilon_1}{\lambda^2}&=A\frac{m^2}{\lambda^2}+2B\frac{m}{\lambda p}+\frac{A}{p^2},\\ B\frac{\epsilon_2}{\lambda^2}&=B\frac{m^2}{\lambda^2}+2A\frac{m}{\lambda p}+\frac{B}{p^2}.\end{aligned}\right\} \tag{11}$$

Eliminating B/A,

$$m^4-m^2(\epsilon_1+\epsilon_2+2\lambda^2/p^2)+(\epsilon_1-\lambda^2/p^2)(\epsilon_2-\lambda^2/p^2)=0. \tag{12}$$

The treatment of these equations is simplified by introducing 'reduced' quantities: $\lambda'=\lambda/(p\sqrt{\epsilon})$; $m'=m/\sqrt{\epsilon}$. According to (5) reflexion occurs for $\lambda'=1$. ϵ stands for the mean dielectric constant ($\epsilon=\frac{1}{2}(\epsilon_1+\epsilon_2)$). $(\epsilon_2-\epsilon_1)/(2\epsilon)=\alpha$ is a measure of the double refraction. $\alpha=0{\cdot}05$ fits rather well the data reported by Friedel (1922).

Equation (12) in its 'reduced' form is

$$m'^4-2m'^2(1+\lambda'^2)+(1-\alpha-\lambda'^2)(1+\alpha-\lambda'^2)=0. \tag{12a}$$

Finally
$$f=\frac{B}{A}=\frac{1-\alpha-m'^2-\lambda'^2}{2m'\lambda'}. \tag{13}$$

Apart from the notation, these formulae were also given by Mauguin (1911).

7. Discussion of the equations

Equation (12) gives two values for m'^2. The larger one, $m_2'^2$, proves to be positive for all values of λ'. The other root, m_1^2, is negative if $1-\alpha<\lambda'^2<1+\alpha$. (For these values of λ'^2 the product of the roots of (12a) is negative.) This implies that m_1' is imaginary in the region where the reflexion occurs according to equation (5). This reflexion will be dealt with in §§ 8 and 9. For m' one of the roots of m'^2 can be chosen. One is inclined to take the positive root since (10) has to represent a wave in the positive direction of the z axis. It will be shown, however, in §§ 7 and 8 (footnotes) that m_1' has to be negative when $\lambda'^2>1+\alpha$. A positive value would give a wave in the wrong direction.

For part of the discussion we want explicit expressions for m_1' and m_2'; these are obtained by expanding in series:

$$\left.\begin{aligned} m_1'&=1-\lambda'-\frac{\alpha^2}{8\lambda'(1-\lambda')}+\ldots,\\ m_2'&=1+\lambda'+\frac{\alpha^2}{8\lambda'(1+\lambda')}+\ldots.\end{aligned}\right\} \tag{14}$$

For $\lambda'^2<\alpha$ and $\lambda'\approx 1$ other expansions should be used. Each value of m corresponds to a normal wave with an ellipticity B/A which is obtained from (13). B_2/A_2 is always real and negative, corresponding to an electrical vector which rotates in clockwise direction. B_1/A_1 is real and negative except for the region of total reflexion (see Fig. 1). The values of B/A and m' are given graphically in Fig. 4.

The first wave in the region $|1-\lambda'^2|<\alpha$ requires special discussion. Substituting in (9) an imaginary value of m', $m'=-i\mu$, where μ stands for a real number, one obtains

$$\left.\begin{aligned} E_\xi&=A_1 e^{-2\pi\mu z/\lambda}\cos 2\pi t/T,\\ E_\eta&=iB_1 e^{-2\pi\mu z/\lambda}\cos 2\pi t/T.\end{aligned}\right\} \tag{15}$$

It follows from (13) that iB/A is real in this region so that the resulting wave is linearly polarized. Its amplitude decreases by the exponential factor and its phase does not depend on z.

Not too much stress should be laid upon the waves *in* the crystal since these waves are unobservable, in contrast to the waves that leave the crystal.

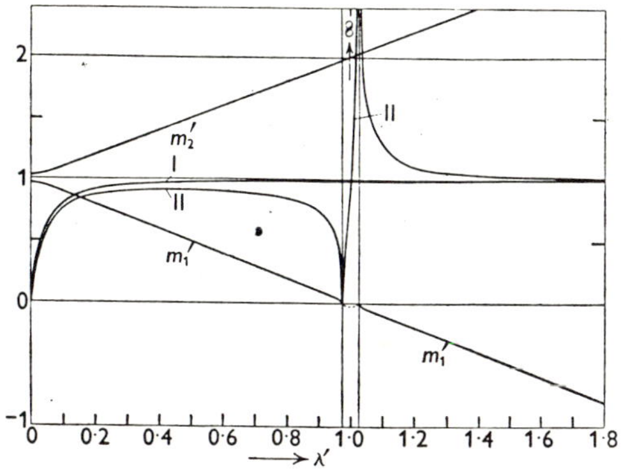

Fig. 4. Curves I and II, the ellipticities B_1/A_1 and $-A_2/B_2$ of the two normal waves. m_1' and m_2', the 'refractive indices' (see text). Vertical lines at $\lambda'=1\pm\frac{1}{2}\alpha$. Arbitrarily α was chosen as 0·05.

8. The rotatory power

(A) Mauguin (1911) confined himself to the case $p\gg\lambda$. The two normal waves are nearly linearly polarized since $B_1/A_1\approx A_2/B_2\approx 0$. The first wave is polarized along the ξ axis, the second wave along the η axis. Since the ξ–η system rotates relative to a fixed system of co-ordinates, the planes of polarization also rotate. The sense of rotation is the same as the sense of rotation of the cover-glass (see §5). This rotation is superimposed on the normal phenomena of double refraction, the refractive indices m_1 and m_2 (10) having the 'normal' values $\sqrt{\epsilon_1}$ and $\sqrt{\epsilon_2}$.

(B) *Larger values of* λ'. Here the normal waves are nearly circularly polarized with opposite sign of rotation and a *different* velocity (see §6). This brings about a rotation of the plane of polarization of a linearly polarized beam which can be considered as resulting from these normal waves. The direction of the resulting vector is the bisector of the angle between simultaneous positions of the two rotating vectors. Calling the angles of these three vectors with the x axis ψ, ψ_1 and ψ_2 respectively, the rotatory power is

$$\frac{d\psi}{dz}=\frac{1}{2}\left(\frac{\partial\psi_1}{\partial z}+\frac{\partial\psi_2}{\partial z}\right)_t.$$

Now $(\partial\psi/\partial z)_t$ consists of two parts: its rotation in the ξ–η system and the rotation of this system itself. One finds

$$\frac{\partial\psi_2}{\partial z}=\frac{2\pi}{p}(1-m_2'/\lambda') \quad\text{and}\quad \frac{\partial\psi_1}{\partial z}=\frac{2\pi}{p}(1+m_1'/\lambda').$$

Inserting m' from (14) this becomes

$$\left.\begin{aligned}\frac{\partial\psi_1}{\partial z}&=\frac{2\pi}{p\lambda'}\left(1-\frac{\alpha^2}{8\lambda'(1-\lambda')}+\ldots\right),\\ \frac{\partial\psi_2}{\partial z}&=\frac{2\pi}{p\lambda'}\left(1+\frac{\alpha^2}{8\lambda'(1+\lambda')}+\ldots\right).\end{aligned}\right\}\qquad(16)$$

Unless $\lambda'\approx 1$, $\partial\psi_1/\partial z$ approximately equals $2\pi\sqrt{\epsilon}/\lambda$. This has a simple physical meaning. When we consider the simultaneous positions of the rotating vectors for different values of z, they are arranged in a screwlike way. A vector which rotates (in the course of time) in a clockwise direction will make a left-turning screw. So the first wave should make a right-turning screw, its pitch being one wave-length ($\lambda/\sqrt{\epsilon}$). Indeed, (16) gives the correct sign and magnitude.* The rotatory power comes out as

$$\frac{\partial\psi}{dz}=-\frac{2\pi}{p}\frac{\alpha^2}{8\lambda'^2(1-\lambda'^2)}.\qquad(17)$$

The factor $(1-\lambda'^2)$ causes the remarkable phenomena described in § 1. For $\lambda'^2<(1-\alpha)$ the electrical vector rotates in anticlockwise direction since $d\psi/dz$ is negative. When the wave moves towards the observer he sees a clockwise rotation; indeed, in Fig. 1 it is seen that the rotatory power should be right-handed for $\lambda'<1$. For $\lambda'>1$ the sign changes because of the factor $(1-\lambda'^2)$ in (17). The magnitude of $d\psi/dz$ is also of the right order. At the edges of the region R we know that $1-\lambda'^2=\alpha$. According to (17), the rotatory power becomes $\dfrac{d\psi}{dz}=\dfrac{2\pi}{p}\dfrac{\alpha}{8}$. Inserting $p=0{\cdot}4\mu$ and $\alpha=0{\cdot}05$, which gives a reasonable band of reflexion (see Fig. 1), the rotatory power becomes 16 revolutions per mm.

Because of the factor λ'^2 the decrease of the rotatory power is most pronounced at the long-wave-length side of the reflexion band; this agrees with the experimental results.

Up to now we have neglected the fact that the waves are not exactly circularly polarized. This is unimportant, however, since it will be shown in §§8 and 9 that the wave which leaves the crystal is nearly circularly polarized. Furthermore, the amplitude of the first wave is smaller than the amplitude of the second wave, at least close to R. This makes the resulting wave elliptically polarized. The rotatory power is now a rotatory power of the long axis. At the edges of the region R the first wave has an intensity zero; this explains why the rotatory power disappears. Even when the intensity of the first wave does not vanish completely (when the layer of crystal is thin) the rotatory power will disappear since the emerging wave results then from a large number of internal reflexions (see § 11).

If the reflexion is neglected a pile of doubly refracting sheets still shows rotatory power

$$\frac{d\psi}{dz}=-\frac{2\pi\alpha^2}{8p\lambda'^2}.$$

* It should be remarked here that $\partial\psi_1/\partial z$ would come out with the wrong sign if for m_1' a positive value had been chosen in the region $\lambda'>1$.

This is, in our notation, a formula given by Pockels (1906, p. 291) and simplified by supposing that the angle ϕ between subsequent layers is small. It differs from our formula (17) just by the important factor $(1-\lambda'^2)$ in the denominator.

Finally, it can be remarked that the formula does not apply to substances like quartz which have a screw axis on an atomic scale. Inserting reasonable values for α and p, one arrives at very small rotatory powers. Moreover $d\psi/dz$ would be proportional to λ^{-4}, whereas it should be proportional to λ^{-2}. These rotatory powers have already been explained in a quite different way.

9. Calculation of the reflexion at the boundaries of a liquid crystal

On both sides of the boundaries E and H must be the same; this gives again the relations between incident and reflected waves. H can be expressed in terms of E by combining (10), (8) and (2).

One finds

$$\left.\begin{aligned} H_\xi &= -iA[\lambda/p+mf] \\ H_\eta &= A[m+(\lambda/p)f] \end{aligned}\right\} \exp[2\pi i(t/T-mz/\lambda)]. \quad (18)$$

These formulae become more comprehensible when the factor $[m+(\lambda/p)f]=(\sqrt{\epsilon})(m'+\lambda'f)$ is considered in more detail. It can be shown that $(m'+\lambda'f)$ is nearly unity, except for the first wave in the region $\lambda'\approx 1$. This means that $m+(\lambda/p)f$ is nearly the normal refractive index. Furthermore, $\lambda'+m'f=\dfrac{f}{\lambda'f+m'}$. (This can be proved from (12) and (13).) Denoting $\lambda'f+m'$ by q, (18) reduces to

$$H_\xi = -\frac{\sqrt{\epsilon}}{q}E_\eta \quad \text{and} \quad H_\eta = q\sqrt{\epsilon}\,E_\xi. \quad (18a)$$

The only difference from the corresponding formulae for a normal substance (see (19)) is the factor q which is exactly unity for normal substances. These formulae demonstrate once more that m does not play the role of the refractive index.*

A. *The boundary glass–crystal*

The components of the incident and the reflected wave in the cover-glass are represented by

$$\left.\begin{aligned} &\left.\begin{aligned} E_x&=e_x \\ E_y&=ie_y \end{aligned}\right\} \exp[2\pi i(t/T-n_0z/\lambda)] \qquad \begin{aligned} H_y&=n_0E_x, \\ H_x&=-n_0E_y, \end{aligned} \\ &\left.\begin{aligned} E'_x&=e'_x \\ E'_y&=ie'_y \end{aligned}\right\} \exp[2\pi i(t/T+n_0z/\lambda)] \qquad \begin{aligned} H'_y&=-n_0E'_x, \\ H'_x&=n_0E'_y. \end{aligned} \end{aligned}\right\} \quad (19)$$

* Formula (18a) enables the calculation of the Poynting vector $\mathbf{P}=\frac{c}{4\pi}[E, H]$ which gives the direction and the magnitude of the propagation of energy. $\mathbf{P}$ is zero for the first wave in the region of total reflexion, since the imaginary value of q causes a phase shift of 90° between E and H. The sign of P depends on the sign of m, chosen in § 7. It was anticipated there that m'_1 had to be chosen negative for $\lambda'^2>(1+\alpha)$. A positive value would have given a negative P; this means that the wave would propagate in the direction of the negative z axis.

It has been explained already in § 3 that the amplitudes e may be complex in order to account for phase shifts. Since the waves turn out to be elliptically polarized we have denoted the y amplitudes by ie_y so that e_y will in general be real.

In order to avoid the normal reflexion at the boundary we will suppose n_0 of the glass to be equal to $\sqrt{\epsilon}$ of the crystal. The boundary conditions for E and H can now be written as

$$\left.\begin{aligned} e_x+e'_x&=A, & e_y+e'_y&=fA, \\ e_x-e'_x&=qA, & e_y-e'_y&=\frac{f}{q}A. \end{aligned}\right\} \quad (20)$$

It is supposed here that the incident wave is chosen so that only *one* normal wave in the crystal results.

The equations are solved by

$$\left.\begin{aligned} 2e_x&=A(1+q) \\ 2e_y&=Af(1+q)/q \\ 2e'_x&=A(1-q) \\ 2e'_y&=-Af(1-q)/q \end{aligned} \quad \text{and} \quad \begin{aligned} \frac{e_x}{e_y}&=-\frac{e'_x}{e'_y}=\lambda'+\frac{m'}{f}, \\ \frac{e'_x}{e_x}&=-\frac{e'_y}{e_y}=\frac{1-q}{1+q}. \end{aligned}\right\} \quad (21)$$

B. *The boundary crystal–glass*

This can be treated in a similar way. It should be noted that m, f and q for the reflected wave change sign (see (13)). The first important result is that the first normal wave in the crystal gives rise to a reflected wave which also consists of the first normal wave only. The same holds for the second wave. Therefore we can confine ourselves to *one* incident normal wave. The results are

$$A'=-\frac{1-q}{1+q}A, \quad e''_x=\frac{2q}{1+q}A, \quad e''_y=f\frac{2}{1+q}A. \quad (22)$$

Here A' stands for the amplitude of the reflected wave; e''_x and e''_y represent the amplitudes of the outgoing wave in the glass.

10. Reflexion at the boundaries (discussion)

A. *The polarization of the waves*

According to (21) and (22) $\dfrac{e_x}{e_y}=\dfrac{e''_x}{e''_y}=\lambda'+\dfrac{m'}{f}$. When $\lambda'>\alpha$ (i.e. the region we are interested in) this is approximately equal to $1+\frac{1}{2}\alpha/\lambda$ for the first wave, or $-1+\frac{1}{2}\alpha/\lambda$ for the second wave ((12) and (13)). This means that the waves in the glass are nearly circularly polarized. The waves in the glass which correspond to the first wave in the crystal and which move in the direction of the incident light have an anticlockwise rotating electrical vector. The deviations from circularity are in opposite directions for both waves, but they are very small whereas these deviations for the waves in the crystal may be very large (see Fig. 4). It is easily found from the formulae that the sense of the rotation is reversed by the reflexion.

B. *The phase relations*

Except for the first wave in the region $|1-\lambda'^2|<\alpha$, the 'refractive index' q is real. According to (21) and (22) this means that the ratios e_x'/e_x, etc., are real, so that there is no phase shift. In the region of total reflexion, however, the first wave has an imaginary q and $e_x'/e_x=(1-q)/(1+q)$ is complex. If $\arctan|q|=\phi$, there is a phase shift of 2ϕ between the incident and the reflected wave. Of course this phase shift is of no direct importance in the measurements. It is only a serious complication of the calculations when one does not make use of complex amplitudes.

C. *The intensity of the reflected wave*

According to (21) and (22) the reflexion is governed by the factor $(1-q)/(1+q)$. It can be shown that $(1-q)$ is approximately equal to $\frac{1}{2}\alpha/m$. This approximation does not hold for the first wave in the region $\lambda'\approx 1$, where $m'\approx 0$; this region will be considered separately. Since α is very small, and $m'\approx 1$ or even larger, the amplitude of the reflected wave is at most $\frac{1}{4}\alpha$ times the amplitude of the incident wave. For very large wave-lengths, λ', the reflexion goes to zero since $m'\approx 1\pm\lambda'$ (see (14)). This is reasonable since, for a long wave-length, only the mean refractive index, n_0, is of interest. We have chosen n_0 equal to the refractive index of the cover-glass so that the normal reflexion vanishes. The intensity of the reflected wave is proportional to the square of the amplitude. Since the latter is small the former can be neglected completely.

D. *The reflexion of the first wave in the region* $\lambda'\approx 1$

For $|1-\lambda'^2|<\alpha$, q is imaginary. The absolute value of $e_x'/e_x=(1-q)/(1+q)$ equals unity. This means that the wave is completely reflected. For $\lambda'^2<(1-\alpha)$ or $\lambda'^2>(1+\alpha)$, but not too far from $\lambda'=1$, the reflexion is still appreciable (see Fig. 1).

Of course both reflexions at the upper and the lower boundary of the crystal have to be taken into account, as well as multiple reflexions of the wave in the crystal. This will be done in the next section.

11. Reflexion at thin layers

According to § 10 we can confine ourselves to the first wave with an anticlockwise rotating electrical vector and the region $\lambda'\approx 1$. We have found already that all reflected waves have an electrical vector which rotates clockwise. Generally, however, there is a phase difference. Beyond the region of total reflexion, this is a consequence of the difference in optical path, d, traversed. In the region of total reflexion the phase difference results from the reflexion. Moreover, there is a difference in amplitude. In the region of total reflexion this results from a decrease of amplitude *in* the layer, beyond this region it is caused by reflexion at the boundaries. When the amplitude of the incident circularly polarized wave e_x equals unity, the amplitude of the first reflected wave e_x' equals $(1-q)/(1+q)$, whereas $A_1=2/(1+q)$ (equation (22)). At the second boundary this amplitude is $A_1\exp[-2\pi i m_1' d\lambda^{-1}\sqrt{\epsilon}]=A_1 s$. Here s represents the absorption when m_1' is imaginary, and the phase shift when m_1' is real. Each of the successive waves which leave the crystal is obtained from the foregoing one by multiplication by $\{s(1-q)/(1+q)\}^2$. Summing the geometric series the total amplitude of the reflected wave is:

$$R=\frac{(1-q^2)(1-s^2)}{(1+q)^2-s^2(1-q)^2}. \qquad (23)$$

R is complex. Since we are only interested in the amplitude, we consider its absolute value.

(1) *Beyond the region of total reflexion*

Here q is real and s^2 is complex. Therefore s^2 can be represented by $s^2=\cos 2\delta+i\sin 2\delta$, where δ stands for the phase shift in the layer d, or $\delta=2\pi m_1' d\lambda^{-1}\sqrt{\epsilon}$. The intensity R^2 of the reflected wave is then

$$R^2=\frac{(1-q^2)^2\sin^2\delta}{(1-q^2)^2\sin^2\delta+4q^2}. \qquad (24)$$

The intensity R^2 is thus a periodic function of the thickness d. When $\delta=0°$, 180°, etc., the reflected wave vanishes; R^2 reaches a maximum for $\delta=90°$, 270°, etc.

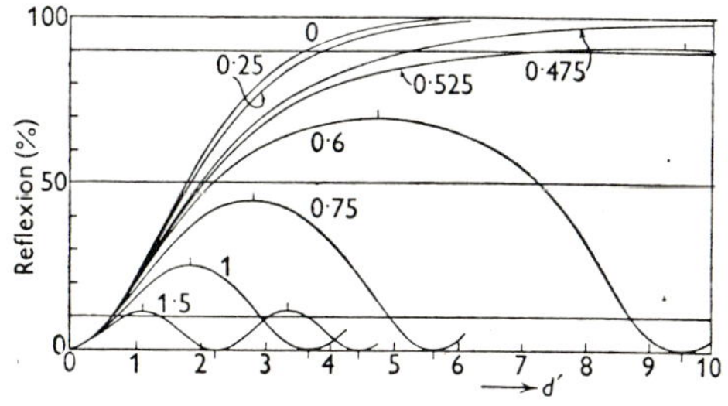

Fig. 5. The reflexion as a function of the reduced thickness $d'=2\pi d\lambda^{-1}\alpha\sqrt{\epsilon}$. The wave-length is $\lambda'=1-m\alpha$, m being given in the figure. For $m>0{\cdot}5$ the curves are periodic; the maxima and minima have been marked by vertical lines.

The relation (24) is represented in Fig. 5. It is seen that the fluctuations become slower when λ' approaches unity. This also follows from the simple reasoning of § 4. When λ' is close to the right value, the deviation makes itself perceptible only when the wave is reflected on a large number of layers.

(2) *Inside the region of total reflexion*

Here s is real and q becomes imaginary (q^2 negative). One finds now

$$R^2=\frac{(1-s^2)^2(1-q^2)^2}{(1-s^2)^2(1-q^2)^2-16s^2q^2}. \qquad (24)$$

This function is also represented in Fig. 5. The first parts of all curves are seen to be nearly identical and can be derived from the formulae that the amplitude R for thin layers equals $\pi\alpha d\lambda^{-1}\sqrt{\epsilon}$. This formula was derived in § 4 in an elementary way (equation (6)).

12. Numerical calculations

Strictly speaking numerical calculations can be performed only if a special value of α is chosen. The parameter p is eliminated by making use of the reduced wave-length $\lambda' = \lambda/(p\sqrt{\epsilon})$. For small values of α, however, one can find very good approximations, in which α is also eliminated. For these small values of α the reflexion is symmetrical at both sides of $\lambda' = 1$; therefore we confine ourselves to $\lambda' \leqslant 1$. Introducing Δ, so that $\lambda'^2 = 1 - \alpha(1+\Delta)$, one finds by expanding in series:

$$m_1'^2 \approx \left(\frac{\alpha}{2}\right)^2 (2\Delta + \Delta^2), \quad f_1 \approx \sqrt{\frac{\Delta}{2+\Delta}}, \quad q^2 = \frac{\Delta}{2+\Delta}.$$

In the region of total reflexion Δ varies from -1 to 0, so that m_1' and f_1 are imaginary. In the expression for s (see § 10) the exponent $2\pi i m_1' d\lambda^{-1}\sqrt{\epsilon}$ can be 'normalized' by introducing $d' = 2\pi\alpha d\lambda^{-1}\sqrt{\epsilon}$. Here d' will be called the reduced thickness. The exponent becomes now $d'\sqrt{(2\Delta+\Delta^2)}$. So there is only one parameter Δ left.

13. The planes of Grandjean

The phenomena predicted in the foregoing sections do not include the periodic bright and dark lines described in § 1 (5). It seems to us, therefore, that these lines are caused by periodical disturbances of the screwlike structure. It is probable that the orientation of the molecules in the two boundaries is prescribed by the structure of the surfaces—especially when the wedge consists of mica, which is often used. When the pitch, f, of the screw exactly fits the width, w, of the wedge the screwlike structure can develop. This requires that w equals $\frac{1}{2}kp$, or perhaps kp (k integer). Of course there are other parts of the wedge where the structure does not fit in w, so that the normal arrangement is disturbed and dark lines appear between crossed nicols. This hypothesis also explains why the planes of Grandjean are less pronounced when a wedge of glass is used since, generally, glass will not show these preferred directions. It is important to know, however, that the study of liquid crystals has often revealed the existence of preferred orientations even on glass (see Mauguin, 1911).

If this explanation is correct the planes of Grandjean should not be visible in light containing only the second normal wave, which is unaffected even when the screwlike structure is present. This has in fact been found. The experiments should, however, be repeated in monochromatic light.

14. Comparison with the experiments

In the course of the discussions it has been shown already that qualitatively the predictions of the theory agree with the experimental results summarized in § 1. No complete data are available to test the theory quantitatively. The measurements would not be difficult to make, however, since the theory contains only three constants, p, α and the refractive index n, which can be determined in various ways. The quantity p is determined from the wave-length λ_0 of the central part of the reflexion band ($pn = \lambda_0$). The wave-lengths can be reduced to λ' by dividing them by λ_0 ($\lambda' = \lambda/\lambda_0$). The quantity α can be determined from the width of the band (see Figs. 1 and 5). Then the rotatory power and the intensity of the reflected beam can be calculated as a function of λ; they depend only on α (see Fig. 5). The interesting fluctuations, shown in Fig. 5, have not yet been observed.

Finally, α appears in the ellipticity of the reflected beam. When α is not too small the ellipticity can be determined, as well as the directions of the axes. The latter are interesting with respect to the occurrence of fixed orientations on the cover-glass mentioned in § 12.

We have confined ourselves to perpendicular incidence. Oblique incidence may give rise to new phenomena which perhaps have been observed already though they still lack an explanation.

I express my thanks to Hn. de Vries (de Vries & Backer, 1950), who made some new cholesterol derivates which drew our attention to this class of liquid crystals, and especially to Dr W. G. Perdok, who made some orientating observations which afterwards proved to agree with the descriptions given by Friedel (1922). At present Dr Perdok is testing the theory and its basic assumptions quantitatively.

Finally, I thank Prof. F. Zernike and Prof. H. Brinkman for some improvements in the text of the manuscript.

References

Friedel, G. (1922). *Ann. Phys., Paris*, (9), **18**, 273.
Mauguin, C. (1911). *Bull. Soc. franç. Minér.* **34**, 71.
Pockels, F. C. A. (1906). *Lehrbuch der Kristalloptik.* Leipzig and Berlin: Teubner.
Vries, Hn. de & Backer, H. J. (1950). *Rec. Trav. chim. Pays-Bas*, **69**, 759.

Molecular Crystals and Liquid Crystals. 1970. Vol. 11, pp. 395–405

Reflection and Transmission by Single-Domain Cholesteric Liquid Crystal Films: Theory and Verification†

DWIGHT W. BERREMAN and TERRY J. SCHEFFER

Bell Telephone Laboratories, Incorporated
Murray Hill, New Jersey

Received October 19, 1970

Abstract—We have developed a fast and essentially exact numerical technique for computing propagation, reflection and transmission of light by a flat layer of any linear optical medium in which the dielectric tensor varies only in a direction normal to the surfaces. Using this technique with Oseen's spiraling-dielectric-tensor model of a single domain in a cholesteric liquid crystal, we predicted triplet Bragg reflection bands of both first and higher orders for light incident obliquely on thin films, similar to the triplet bands that Taupin predicted by a different technique for semi-infinite samples. We have observed the first and second order Bragg reflection bands for light incident at 45 degrees on single-domain cholesteric films between two glass prisms. The films used were mixtures of 4,4′-Bis(*n*-hexyloxy)azoxybenzene, which is nematic at about 100 °C, and dextro-4,4′-Bis(2-methylbutoxy)azoxybenzene which is asymmetric and causes the cholesteric spiral twist in the mixture. Adjustment of parameters in a general spiraling ellipsoid model to fit the data shows that the dielectric ellipsoids of such films are approximately prolate spheroids with the major axis normal to the spiral axis, as hypothesized by Oseen and Taupin. In mixtures having a pitch of 0.764 microns, for example, the two unlike principal values of the dielectric tensor are approximately 3.060 and 2.430 for blue light around the second order triplet. Additional Bragg reflection bands, which we predicted if no major axis of the dielectric ellipsoid were parallel to the spiral axis, were not observed.

1. Mathematical Technique

We have used a 4×4 matrix formulation of the electromagnetic wave equations in stratified media to compute the reflectance and transmittance of single-domain cholesteric liquid crystal films.(1) Our technique is basically equivalent to the 4×4 matrix technique first described by Teitler and Henvis,(2) and applied by them to

† Presented at the Third International Liquid Crystal Conference, Berlin, August 24–28, 1970.

finite layers of homogeneous anisotropic media. We have found that the method can easily be extended to the numerical solution of problems involving media with continuously varying anisotropic dielectric properties.

Prior to our publication,[1] Taupin[3] found an entirely different technique for computing some of the optical properties of certain models of single domain cholesteric liquid crystals with obliquely incident light using truncated infinite matrices. Recently Dreher *et al.*[4] have found propagation eigenvalues for oblique rays in such crystals using a single fourth-order differential equation. We believe the generality and simplicity of the 4 × 4 matrix technique makes it a useful alternative to Taupin's method or the method of Dreher *et al.* for computing optical properties of cholesteric liquid crystals.

2. Propagation in Stratified Media

When stratified, nonmagnetic, dielectric media carry electromagnetic waves of the form

$$\psi(z) \exp(ikx - i\omega t),$$

Maxwell's equations can be reduced to the matrix form

$$\frac{\partial}{\partial z}\begin{bmatrix} E_x \\ iH_y \\ E_y \\ -iH_x \end{bmatrix} = \frac{\omega}{c}\begin{bmatrix} \left(-i\dfrac{kc\epsilon_{xz}}{\omega\epsilon_{zz}}\right) & \left[1-\dfrac{1}{\epsilon_{zz}}\left(\dfrac{kc}{\omega}\right)^2\right] & \left(-i\dfrac{kc}{\omega}\dfrac{\epsilon_{yz}}{\epsilon_{zz}}\right) & 0 \\ \left(-\epsilon_{xx}+\dfrac{\epsilon_{xz}^2}{\epsilon_{zz}}\right) & \left(-i\dfrac{kc}{\omega}\dfrac{\epsilon_{xz}}{\epsilon_{zz}}\right) & \left(\dfrac{\epsilon_{xz}\epsilon_{yz}}{\epsilon_{zz}}-\epsilon_{xy}\right) & 0 \\ 0 & 0 & 0 & 1 \\ \left(\dfrac{\epsilon_{xz}\epsilon_{yz}}{\epsilon_{zz}}-\epsilon_{xy}\right) & \left(-i\dfrac{kc}{\omega}\dfrac{\epsilon_{yz}}{\epsilon_{zz}}\right) & \left[\dfrac{\epsilon_{yz}^2}{\epsilon_{zz}}-\epsilon_{yy}+\left(\dfrac{kc}{\omega}\right)^2\right] & 0 \end{bmatrix}\begin{bmatrix} E_x \\ iH_y \\ E_y \\ -iH_x \end{bmatrix}$$

or

$$\frac{\partial}{\partial z}\psi(z) = \frac{\omega}{c}\mathscr{D}(z)\psi(z).$$

We shall call $\mathscr{D}(z)$ the differential propagation matrix. When $\mathscr{D}(z)$ does not vary appreciably over an interval h, an integral of this equation is

$$\begin{aligned}\psi(z+h) &= \mathbf{P}(z,h)\psi(z) = \exp[\mathscr{D}(z)(h\omega/c)]\psi(z) \\ &= [\mathbf{1} + \mathscr{D}(z)(h\omega/c) + \mathscr{D}(z) : \mathscr{D}(z)(h\omega/c)^2/2! + \ldots]\psi(z).\end{aligned}$$

We shall call $\mathbf{P}(z, h)$ the local propagation matrix.

Now consider larger intervals of length $l = mh$, where the total variation of $\mathscr{D}(z)$ is large over l, but small over each of the m subintervals, h. We may write a general propagation matrix $\mathbf{F}(z, l)$ such that

$$\psi(z + l) = \mathbf{F}(z, l)\psi(z).$$

An obvious approximation for $\mathbf{F}(z, l)$ is

$$\mathbf{F}(z, l) \approx \mathbf{P}(z + l - h, h) : \mathbf{P}(z + l - 2h, h) : \ldots \mathbf{P}(z + h, h) : \mathbf{P}(z, h).$$

However, for most practical problems another approximation for $\mathbf{F}(z, l)$ converges faster. From the symmetry of the physical problem we know that

$$\mathbf{P}(z, h) = \mathbf{P}^{-1}(z, -h) \approx \mathbf{P}^{-1}(z + h, -h).$$

If this expression is substituted for alternate terms in the preceding product series for $\mathbf{F}(z, l)$, we obtain the following more symmetrical expansion, assuming the number of subintervals, m, is even.

$$\mathbf{F}(z, l) \approx \mathbf{P}(z + l - h, h) : \mathbf{P}^{-1}(z + l - h, -h) : \ldots \mathbf{P}(z + h, h) : \mathbf{P}^{-1}(z + h, -h).$$

If ϵ is periodic with period l, then

$$\mathscr{D}(z + l) = \mathscr{D}(z),$$
$$\mathbf{P}(z + l, h) = \mathbf{P}(z, h)$$

and

$$\psi(z + Nl) = \mathbf{F}(Nl)\psi(z).$$
$$= \mathbf{F}^{N}(l)\psi(z)$$

3. Computing Reflectance and Transmittance

Let the subscript i denote incident, r reflected and t transmitted field components for beams in isotropic media separated by a flat stratified layer of thickness T. Let the first medium, which contains the incident and reflected rays, have optical dielectric constant ϵ_1, and let the last medium have optical dielectric constant ϵ_2. Let the angle of incidence within the first medium be θ_1 (see Fig. 1). Snell's law gives

$$\epsilon_1^{1/2} \sin \theta_1 = \epsilon_2^{1/2} \sin \theta_2.$$

The following 6 relations are easily obtained from Maxwell's equations

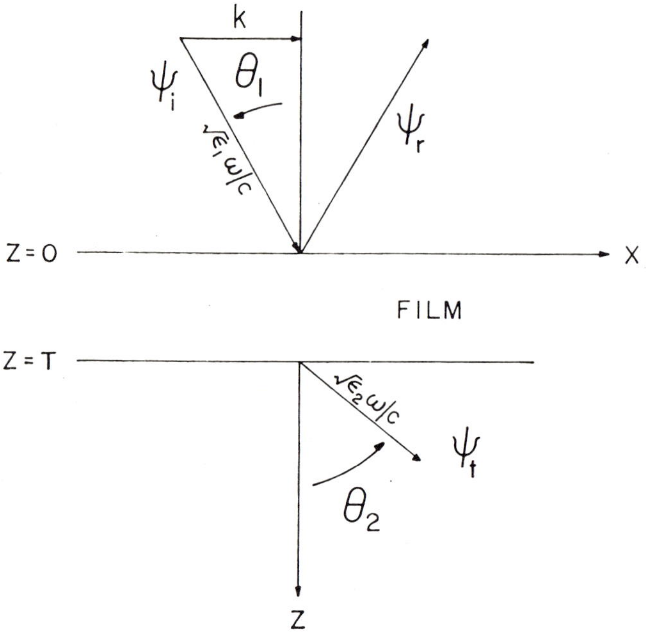

Figure 1. Illustration of variables used in computing the reflectance and transmittance of a stratified film of thickness T.

in isotropic media

$$\begin{aligned}
(H_y/E_x)_i &= -(H_y/E_x)_r = \epsilon_1^{1/2}/\cos\theta_1 \\
(H_x/E_y)_i &= -(H_x/E_y)_r = \epsilon_1^{1/2}\cos\theta_1 \\
(H_y/E_x)_t &= \epsilon_2^{1/2}/\cos\theta_2 \\
(H_x/E_y)_t &= \epsilon_2^{1/2}\cos\theta_2
\end{aligned}$$

Matching fields at the layer surfaces gives the matrix equation

$$\psi_t(T) = \mathbf{F}(T)\,(\psi_i(0) + \psi_r(0))$$

which can now be expressed as 4 linear equations in 6 field variables, such as E_{xi}, E_{yi}, E_{xr}, E_{yr}, E_{xt} and E_{yt}. Given E_{xi} and E_{yi}, we can compute the remaining 4, which give reflectance and transmittance of the combined layer and two interfaces.

4. Optical Model of a Cholesteric Liquid Crystal

A very simple optical model of a perfectly ordered cholesteric liquid crystal was first studied by Oseen[5] and later by Hl. de

Vries.[6] They only investigated light propagated normal to the x, y plane. A slight generalization of their model, which appears to describe correctly the samples of real liquid crystals that we studied, is given by the following dielectric tensor.

$$\boldsymbol{\varepsilon} = \begin{bmatrix} \bar{\epsilon} + \delta \cos 2\beta z & \delta \sin 2\beta z & 0 \\ \delta \sin 2\beta z & \bar{\epsilon} - \delta \cos 2\beta z & 0 \\ 0 & 0 & \epsilon_3 \end{bmatrix}$$

The value of ϵ_3 was irrelevant to Oseen and de Vries' investigations because they only found solutions for normally incident light. Taupin[3] recently described theoretical solutions for light obliquely incident on semi-infinite samples, assuming that $\epsilon_3 = \bar{\epsilon} - \delta$, which appears to be at least approximately the correct relationship for our samples.

Our model gives a differential propagation matrix

$$\mathscr{D}(z) = \mathscr{D}_0 + \mathscr{D}_2(z),$$

where

$$\mathscr{D}_0 = \begin{bmatrix} 0 & 1 - (kc/\omega)^2/\epsilon_3 & 0 & 0 \\ -\bar{\epsilon} & 0 & 0 & 0 \\ 0 & 0 & 0 & 1 \\ 0 & 0 & -\bar{\epsilon} + (kc/\omega)^2 & 0 \end{bmatrix}$$

and

$$\mathscr{D}_2(z) = \begin{bmatrix} 0 & 0 & 0 & 0 \\ -\delta \cos 2\beta z & 0 & -\delta \sin 2\beta z & 0 \\ 0 & 0 & 0 & 0 \\ -\delta \sin 2\beta z & 0 & +\delta \cos 2\beta z & 0 \end{bmatrix}$$

In this model, dielectric ellipsoids spiral about the z axis with pitch $2\pi/\beta$, but the period of the periodic $\mathscr{D}(z)$ matrix is $l = \pi/\beta$ (see Fig. 2). If the ellipsoids were tilted, $\boldsymbol{\varepsilon}$ would have no zeros and $\mathscr{D}$ would contain sin (βz) and cos (βz) terms, so that the period would be $l = 2\pi/\beta$. We found solutions for this case. Additional strong Bragg reflection bands corresponding to the longer fundamental period appear in this case. These additional bands were not observed

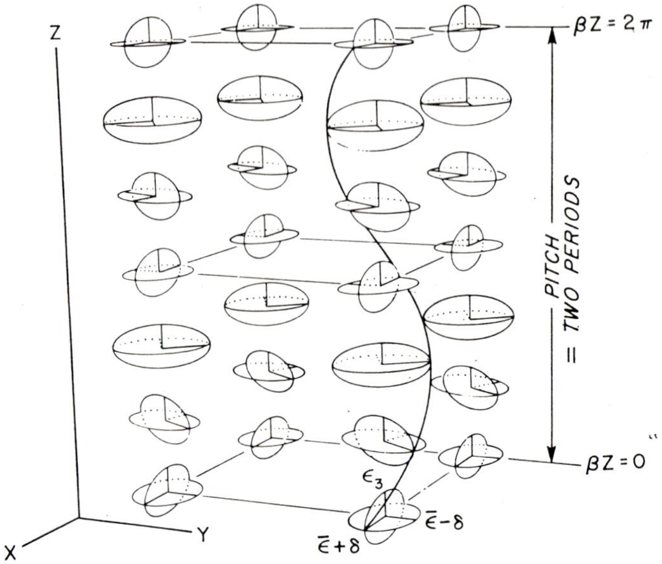

Figure 2. Spiraling dielectric ellipsoids in Oseen's optical model of a cholesteric liquid crystal.

experimentally. Hence the dielectric ellipsoids must not be tilted in our sample.

If the alternating product series expression is used to compute $\mathbf{F}(l)$, a rather rough approximation for $\mathbf{P}(z, h)$ may be used without introducing excessive cumulative errors in $\mathbf{F}$. With the generalized Oseen model, a satisfactory approximation is

$$\mathbf{P}(z, h) \approx \mathbf{P}_0(h) + (h\omega/c)\mathscr{D}_2(z).$$

The z-independent term $\mathbf{P}_0(h)$ is the propagation matrix corresponding to the invariant part, $\mathscr{D}_0$, of the differential propagation matrix. It may be written in the following exact, closed form.

$$\mathbf{P}_0(h) = \begin{bmatrix} \cos(abh\omega/c) & (a/b)\sin(abh\omega/c) & 0 & 0 \\ (-b/a)\sin(abh\omega/c) & \cos(abh\omega/c) & 0 & 0 \\ 0 & 0 & \cos(vh\omega/c) & (1/v)\sin(vh\omega/c) \\ 0 & 0 & (-v)\sin(vh\omega/c) & \cos(vh\omega/c) \end{bmatrix}$$

where

$$a = [1 - (kc/\omega)^2/\epsilon_3]^{1/2}$$
$$b = (\bar{\epsilon})^{1/2}$$

and

$$v = [\bar{\epsilon} - (kc/\omega)^2]^{1/2}$$

We computed the reflectance spectrum for obliquely incident plane wave radiation interacting with a system obeying Oseen's optical model. Figure 3 (bottom) shows the first and second order Bragg reflection bands that we predicted for certain principal values of the dielectric tensor.

5. Experiment

A straightforward experiment to verify our calculations would be to measure the reflectivity of a single domain in a cholesteric liquid crystal where the helicoidal axis is uniformly perpendicular to the film surface over the whole area of the light beam. To our knowledge, there have been no prior reflectivity studies made on single-domain cholesteric liquid crystal systems.

The cholesteric films that Fergason(7) and Adams, Haas, and Wysocki(8) used in their optical experiments were not single domain systems. To explain the reflectivity of these films Fergason(7) assumed a distribution in orientation of small Bragg scattering domains embedded in a matrix of constant refractive index. Experiments of Adams, Haas, and Wysocki(8) have shown that there is also a variation in the Bragg spacing, making it impractical to measure the distribution in orientation of these domains. Without knowledge of this orientational distribution, only qualitative comparison of experiments with the single-domain optical theories of Oseen,(5) de Vries,(6) and others(3,4,9) can be made.

We have used a single-domain cholesteric film in which the helicoidal axis is uniformly perpendicular to the film surface over regions of a square centimeter or more. Our cholesteric film is a binary mixture of non-mesomorphic dextro-4,4′-Bis(2-methylbutoxy)azoxybenzene (2 MBAB) and nematic 4-4′-Bis(*n*-hexyloxy)azoxybenzene.(10) We can vary the pitch of the resulting cholesteric mesophase from infinity to 0.24 μ by increasing the mole fraction of 2 MBAB in the mixture from 0 to 85%, which is the upper limit for

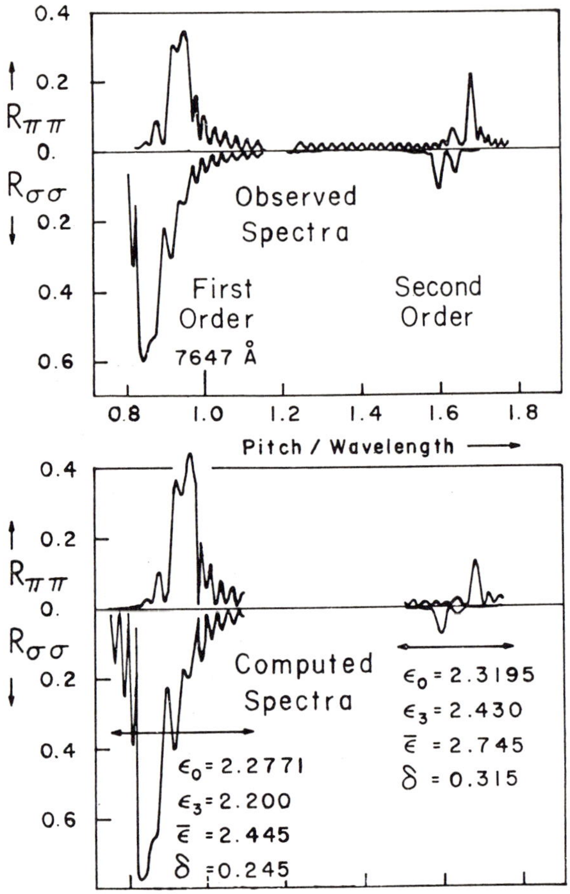

Figure 3. First and second order reflectance spectra of a cholesteric liquid crystal film 15 pitch lengths or 11.47 μ thick, confined between two glass prisms of optical dielectric constant ϵ_0. Light beam is incident at 45 degrees. Polarizer and analyzer were parallel to the plane of incidence for $R_{\pi\pi}$ and normal to it for $R_{\sigma\sigma}$ measurements. Mole fraction of 2 MBAB is 0.45 and temperature is 88 °C. Small oscillations are interference fringes from the two film-prism interfaces.

the existence of a pure cholesteric phase. The pitch shows only a small temperature dependence, decreasing only a few tenths of a percent for each centigrade degree increase in temperature. The mesomorphic range depends upon the fraction of 2 MBAB in the mixture, but all ranges fall between 42 and 130 °C.

The experimental arrangement that we used for measuring Bragg reflection from this cholesteric film is shown in Fig. 4. We directed an obliquely incident, plane-polarized, monochromatic beam of parallel light at the liquid crystal film contained between the faces of two 36 × 25 × 25 mm right angle glass prisms. We observed that the reflected light from the film always emerged at the specular angle. The reflected beam passed through an analyzer and was focused on a photomultiplier detector. We measured the reflected intensities for both sigma and pi polarized radiation. For pi polarized radiation both the polarizer and analyzer were oriented so that the electric field vector of the radiation passing through them would be in the plane defined by the incident and reflected beam. The polarizer and analyzer were rotated 90 degrees from this position for sigma polarized radiation. The analyzer was needed because it eliminated the dependence of detector sensitivity on polarization. We compared the reflected intensities with 100% reflection values that we measured in a separate experiment by using only the lower prism so the light beam would be totally internally reflected.

We made permanent spacers for the sample by evaporating chromium at three spots on the lower prism face, electroplating a thicker layer of gold on them and then polishing the gold to the desired thickness. The sample thickness was determined at the time of the experiment by measuring the wavelengths of a series of interference minima observed at normal incidence within an air bubble trapped in the film. We calculated the cholesteric pitch by measuring the spacing of the Grandjean-Cano[11,12] discontinuities produced by placing some of the sample between a convex lens surface of known curvature and a flat glass plate, both of which had been rubbed in one direction with lens tissue to insure a well-oriented sample.

We defined the molecular orientation at the surfaces of the liquid crystal film by rubbing the prisms on lens tissue in the x-direction. (See Fig. 4.) We introduced the sample in its isotropic phase between the heated glass prisms and then allowed the system to cool until the sample passed to its cholesteric phase. The desired temperature was held within a degree centigrade or so by means of a simple thermostatic oven arrangement. Before starting the reflection measurements we sheared the film by moving the upper prism back and forth several times in the x direction to obtain the Grandjean plane texture.

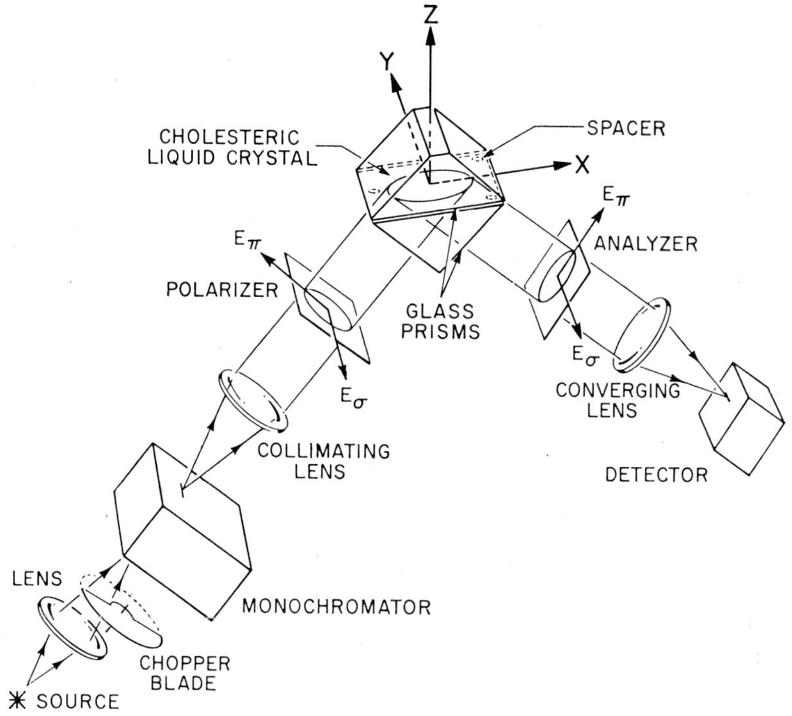

Figure 4. Apparatus for measuring reflectance of oblique rays by a liquid crystal film.

Our sample contained two or three parallel Grandjean discontinuities about 1 cm apart because it was slightly wedge-shaped. We took reflectivity measurements over a single domain region of uniform pitch by positioning a mask so that only light reflected from a $2 \times \frac{1}{2}$ mm rectangular area half way between two discontinuities and parallel to them was allowed to strike the detector.

The measured reflectivity curves for our cholesteric system are shown in Fig. 3 (top).

6. Conclusions

The mathematical method outlined here was used to generate reflectance spectra for liquid crystal films having the known thickness and pitch of our samples. By adjusting $\bar{\epsilon}$, δ and ϵ_3 we were able to fit frequencies of variations in reflectance quite closely. (See Fig. 3, top.) The method can easily be used for models with dielectric tensors that vary in a more complicated way. Our experimental data did not show additional Bragg reflection bands or other spectral features that we predicted if the principal axis ϵ_3 were not parallel to the z axis. We found that $\epsilon_3 \approx \bar{\epsilon} - \delta$; that is, that the dielectric tensor ellipsoid is (at least approximately) a prolate spheroid, as assumed by Taupin.[3] Our computations show that the second order Bragg reflection band for an oblate spheroid would be much less symmetric about the central component of the band (which is common to both π and σ polarized radiation) than that computed for a prolate spheroid.

We are in doubt as to whether the discrepancy between the computed and the measured intensities shown in Fig. 3 is significant. The first order Bragg reflection band appears to be somewhat weaker and the second somewhat stronger than predicted. Altering the initial and final azimuth of the dielectric ellipsoid or its principal values only made the fit worse. Assuming an error in pitch measurement did not help either. Thin regions near the surfaces with anomalous dielectric properties might account for the discrepancy. However, there might have been an experimental error due to difficulty in getting the same alignment of the sample and reference beams.

Acknowledgement

We wish to acknowledge the capable assistance of F. C. Unterwald in the optical measurements.

REFERENCES

1. Berreman, D. W. and Scheffer, T. J., *Phys. Rev. Lett.* **25**, 577 (1970).
2. Teitler, S. and Henvis, B. W., *J. Opt. Soc. Amer.* **60**, 830 (1970).
3. Taupin, D., *J. Phys. (France)* **30**, C4–32 (1969).
4. Dreher, R., Meier, G. and Saupe, A. O., *Third Internat. Liq. Cryst. Conf.*, Berlin, 1970 (to be published).
5. Oseen, C. W., *Trans. Faraday Soc.* (GB) **29**, 833 (1933).
6. de Vries, Hl., *Acta Cryst.* (Internat.) **4**, 219 (1951).
7. Fergason, J. L., *Mol. Cryst. and Liq. Cryst.* **1**, 293 (1966).
8. Adams, J. E., Haas, W., and Wysocki, J., *J. Chem. Phys.* **50**, 2458 (1969).
9. Conners, G. H., *J. Opt. Soc. Amer.* **58**, 875 (1968).
10. Sackmann, E., Meiboom, S. and Snyder, L. C., *J. Amer. Chem. Soc.* **89**, 5981 (1967).
11. Grandjean, F., *CR Acad. Sci. (France)* **172**, 71 (1921).
12. Cano, R., *Bull. Soc. Franc. Mineral. Crist.* **91**, 20 (1968).

Mol. Cryst. Liq. Cryst., 1987, Vol. 143, pp. 17-29
Photocopying permitted by license only

LIGHT PROPAGATION IN ANISOTROPIC STRATIFIED MEDIA IN THE QUASI ADIABATIC LIMIT

P ALLIA[*]°, C. OLDANO[+]° and L. TROSSI[+]
[+] Dipartimento di Fisica, Politecnico, Torino, Italy
* I.E.N. G. Ferraris, Torino, Italy
° GNSM-CNR and CISM-MPI, Torino, Italy

Abstract We propose a new perturbative treatment of the equations for electromagnetic wave propagation in the adiabatic limit in liquid crystals where the director is continuously rotating through the sample with an arbitrary law. In this approach a simple, 2x2 Jones matrix is obtained by proper transformation of the conventional 4x4 Berreman matrix. An approximate propagation equation is deduced for the Jones vector giving the polarization state of the wave propagating in the forward direction. The results obtained by using this equation are compared with the exact ones and the limits of validity of the present approach are discussed.

INTRODUCTION

Several methods have been proposed in order to solve the general problem of the propagation of electromagnetic waves in optically anisotropic stratified media where the director rotates continuously within the sample thickness, according to an arbitrary rotation law. In some particular cases of great physical interest, as for instance in N^* and S_c^* liquid crystals, the rotation of the director is uniform. In these cases, exact analytical solutions of the Maxwell's equations for oblique incidence of light may be obtained[1-2]. In the general case, however, only numerical solutions of these equations may be found, by exploiting for instance the so-called Berreman's 4x4 matrix method[3]. However, the numerical methods provide exact results but are hardly able to give a physical insight into the features of the problem. As a consequence, many efforts have been recently made in order to give simple, approximate solutions to the Maxwell's equations in particular conditions[4]. A simple, but very interesting case occurs when the director rotates very slightly over distances of the order of the wavelength of the incident light, λ. This is termed the quasi-adiabatic limit. In N^* liquid crystals and for light propagating along the helix axis the quasi-adiabatic limit may be expressed by the condition $p(n_e - n_o) \gg \lambda$, p being the pitch, n_e and n_o the extraordinary and ordinary refractive indices.

The quasi-adiabatic limit is of interest in many optical devices, as for instance the twisted nematic cell. For this problem the Berreman's matrix method requires very long computations. As long as the polarization properties of the transmitted beam are concerned, a simple approximation, good enough to any practical purpose, may be obtained by neglecting the two backward-propagating solutions of the Maxwell's equations cast in the Berreman's form. As a consequence, it is possible to adopt a formalism much simpler than the one involved in the treatment of the 4x4 Berreman's matrix, which may in fact be properly approximated by a reducible matrix. This allows one to study the changes of the polarization state of the incident light propagating within the medium by means of a 2x2 Jones matrix[5]. While in the rigorous adiabatic limit the polarization plane of an incident beam of linearly polarized light remains always parallel (or perpendicular) to the rotating director, in the quasi adiabatic limit a small component of opposite polarization is present.

The aim of this work is to write explicitly the Jones matrix in the case of normal incidence of light and for any type of stratified media. This is done starting from the Berreman's matrix and making use of some proper transformations followed by a suitable perturbative treatment. The results obtained by means of the Jones matrix method will be considered together with the ones obtained by a numerical solution of the Maxwell's equations, in order to establish, by direct comparison, the conditions defining the long wavelength region where the Jones matrix treatment is suitable. This is equivalent to determine a maximum acceptable value for the rotation of the director over one wavelength, or, in periodic structures, to seek for the minimum acceptable value of the ratio p/λ.

THEORY

Let us consider a sample of an optically anisotropic medium. The sample is in the shape of an infinite slab of thickness d, where the dielectric tensor continuously rotates within the sample thickness, forming the angles $\theta(z)$, $\varphi(z)$ with respect to a fixed reference system (x,y,z). The slab is parallel to the (x,y) plane (see Figure 1).

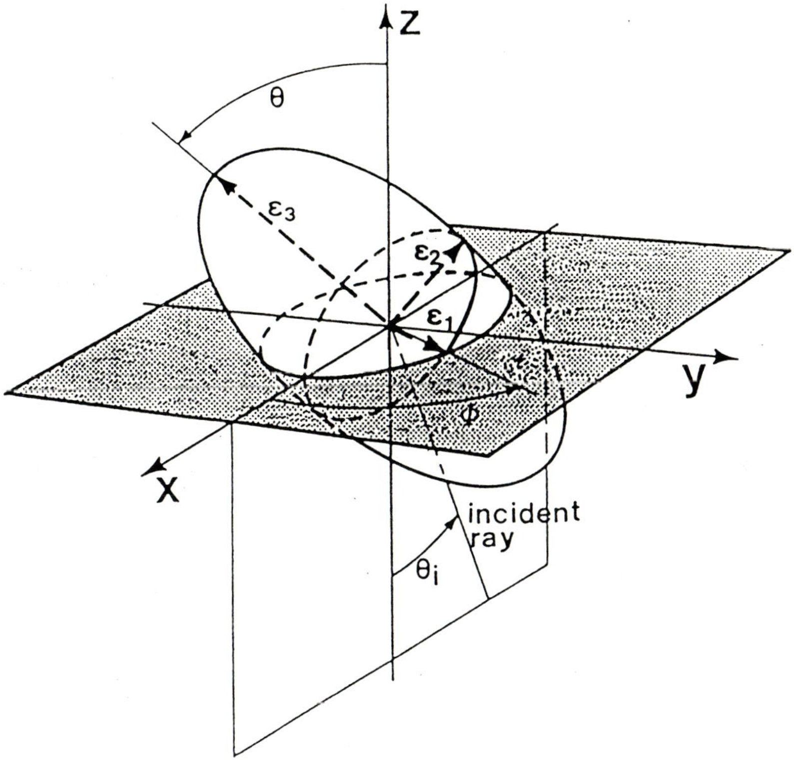

FIGURE 1. Euler angles of the dielectric tensor principal axes (1,2,3) with respect to the sample (x,y,z) frame (after Berreman[6]).

The Berreman's equations are written in the canonical form as

$$d\boldsymbol{\Psi}/dz=(i\omega/c)\boldsymbol{\Delta}\boldsymbol{\Psi} \qquad (1)$$

where

$$\boldsymbol{\Psi}=\begin{pmatrix} E_x \\ H_y \\ E_y \\ -H_x \end{pmatrix}$$

Δ is the Berreman's matrix, which for normal incidence of light reduces to

$$\Delta = \begin{pmatrix} 0 & 1 & 0 & 0 \\ \Delta_{21} & 0 & \Delta_{23} & 0 \\ 0 & 0 & 0 & 1 \\ \Delta_{23} & 0 & \Delta_{43} & 0 \end{pmatrix} \quad (2)$$

where

$$\Delta_{21} = \cos 2\varphi(\varepsilon_{11} - \varepsilon_{22})/2$$

$$\Delta_{23} = \sin 2\varphi(\varepsilon_{11} - \varepsilon_{22})/2$$

$$\Delta_{43} = \cos 2\varphi(\varepsilon_{11} - \varepsilon_{22})/2 + (\varepsilon_{11} + \varepsilon_{22})/2 \quad (3)$$

$$\varepsilon_{11} = \varepsilon_1 \qquad \varepsilon_{23} = (\varepsilon_2 - \varepsilon_3)\sin\theta\cos\theta$$
$$\varepsilon_{22} = \varepsilon_2\cos^2\theta + \varepsilon_3\sin^2\theta$$
$$\varepsilon_{33} = \varepsilon_2\sin^2\theta + \varepsilon_3\cos^2\theta$$

ε_1, ε_2, ε_3 being the principal values of the dielectric tensor. The Berreman's equation is first expressed in a reference system (x_R, y_R, z_R) rigidly following the rotation of the director around the z-axis, expressed by the azimuth φ.

The transformation matrix is

$$R(z) = \begin{pmatrix} \cos\varphi(z) & 0 & -\sin\varphi(z) & 0 \\ 0 & \cos\varphi(z) & 0 & -\sin\varphi(z) \\ \sin\varphi(z) & 0 & \cos\varphi(z) & 0 \\ 0 & \sin\varphi(z) & 0 & \cos\varphi(z) \end{pmatrix} \quad (4)$$

and the transformed Berreman's matrix becomes

$$\boldsymbol{\Delta}_R = R^{-1}\boldsymbol{\Delta}\,R = \begin{pmatrix} 0 & 1 & 0 & 0 \\ \varepsilon_{11} & 0 & 0 & 0 \\ 0 & 0 & 0 & 1 \\ 0 & 0 & \varepsilon_{22}-\frac{\varepsilon_{23}^2}{\varepsilon_{33}} & 0 \end{pmatrix} \tag{5}$$

The Berreman's equations may be written

$$d\boldsymbol{\Psi}_R/dz = (i\boldsymbol{\omega}/c)H_R\,\boldsymbol{\Psi}_R \tag{6}$$

where, setting $a=(ic/\boldsymbol{\omega})d\boldsymbol{\varphi}/dz$,

$$H_R = \begin{pmatrix} 0 & 1 & -a & 0 \\ \varepsilon_{11} & 0 & 0 & -a \\ a & 0 & 0 & 1 \\ 0 & a & \varepsilon_{22}-\frac{\varepsilon_{23}^2}{\varepsilon_{33}} & 0 \end{pmatrix} \tag{7}$$

and $\boldsymbol{\Psi}_R = R^{-1}.\boldsymbol{\Psi}$

The following step consists in a second change of matrix representation, by introducing a more convenient basis whose eigenvectors are linearly polarized waves propagating within the medium in the forward and backward directions.

In the new basis, the Berreman's equations take the form:

$$d\Psi_T/dz = H_T \Psi_T \tag{8}$$

where and

$$H_T = i\omega/c \left[T H_R T^{-1} + (ic/\omega) T \left(dT^{-1}/dz \right) \right]$$

the transformation matrix T is defined as:

$$T = \begin{pmatrix} \sqrt{n_1} & 1/\sqrt{n_1} & 0 & 0 \\ 0 & 0 & -\sqrt{n(\theta)} & 1/\sqrt{n(\theta)} \\ \sqrt{n_1} & -1/\sqrt{n_1} & 0 & 0 \\ 0 & 0 & \sqrt{n(\theta)} & -1/\sqrt{n(\theta)} \end{pmatrix} \tag{9}$$

where $n_1 = \sqrt{\varepsilon_{11}}$; $n(\theta) = \left[\varepsilon_2 . \varepsilon_3 / (\varepsilon_2 \sin^2\theta + \varepsilon_3 \cos^2\theta) \right]^{1/2}$

We are concerned with the polarization properties of the transmitted light in the quasi-adiabatic limit. As a consequence, it is possible to assume that, in a first approximation, the backward-propagating waves are negligible since they give rise only to second or higher order effects. By explicitly considering the waves propagating in the forward direction, the transformed Berreman's matrix H_T may be reduced to a 2x2 Jones matrix by keeping only the matrix elements of the first (upper left) block. The Jones matrix may be cast in the very simple form

$$J(z) = \begin{pmatrix} k_1 & -i\eta(\theta) d\varphi/dz \\ i\eta(\theta) d\varphi/dz & k(\theta) \end{pmatrix} \tag{10}$$

where $k_1 = (\omega/c) n_1$; $k(\theta) = (\omega/c)\, n(\theta)$;
$\eta(\theta) = (\sqrt{n_1/n(\theta)} + \sqrt{n(\theta)/n_1})/2$

Notice that J(z) is an hermitean matrix, and generally the extraordinary refractive index, n, is a function of $\theta(z)$. The solution of the Jones equation $d\Psi_T/dz = iJ(z)\Psi_T$ is equivalent to the search for the evolution operator U(z) defined as $\Psi_T(z)=U(z)\Psi(0)$ (z=0 corresponding to the first surface of the slab). U(z) satisfies to the equation

$$dU/dz=iJ(z)\,U(z) \tag{11}$$

Notice that U(z) is an unitary matrix (since J is hermitean). In the general case of arbitrary dependence of both and on z, it is convenient to solve perturbatively the Jones equation by setting $J=J_o+V$, where

$$J_o=\begin{pmatrix} k_1 & 0 \\ 0 & k(\theta) \end{pmatrix};\ V=\begin{pmatrix} 0 & -i\eta(\theta)d\varphi/dz \\ i\eta(\theta)d\varphi/dz & 0 \end{pmatrix} \tag{12}$$

The equation (11) may be solved either numerically (by making use of an algorithm much simpler than the one required by the full 4x4 Berreman's matrix), or using a standard perturbative treatment. In both cases, use is made of the interaction matrix formalism, in which V plays the role of the perturbative term.

The evolution matrix U_o, corresponding to the case V=0 is

$$U_o=\begin{pmatrix} \exp(ik_1z) & 0 \\ 0 & \exp\left\{i\int_0^z k\left[\theta(z')\right]dz'\right\} \end{pmatrix} \tag{13}$$

The interaction matrix U_I, to the first perturbative order, is

$$U_I(z)=\begin{pmatrix} 1 & u(z) \\ -u^*(z) & 1 \end{pmatrix} \tag{14}$$

where

$$u(z)=\int_0^z \exp\left\{-i\int_0^{z'}\left[k_1 z''-k(z'')\right]dz''\right\}\eta(z')\frac{d\varphi}{dz'}dz' \text{ and}$$

$u^*(z)$ is the complex conjugate of $u(z)$.

As a consequence, the evolution matrix $U(z)=U_o\cdot U_I$ may be written as

$$U(z)=\begin{pmatrix} \exp(ik_1z) & \exp(ik_1z)u(z) \\ -\exp\left[i\int_0^z k(z')dz'\right]u^*(z) & \exp\left[i\int_0^z k(z')dz'\right] \end{pmatrix} \tag{15}$$

In N^* and S^*_c liquid crystals, as well as in the twisted nematic cell, the formalism can be simplified, since the Jones matrix is independent of z, $d\varphi/dz$ and being both constants in these cases. By setting $k(\theta)=\text{const}=k_2$, the characteristic equation for the wave vectors of the forward propagating modes is simply

$$(k_1-k)(k_2-k)-\eta^2q^2=0 \tag{16}$$

where $q=d\varphi/dz$; $\eta(\theta)=\text{const}=\eta$

Eq. (16) admits the solutions

$$k=(k_1+k)/2 \pm (k_1-k_2)/2\left[1+(4\,\eta^2 q^2/(k_1-k_2)^2\right]^{1/2} \qquad (17)$$

In the quasi-adiabatic limit, the wave vectors become

$$\begin{aligned} k_I &= k_1 + \eta^2 q^2/(k_1-k_2) \\ k_{II} &= k_2 - \eta^2 q^2/(k_1-k_2) \end{aligned} \qquad (18)$$

The corresponding eigenfunctions are ellyptically polarized waves, orthogonal to each other. The ratio between the minor and the major axis of the ellypse is $\eta q/(k_1-k_2) \ll 1$.

We mention here the fact that in all cases where $k(\theta)$ is varying within a restricted interval around a given value $\bar{k}$, it is convenient to set in Eq.(12), $J_o = \begin{pmatrix} k_1 & 0 \\ 0 & \bar{k} \end{pmatrix}$.

As a consequence, in this case the Eqs.(13-15) can be cast in a much simpler form. This approach will be further discussed in a forthcoming paper.

A simple application

We will check the validity of the approach proposed in the previous Section in a particularly simple case. Let us consider an uniaxial cholesteric liquid crystal of pitch p, with ordinary and extraordinary refractive indexes in the ratio $n_o/n_e=9/11$. We seek for solutions of the Jones equations (eq.11) for normally incident light. The results are expressed in the rotating reference system (x_R, y_R, z_R).

The relative difference $\Delta k_i/k_i$ (i=I,II) between exact and approximate wavevectors of both forward-propagating modes are reported in Figure 2.

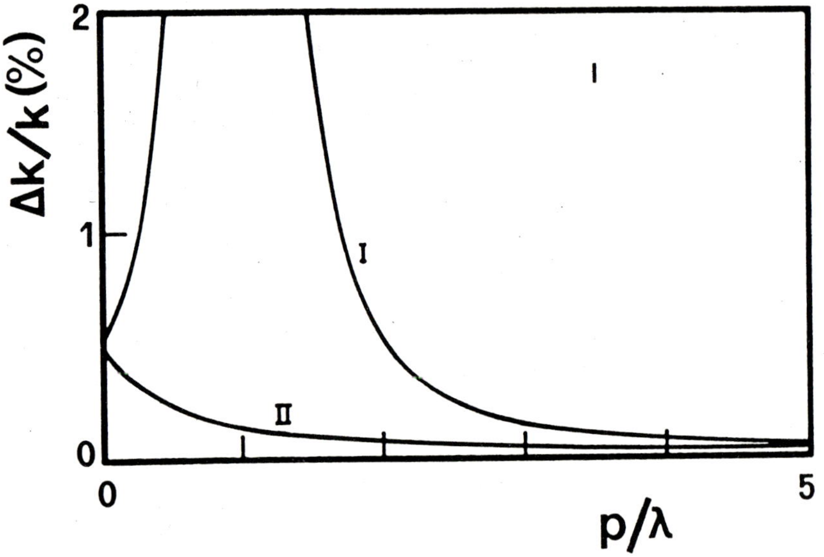

FIGURE 2. Relative differences between exact and approximate wavevectors of the eigenmodes for light propagating in cholesteric L.C.

The approximate results are calculated from Eq.16, the exact solutions are the ones found in any text book on L.C.[7]. It is clearly seen that the approximation proposed in the present paper becomes and remains very good for wavelengths even smaller than the ones predicted by the theory ($\lambda \ll p\,\Delta n/n$), i.e., of the order of the pitch ($\lambda \simeq p$).

The large discrepancy between exact and approximate values of k_I, appearing in Figure 2 in the region $\lambda \simeq p$, is clearly related to the existence of a reflection band for the first mode in N^* liquid crystals at normal incidence. This band is obviously not accounted for by the Jones matrix formalism, which loses entirely its validity in the region $p \cong \lambda$.

The proposed formalism has been tested in several other cases, always for normal incidence of light either by exploiting the first-order perturbative formula (eq.15) or by solving the equation for U(z) numerically.

We may summarize the results as follows: for a L.C. with $n \simeq 0.2$ the numerical solutions of eq.(11) give the correct polarization state (within $\pm$ 0.5%) when the variations of the director typically involves a rotation of one radians over one wavelength.

This means that the backward components may be totally neglected in most of the cases of practical interest. The first-order perturbative approach provides instead solutions which are correct when one full turn (2π) of the rotating director occurs over a distance much greater than $\lambda / \Delta n$, n being the difference between ordinary and extraordinary refractive indexes. In other words, the first-order perturbative approach gives acceptable results in a wavelength region more restricted than the one where the numerical solution of eq.11 is already valid.

REFERENCES

1. R. Dreher and G. Meier, Phys. Rev. A8 1616 (1973)
2. C. Oldano, E. Miraldi and P. Taverna Valabrega, Phys. Rev. A27, 3291 (1983); C. Oldano, P. Allia and L. Trossi, J. Phys. (Paris) 46, 573 (1985); M.A. Peterson,

Phys. Rev. A27 520 (1983)

3. D.W. Berreman, J. Opt. Soc. Am. 62, 502 (1971)
4. A. Schunhofer, H.G. Kuball and C. Puebla, Chem. Phys. 76 453 (1983); P. Allia, C. Oldano and L. Trossi, J. Opt. Soc. Am. B3, 424 (1986); see also S. Chandrasekhar, Liquid Crystals, Cambridge University Press, Cambridge (1977)
5. R.C. Jones, J. Opt. Soc. Am. 31, 488 (1941); J. Opt. Soc. Am. 31, 499 (1941); J. Opt. Soc. Am. 32, 486 (1942)
6. D.W. Berreman, Mol. Cryst. Liq. Cryst. 22, 175 (1973)
7. P.G. de Gennes, The Physics of Liquid Crystals, Clarendon Press, Oxford (1974).

Pseudo-Stokes parameter representation of light propagation in layered inhomogeneous uniaxial media in the geometric optics approximation

E. Santamato

Dipartimento di Fisica Nucleare, Struttura della Materia e Fisica Applicata, Pad. 20, Mostra d'Oltremare, 80125 Napoli, Italy

Y. R. Shen

Department of Physics, University of California, Berkeley, California 94720

Received June 2, 1986; accepted October 5, 1986

We study the propagation of normally incident, elliptically polarized light in layered inhomogeneous uniaxial structures by using the geometric optics approximation (GOA). We show that, in the GOA, light propagation can be described by the evolution of four real-valued pseudo-Stokes parameters. These parameters are linearly related to the familiar Stokes parameters and obey a simple first-order differential equation. Conservation of energy and angular momentum is preserved under this approximation.

INTRODUCTION

Electromagnetic wave propagation in inhomogeneous and anisotropic media is often difficult to analyze; approximate solutions are therefore of great significance. Thus there exist the geometric optics approximation (GOA), the phase-integral method, and the method of perturbation calculation.[1] In most practical cases the properties of the inhomogeneous medium vary slowly over a wavelength, and the GOA is the most suitable analytical method.

The GOA has been widely used in the study of light propagation in photoelastic materials[2] and liquid crystals.[3] Recently, Ong and Meyer presented a general GOA formalism for wave propagation in optically inhomogeneous, planar, uniaxial media and found excellent agreement between the zeroth-order GOA and the exact solution for the case of a periodically bent nematic liquid crystal.[4] The approach of Ong and Meyer, however, is restricted to linearly polarized light in planar structures.

In this work, the GOA is extended to the more general case of elliptically polarized light in a spatially twisted and tilted uniaxial structure. Maxwell's equations for normal incidence are solved asymptotically in the limit of small wavelength in a rigorous way, and the zeroth-order terms are retained. The evolution of the polarization state of the light wave as it propagates in the medium is studied by introducing a set of four real-valued parameters (the pseudo-Stokes parameters) that are linearly related to the usual Stokes parameters. The main advantages of using the pseudo-Stokes parameters are that their evolution equation takes a very simple form and that the polarization state of the optical wave can be represented by a point on the Poincaré sphere, as it does in homogeneous nonabsorbing media. With the aid of the Poincaré sphere, we can then visually grasp the complicated changes of the wave polarization as the wave propagates along its path in an inhomogeneous, locally uniaxial medium.

The GOA approach presented in this paper is compatible with the requirements of energy and angular momentum conservation along the wave propagation direction. A consequence of the angular momentum conservation law is that the incident wave may exert a mechanical torque on the anisotropic sample. This torque is usually very small and may be neglected, but in some materials, such as liquid crystals, this electromagnetic torque may lead to observable effects.[5] The problem of reflection at boundaries is also considered. It is shown that in the GOA conservation of both energy and angular momentum is still satisfied at the reflecting boundaries.

OPTICS IN INHOMOGENEOUS UNIAXIAL MEDIA

We consider an elliptically polarized plane wave of frequency ω impinging at normal incidence onto an inhomogeneous, nonabsorbing, locally uniaxial film of thickness d. The dielectric tensor of the medium has the form

$$\epsilon_{ij} = \epsilon_\perp + (\epsilon_\parallel - \epsilon_\perp)n_i n_j \qquad (i, j = 1, 2, 3), \tag{1}$$

where n_i are the Cartesian components of the unit vector $\hat{n}$, describing the local orientation of the optical axis, $\epsilon_\perp = n_o^2$, $\epsilon_\parallel = n_e^2$, and n_o and n_e are the ordinary and extraordinary refractive indices, respectively. We assume n_o and n_e to be constant, while the direction of $\hat{n}$ may vary in the sample. We take the x_3 axis to be along the surface normal and assume that $\hat{n}$ varies with x_3 only.

At normal incidence, div $\mathbf{D} = 0$ yields

$$D_3 = \epsilon_{31}E_1 + \epsilon_{32}E_2 + \epsilon_{33}E_3 = 0. \tag{2}$$

Equation (2) can be used to eliminate E_3 in the wave equations, which can then be written explicitly as a set of two coupled second-order differential equations for E_1 and E_2:

$$\mathrm{d}^2\mathbf{E}/\mathrm{d}x_3^2 + (k_0^2/\epsilon_{33})\tilde{M}\mathbf{E} = 0, \tag{3}$$

0740-3232/87/020356-04$02.00

where $k_0 = \omega/c$, $\mathbf{E} = (E_1, E_2)$, and $\tilde{M}$ is a rank-two tensor with elements

$$M_{11} = \epsilon_{11}\epsilon_{33} - \epsilon_{13}^2,$$
$$M_{22} = \epsilon_{22}\epsilon_{33} - \epsilon_{23}^2,$$
$$M_{12} = M_{21} = \epsilon_{12}\epsilon_{33} - \epsilon_{13}\epsilon_{23}. \quad (4)$$

An integration of Eq. (3) gives the energy flux conservation along x_3:

$$-\mathrm{Im}(E_1{}^*E_1{}' + E_2{}^*E_2{}') = k_0\langle \wp_3\rangle = \text{const.}, \quad (5)$$

where the prime denotes the derivative with respect to the x_3 coordinate, the star denotes the complex conjugate, and $\langle \wp_3\rangle$ is the x_3 component of the time-averaged Poynting vector.

An electromagnetic wave propagating in an anisotropic medium exerts an average local mechanical torque τ per unit volume on the medium with[5]

$$\tau = (\Delta\epsilon/8\pi)\mathrm{Re}[(\hat{n}\cdot\mathbf{E}^*)(\hat{n}\times\mathbf{E})], \quad (6)$$

where $\Delta\epsilon = \epsilon_\parallel - \epsilon_\perp$. The total torque along $\hat{x}_3$ acting on the medium of thickness d is

$$\mathrm{T}_3 = \mathrm{A}\int_0^{\mathrm{d}} \tau_3(\mathrm{x}_3)\mathrm{dx}_3, \quad (7)$$

where A is the optical-beam cross section. One can show, by using Eqs. (1), (3), (4), and (6), that τ_3 is a total derivative,

$$\tau_3 = -\mathrm{dL}_3/\mathrm{dx}_3, \quad (8)$$

with $L_3 = (1/8\pi k_0{}^2)\mathrm{Re}(\mathrm{E}_2{}^*E_1{}' - E_1{}^*E_2{}')$. Equation (7) then reduces to

$$\mathrm{T}_3 = -\mathrm{A}\Delta\mathrm{L}_3, \quad (9)$$

where $\Delta L_3 = L_3(d) - L_3(0)$. We can identify L_3 as the average angular momentum per unit area and unit time carried by the propagating wave, analogous to the case of wave propagation in vacuum.[6] Equation (8) or (9), therefore, describes the conservation of angular momentum. If the sample were free to rotate around the x_3 axis, T_3 should be equal to the change of the sample angular momentum per unit time.

THE GEOMETRIC OPTICS APPROXIMATION

Let us express $\mathbf{E}$ in Eq. (3) in terms of the eigenvectors $\hat{e}_1$ and $\hat{e}_2$ of the local symmetric matrix $\tilde{M}$:

$$\mathbf{E} = c_1\hat{e}_1\exp(ik_0\phi_1) + c_2\hat{e}_2\exp(ik_0\phi_2). \quad (10)$$

In the coordinates of (x_1, x_2) the unit eigenvectors of $\tilde{M}$ are given by $\hat{e}_1 = (\cos\Phi, \sin\Phi)$ and $\hat{e}_2 = (-\sin\Phi, \cos\Phi)$, with eigenvalues $\lambda_1 = \epsilon_\parallel\epsilon_\perp$ and $\lambda_2 = \epsilon_\perp\epsilon_{33}$, respectively. The angle Φ is the azimuthal angle of the local director $\hat{n}$ of the optical axis, expressed in polar coordinates, viz., $\hat{n} = (\sin\theta\cos\Phi, \sin\theta\sin\Phi, \cos\theta)$. The polar angle θ refers to the tilt angle between $\hat{n}$ and the x_3 axis. Notice that $\hat{e}_1$ and $\hat{e}_2$ depend on x_3, since $\Phi = \Phi(x_3)$. Insertion of Eq. (10) into Eq. (3) yields

$$\sum_{j=1}^{2}\{[(c_j\hat{e}_j)'' + 2ik_0\sqrt{\phi_j{}'}(\sqrt{\phi_j{}'}c_j\hat{e}_j)' - k_0{}^2(\phi_j{}'^2 - \lambda_j/\epsilon_{33})c_j\hat{e}_j]\exp(ik_0\phi_j)\} = 0. \quad (11)$$

The GOA corresponds to the limit when $k_0 = 2\pi/\lambda$ (λ = wavelength) appears as a dominant parameter in the above equation. With $k_0 \to \infty$, we find from Eq. (11) that $\phi_j{}' = \sqrt{\lambda_j/\epsilon_{33}}$ ($j = 1, 2$), or

$$\phi_1 = \int_0^{x_3}[n_on_e/\sqrt{\epsilon_{33}(x)}]\mathrm{d}x + \text{const.}, \quad (12a)$$

$$\phi_2 = n_ox_3 + \text{const.} \quad (12b)$$

We note that ϕ_1 and ϕ_2 are the phases of the extraordinary and ordinary waves, respectively. With ϕ_1 and ϕ_2 given by Eqs. (12a) and (12b), Eq. (11) reduces to a set of two second-order differential equations for the coefficients c_1 and c_2 only. The solution of this set of equations can be found by first expanding c_1 and c_2 into power series of $1/k_0$:

$$c_j = \sum_{n=0}^{\infty} c_j{}^{(n)}k_0{}^{-n}. \quad (13)$$

Insertion of Eq. (13) into Eq. (11) gives

$$\sum_{j=1}^{2}[(c_j{}^{(n-1)}\hat{e}_j)'' + 2i\sqrt{\phi_j{}'}(\sqrt{\phi_j{}'}c_j{}^{(n)}\hat{e}_j)']\exp(ik_0\phi_j) = 0, \quad (14)$$

where, by definition, $c^{(-1)} = 0$.

Equation (14) permits an iterative soluton of $c_j{}^{(n)}$. The zeroth-order GOA corresponds to setting $n = 0$ in Eq. (14). Since $\hat{e}_1{}' = \Phi'\hat{e}_2$ and $\hat{e}_2{}' = -\Phi'\hat{e}_1$ and $\hat{e}_1$ and $\hat{e}_2$ form an orthonormal basis, Eq. (14) in the zeroth-order GOA can be put into the form

$$\sqrt{f}(\sqrt{f}c_1{}^{(0)})' - \Phi'c_2{}^{(0)}\exp(-ik_0\Delta\phi) = 0,$$

$$c_2{}^{(0)\prime} + \Phi'fc_1{}^{(0)}\exp(ik_0\Delta\phi) = 0, \quad (15)$$

where $\Delta\phi = \phi_1 - \phi_2$ and

$$f(x_3) = (n_e/\epsilon_{33}{}^{1/2}) = n_e/(n_e{}^2\cos^2\theta + n_0{}^2\sin^2\theta)^{1/2}. \quad (16)$$

Equations (15) can be written in a more symmetric form by letting

$$g_1 = \sqrt{f}c_1{}^{(0)}\exp(ik_0\Delta\phi/2),$$
$$g_2 = c_2{}^{(0)}\exp(-ik_0\Delta\phi/2). \quad (17)$$

We find that

$$g_1{}' = (i/2)n_ok_0(f-1)g_1 + (\Phi'/\sqrt{f})g_2,$$
$$g_2{}' = -\sqrt{f}\Phi'g_2 - (i/2)n_ok_0(f-1)g_2. \quad (18)$$

We notice that this set of differential equations has the same form as the one considered by Kubo and Nagata.[2] Therefore, by introducing the pseudo-Stokes parameters, defined as $P_0 = |g_1|^2 + |g_2|^2$, $P_1 = |g_1|^2 - |g_2|^2$, $P_2 = 2\,\mathrm{Re}(g_1g_2{}^*)$, and $P_3 = 2\,\mathrm{Im}(g_1g_2{}^*)$, Eqs. (18) can be rewritten in the vectorial form

$$\mathbf{P}' = (\boldsymbol{\omega}\times\mathbf{P}) - P_0\mathbf{T}, \quad (19)$$

where $\mathbf{P} = (P_1, P_2, P_3)$, $\boldsymbol{\omega} = (\omega_1, 0, \omega_3)$, and $\mathbf{T} = (0, T_2, 0)$, with

$$\omega_1 = n_ok_0(f-1),$$
$$\omega_3 = (\sqrt{f} + 1/\sqrt{f})\Phi',$$
$$T_2 = (\sqrt{f} - 1/\sqrt{f})\Phi'. \quad (20)$$

Notice that, by definition, $P_0^2 = P_1^2 + P_2^2 + P_3^2 = \mathbf{P}\cdot\mathbf{P}$, and hence $P_0' = 2\mathbf{P}\cdot\mathbf{P}' = -2(\mathbf{T}\cdot\mathbf{P})P_0$. The quantity f falls in the range $1 \le f \le n_e/n_o$, so that in the usual birefringent media T_2 is very small as compared with both ω_1 (containing the large quantity k_0) and ω_3, and P_0 is approximately constant.

If T_2 in Eq. (19) is neglected, P_0 = constant, and Eq. (19) assumes the simple form of an equation of motion for the precession of a vector with constant length P_0:

$$\mathbf{P}' = \boldsymbol{\omega}\times\mathbf{P}. \tag{21}$$

This then suggests that we can again use the Poincaré sphere to help to visualize the evolution of **P**. By retaining only the dominant terms in Eqs. (5) and (8), we can find also the zeroth-order GOA for the energy flux and angular momentum carried by the optical wave in the medium:

$$(8\pi/n_oc)I \cong |g_1|^2 + |g_2|^2 = P_0 \tag{22}$$

and

$$L_3 = -[n_o(f+1)/(16k_0\sqrt{f}]P_3. \tag{23}$$

Notice that Eq. (22) is consistent with the energy flux conservation law of Eq. (5). Therefore the last term on the right-hand side of Eq. (19) is actually a term pertaining to the higher-order GOA. We can now take Eq. (21) as our basic equation for describing the propagation of elliptically polarized light in an inhomogeneous, uniaxial medium in the zeroth-order GOA.

The polarization state of light, however, is directly described not by our pseudo-Stokes parameters P_j (j = 0, 1, 2, 3) but by the usual Stokes parameters S_j, defined as $S_0 = |E_1|^2 + |E_2|^2$, $S_1 = |E_1|^2 - |E_2|^2$, $S_2 = 2\,\mathrm{Re}(E_1E_2^*)$, $S_3 = 2\,\mathrm{Im}(E_1E_2^*)$. A simple calculation based only on definitions relates S_j to P_j:

$$\begin{aligned} S_0 &= (1/2f)[(f+1)P_0 - (f-1)P_1],\\ S_1 &= (1/2f)[(f+1)P_1 - (f-1)P_0]\cos 2\Phi - (P_2f^{-1/2})\sin 2\Phi,\\ S_2 &= (1/2f)[(f+1)P_1 - (f-1)P_0]\sin 2\Phi + (P_2f^{-1/2})\cos 2\Phi,\\ S_3 &= P_3f^{-1/2}. \end{aligned} \tag{24}$$

The inverses of Eqs. (24) are

$$\begin{aligned} P_0 &= \tfrac{1}{2}[(f+1)S_0 + (f-1)(S_1\cos 2\Phi + S_2\sin 2\Phi)],\\ P_1 &= \tfrac{1}{2}[(f-1)S_0 + (f+1)(S_1\cos 2\Phi + S_2\sin 2\Phi)],\\ P_2 &= (S_2\cos 2\Phi - S_1\sin 2\Phi)f^{1/2},\\ P_3 &= S_3f^{1/2}. \end{aligned} \tag{25}$$

It should be noted that in the limit of low-birefringence media ($n_o \simeq n_e$, $f \simeq 1$), P_j's are related to S_j's by a simple rotational transformation of an angle 2Φ around the x_3 axis. This approximation is usually exploited in the literature.[1,2]

One often likes to introduce the reduced parameters $p_j = P_j/P_0$ (j = 1, 2, 3). These three p_j parameters now describe a point on the Poincaré unit sphere. The motion of the point on the Poincaré sphere, as governed by Eq. (21), can then be used to visualize the evolution of the light polarization.

BOUNDARY EFFECTS

In order to solve Eq. (21), we need the initial values of **P** at the input face $x_3 = 0$ inside the medium. This requires the solution of transmission and reflection at the boundary surface. Let us assume that the sample is immersed in a homogeneous isotropic fluid having a refractive index n_1. Without loss of generality we can choose the x_1 axis along the extraordinary direction on the input face, so that $\phi(0) = 0$. The extraordinary and ordinary wave components obey the Fresnel equations for normal incidence:

$$\begin{aligned} E_1^{\mathrm{R}} &= r_eE_1^{\mathrm{I}}, \quad E_1^{\mathrm{T}} = (1+r_e)E_1^{\mathrm{I}},\\ E_2^{\mathrm{R}} &= r_oE_2^{\mathrm{I}}, \quad E_2^{\mathrm{T}} = (1+r_o)E_2^{\mathrm{I}}, \end{aligned} \tag{26}$$

where the superscripts I, R, and T refer to the incident, reflected, and transmitted waves, respectively, and

$$\begin{aligned} r_e &= (n_1 - n_of_o)/(n_1 + n_of_o) \qquad [f_o = f(0)],\\ r_o &= (n_1 - n_o)/(n_1 + n_o) \end{aligned} \tag{27}$$

are the extraordinary and ordinary Fresnel reflection factors, respectively. Equations (26) and (27) determine the reflected and transmitted fields at the input face in terms of the incident field. The Stokes parameters, and hence the pseudo-Stokes parameters, can be calculated.

A long but straightforward calculation yields for $P_j(0)$

$$\begin{aligned} P_0(0) &= (n_1/n_o)S_0^{\mathrm{I}} - (n_1/2n_o)[(r_e^2 + r_o^2)S_0^{\mathrm{I}} + (r_e^2 - r_o^2)S_1^{\mathrm{I}}],\\ P_1(0) &= (n_1/n_o)S_1^{\mathrm{I}} - (n_1/2n_o)[(r_e^2 - r_o^2)S_0^{\mathrm{I}} + (r_e^2 + r_o^2)S_1^{\mathrm{I}}],\\ P_2(0) &= f_o^{1/2}(1+r_e)(1+r_o)S_2^{\mathrm{I}},\\ P_3(0) &= f_o^{1/2}(1+r_e)(1+r_o)S_3^{\mathrm{I}}, \end{aligned} \tag{28}$$

where we have used Eqs. (25) at $x_3 = 0$ [with $\Phi(0) = 0$] as well as the identities

$$(1 - r_e)/(1 + r_e) = n_of_o/n_1$$

and

$$(1 - r_o)/(1 + r_o) = n_o/n_1.$$

It is worth noting that the first and the last of Eqs. (28) have a direct physical meaning. The Stokes parameters S_0 of both the incident and reflected waves are related to their respective intensities I by $S_0 = (8\pi/n_1c)I$.

Observing that, by definition,

$$S_0^{\mathrm{R}} = |E_1^{\mathrm{R}}|^2 + |E_2^{\mathrm{R}}|^2 = \tfrac{1}{2}[(r_e^2 + r_o^2)S^{\mathrm{I}} + (r_e^2 - r_o^2)S_1^{\mathrm{I}}], \tag{29}$$

the first of Eqs. (28) can be transformed into

$$P_0(0) = (8\pi/n_oc)(I^{\mathrm{I}} - I^{\mathrm{R}}) = (8\pi/n_oc)I^{\mathrm{T}}. \tag{30}$$

Therefore we can identify the constant I in Eq. (22) with the light intensity in the sample. Equation (30) describes the energy flux conservation at the boundary surface.

Analogously, by using the observation that $S_3^{\mathrm{R}} = r_er_oS_3^{\mathrm{I}}$ and using the identity

$$(1 - r_er_o)/[(1 + r_o)(1 + r_e)] = n_o(f_o + 1)/2n_1,$$

the last of Eqs. (28) can be transformed into

$$P_3(0) = [2n_1\sqrt{f_o}/n_o(f_o + 1)](S_3^{\mathrm{I}} - S_3^{\mathrm{R}}). \qquad (31)$$

The Stokes parameters S_3 of both the incident and reflected waves are related to their respective angular momentum fluxes L_3 by $S_3 = -(8\pi k_0/n_1)L_3$. By using the GOA expression of Eq. (23) for L_3 in the sample, Eq. (31) can then be cast into the form

$$L_3^{\mathrm{T}} = L_3^{\mathrm{I}} - L_3^{\mathrm{R}}. \qquad (32)$$

Equation (32) shows that the angular momentum is also conserved at the boundary surface.

We notice that if the index-matching condition $n_1 = n_o$ is satisfied, then $r_o = 0$ and $S_3^R = 0$. In this case $L_3^{\mathrm{T}} = L_3^{\mathrm{I}}$, and L_3 is totally transmitted across the input boundary. This is expected, since the reflected wave must be linearly polarized along the extraordinary direction, and hence it carries no net angular momentum along with it.

Finally, we would mention that Eqs. (24) can also be used to obtain the Stokes parameters of the output wave, once the $P_j(d)$ at $x_3 = d$ are found from Eq. (21).

CONCLUSION

We have shown that the zeroth-order geometric optics approximation can be used to describe the propagation of normally incident, elliptically polarized light in an inhomogeneous, locally uniaxial medium. The approximation corresponds to finding an asymptotic solution of the wave equation in the short-wavelength limit. It is found that a set of pseudo-Stokes parameters, linearly related to the usual Stokes parameters, can be defined to characterize the propagating wave. They obey a simple vectorial differential equation that governs the evolution of these parameters as the wave propagates in the medium. The solution of the equation can be described by the motion of a point on the Poincaré sphere. The result of the geometric optics approximation is sufficiently accurate that conservation of energy and angular momentum still appears to be valid both in the bulk and at the boundary surfaces.

ACKNOWLEDGMENTS

E. Santamoto acknowledges the support of a Consiglio Nazionale delle Ricerche–North Atlantic Treaty Organization fellowship during his visit at the University of California, Berkeley. This work was supported by National Science Foundation Solid State Chemistry grant DMR84-14053.

REFERENCES AND NOTES

1. See, for example, J. R. Wait, *Electromagnetic Waves in Stratified Media* (Pergamon, New York, 1970); L. M. Brekhovskikh, *Waves in Layered Media*, translated by O. Liberman (Academic, New York, 1980).
2. H. K. Aben, Expl. Mech. **5**, 13–22 (1966); H. Kubo and R. Nagata, J. Opt. Soc. Am. **73**, 1719–1724 (1983).
3. B. Ya Zel'dovich, N. V. Tabiryan, and Yu. S. Chilingaryan, Zh. Eksp. Teor. Fiz. **81**, 72–83 (1981) [Sov. Phys. JETP **54**, 32–37 (1981)]; H. L. Ong, Phys. Rev. A **32**, 1098–1105 (1985).
4. H. L. Ong and R. B. Meyer, J. Opt. Soc. Am. A **2**, 198–201 (1985).
5. E. Santamato, B. Daino, M. Romagnoli, M. Settembre, and Y. R. Shen, Phys. Rev. Lett. **57**, 2423 (1986). The radiation torque has been measured in a clever experiment by R. A. Beth, Phys. Rev. **50**, 115–125 (1936).
6. See J. D. Jackson, *Classical Electrodynamics* (Wiley, New York, 1962); J. M. Jauch and F. Rohrlich, *The Theory of Photons and Electrons* (Addison-Wesley, Cambridge, Mass., 1955).

Study of Phase-Matched Normal and Umklapp Third-Harmonic- Generation Processes in Cholesteric Liquid Crystals*

J. W. Shelton† and Y. R. Shen

Department of Physics, University of California, Berkeley, California 94720 *and*
Inorganic Materials Research Division, Lawrence Berkeley Laboratory, Berkeley, California 94720

(Received 4 October 1971)

Using the model of Oseen and de Vries, we show that phase matching of optical third-harmonic generation can be achieved in a cholesteric liquid crystal with the help of the lattice momentum. Many different collinear phase-matching conditions exist. In some cases, the phase-matched third harmonic is generated in the same direction as the fundamental, and in some other cases, it is generated in the opposite direction. In many other cases, phase-matched third-harmonic generation requires the simultaneous presence of fundamental waves propagating in opposite directions. Analogous to electron-electron interaction in a periodic lattice, these processes can be identified as coherent, normal, and umklapp third-harmonic-generation processes. Experiments using a mode-locked Nd laser as the fundamental source verify the existence of most of the predicted phase-matching conditions. Our results agree well with the theoretical predictions.

I. INTRODUCTION

Phase matching in nonlinear optical processes has long been a subject of interest. It not only has led to the successful development of useful nonlinear optical devices, but it has also helped in gaining information about linear and nonlinear optical properties of various materials. The usual technique of phase matching is to compensate the color dispersion by either linear birefringence,[1] or circular birefringence,[2] or anomalous dispersion of a dye.[3] However, we have recently demonstrated that phase matching can also be achieved in a periodic medium through compensation of the color dispersion by the lattice momentum.[4] The particular periodic medium we deal with is the cholesteric liquid-crystalline material characterized by helical structure.[5]

In recent years, liquid crystals have become a field of interest for many research workers. While there is a large amount of work reported on their linear optical properties, reports on their nonlinear optical properties have been extremely rare. It was thought that since liquid crystals have large inherent birefringence, phase matching of harmonic generation would be possible in these materials and they could then be used as effective harmonic generators.[6] Experimental investigations of second-harmonic generation in nematic, smectic, and cholesteric liquid crystals have been made,[6-9] but there is no evidence of any discernible second-harmonic signal, suggesting that the molecular arrangement in these materials has an over-all inversion symmetry.

Third-harmonic generation in liquid crystals is clearly not forbidden by symmetry. It has in fact been observed in a number of liquid crystals by Goldberg and Schnur.[8,9] However, their attempt to achieve phase matching of third-harmonic generation in cholesteric liquid crystals has not been successful. Recently, using the model of Oseen[10] and de Vries[11] for cholesteric liquid crystals, we have been able to predict the helical pitch values at which collinear phase matching of third-harmonic generation would occur, and by tuning the pitch with temperature, have then been successful in observing the phase-matching peaks as predicted.[4] In this paper, we would like to give a complete account of our work on the problem. We show that for third-harmonic generation in a cholesteric liquid crystal, there could exist 15 different collinear phase-matching conditions. We can divide them into three general classes. In the first class, both the fundamental and the phase-matched third-harmonic waves are propagating in the same direction as we normally expect. In the second class, the phase-matched third harmonic is generated in a direction opposite to the fundamental. In the third class, the phase-matched third harmonic is generated only when the fundamental waves are propagating simultaneously in both forward and backward directions. It is obvious from the requirement of momentum conservation that in the last two cases, phase matching can only be achieved if the momentum mismatch between the fundamental and the third harmonic can be absorbed by the medium. In fact, in most cases here, phase matching is achieved through compensation of color dispersion by the lattice momentum of the periodic medium as we shall see later.

In Sec. II, we give a brief review of the linear optical properties of a cholesteric liquid crystal. We point out that propagation of light waves along the helical axis in such a medium is analogous to propagation of electron waves in a one-dimensional periodic lattice. What we have here are Bloch photons instead of Bloch electrons. In Sec. III, we describe the theory of third-harmonic generation in a cholesteric liquid crystal. We show that collinear phase matching is possible with the help of the lattice momentum for many different mode combinations. Clearly, such phase-matched third-harmonic-generation processes fall into the category of nonlinear Bragg diffraction.[12] In analogy to umklapp processes of electrons in solids,[13] they can also be called coherent, optical, umklapp processes. In Sec. IV, we describe the experimental arrangement and show that the results agree with theoretical predictions. Finally, in Sec. V, we discuss the effects of many experimental difficulties on the observed phase-matched third-harmonic generation.

II. THEORY OF LINEAR OPTICAL WAVE PROPAGATION IN CHOLESTERIC LIQUID CRYSTAL

Liquid crystals are generally composed of long, anisotropic organic molecules.[5] The so-called nematic structure has long-range order in molecular orientation with the long molecular axes aligned more or less parallel to one another, although the molecules are fairly free to translate and to rotate about their long axes. The cholesteric structure is simply a twisted version of the nematic. It is formed by twisting the nematic structure about an axis normal to the molecular alignment so as to have an over-all helical structure. The helical pitch is a function of composition, temperature, fields, and other external perturbations, and usually varies from ± 0.2 μm to essentially infinity. Because of their helical structure, cholesteric liquid crystals have some interesting optical properties. In particular, the optical Bragg reflection gives rise to their grating characteristics. We shall see in Sec. III that these materials also have interesting nonlinear optical properties arising from nonlinear optical Bragg reflection.

The theory of linear wave propagation in a cholesteric liquid crystal has been well developed by Oseen[10] and de Vries.[11] They assume that a cholesteric liquid crystal can be treated as a twisted birefringent medium characterized by a dielectric tensor $\overleftrightarrow{\epsilon}(z)$ periodic in z:

$$\overleftrightarrow{\epsilon}(z)=\begin{pmatrix} \bar{\epsilon}[1+\alpha\cos(4\pi z/p)] & \bar{\epsilon}\alpha\sin(4\pi z/p) & 0 \\ \bar{\epsilon}\alpha\sin(4\pi z/p) & \bar{\epsilon}[1-\alpha\cos(4\pi z/p)] & 0 \\ 0 & 0 & \epsilon_\eta \end{pmatrix}, \tag{1}$$

where

$$\bar{\epsilon}=(\epsilon_\xi+\epsilon_\eta)/2, \qquad \alpha=(\epsilon_\xi-\epsilon_\eta)/2\bar{\epsilon}.$$

ϵ_ξ and ϵ_η are the principal dielectric constants in the directions parallel and perpendicular to the molecular alignment, respectively, and p is the helical pitch. Note that in this model, the dielectric tensor $\overleftrightarrow{\epsilon}(z)$ has a period of $\frac{1}{2}p$ rather than p.

Although the general solution for light propagation in any direction is available,[14] we shall consider here only propagation along the helical (z) axis. The corresponding wave equation for monochromatic light is given by

$$\left(\frac{\partial^2}{\partial z^2}+\frac{\omega^2}{c^2}\overleftrightarrow{\epsilon}(z)\right)\cdot\vec{E}(z)=0\ . \tag{2}$$

This equation with a position-dependent dielectric constant is most easily solved by first applying to the equation a rotational transformation

$$R(\theta=2\pi z/p)=\begin{pmatrix} \cos\theta & \sin\theta & 0 \\ -\sin\theta & \cos\theta & 0 \\ 0 & 0 & 1 \end{pmatrix}\ . \tag{3}$$

Physically, the rotational transformation is to untwist the twisted helical structure, so that in the rotating frame the medium now appears to be a simple birefringent material with a dielectric tensor $\overleftrightarrow{\epsilon}_T=\mathrm{R}:\overleftrightarrow{\epsilon}(z):\mathrm{R}^{-1}$ independent of z. This technique is analogous to the rotational-transformation technique used in magnetic resonance.[15] After the transformation, Eq. (2) becomes

$$\left[\frac{\partial^2}{\partial z^2}+\frac{4\pi}{p}\overleftrightarrow{\sigma}\frac{\partial}{\partial z}-\left(\frac{2\pi}{p}\right)^2+\frac{\omega^2}{c^2}\overleftrightarrow{\epsilon}_T\right]\cdot\vec{E}_T=0\ , \tag{4}$$

where

$$\overleftrightarrow{\sigma}=\begin{pmatrix} 0 & -1 & 0 \\ 1 & 0 & 0 \\ 0 & 0 & 0 \end{pmatrix}, \qquad \overleftrightarrow{\epsilon}_T=\begin{pmatrix} \epsilon_\xi & 0 & 0 \\ 0 & \epsilon_\eta & 0 \\ 0 & 0 & \epsilon_\eta \end{pmatrix}\ .$$

The above equation can now be solved readily. Let

$$\vec{E}_T=\sum_{j=\pm}(\mathcal{E}_\xi\hat{\xi}+\mathcal{E}_\eta\hat{\eta})_j\,e^{ik_jz-i\omega t}\ , \tag{5}$$

where $\hat{\xi}=\hat{x}\cos(2\pi z/p)+\hat{y}\sin(2\pi z/p)$ and $\hat{\eta}=-\hat{x}\times\sin(2\pi z/p)+\hat{y}\cos(2\pi z/p)$ are unit vectors parallel and perpendicular to the molecular alignment in the plane at z, respectively. We then find

$$\begin{aligned} &k_\pm^{(\omega)}=(\omega\bar{\epsilon}^{1/2}/c)m_\pm\ ,\\ &(m_\pm)^2=(\lambda'^2+1)\pm(4\lambda'^2+\alpha^2)^{1/2}\ ,\\ &\lambda'=2\pi c/\omega p\bar{\epsilon}^{1/2}\\ &(\mathcal{E}_\eta/\mathcal{E}_\xi)_\pm\equiv if_\pm=i2m_\pm\lambda'/[m_\pm^2+\lambda'^2+(\alpha-1)]\ . \end{aligned} \tag{6}$$

The magnitudes of $\mathcal{E}_{\xi+}$ and $\mathcal{E}_{\xi-}$ are determined by the boundary conditions. The Poynting vectors for the two modes are

$$\vec{S}_\pm=(|\mathcal{E}_\pm|^2c\bar{\epsilon}^{1/2}/2\pi)\frac{\mathrm{Re}[q+|f|^2/q]_\pm}{1+|f_\pm|^2}\hat{z}\ , \tag{7}$$

where

$$q_\pm=(m_\pm/\bar{\epsilon}^{1/2})-\lambda'f_\pm$$

and

$$|\mathcal{E}_\pm|^2\equiv(|\mathcal{E}_\xi|^2+|\mathcal{E}_\eta|^2)_\pm\ .$$

It can be shown that for the minus mode in the region of $\lambda'^2 > 1 + |\alpha|$, a forward propagating wave (with $\vec{S}_-$ in the $+\hat{z}$ direction) actually corresponds to $m_- < 0$ (indicating a negative phase velocity in the rotating frame). We also remark that here two modes are generally not orthogonal since $\vec{E}_{T+} \cdot \vec{E}_{T-} \neq 0$. This nonorthogonal property does not affect the linear wave propagation, but affects the nonlinear wave propagation slightly as we shall see in Sec. III.

The above solutions describe the characteristics of linear wave propagation in a cholesteric liquid crystal along the helical axis. In particular, if $|\lambda'^2 - 1| < |\alpha|$, then m_- is purely imaginary, and the corresponding wave should be totally reflected. We realize that $\lambda' \approx 1$ or $\lambda = 2\pi c/\omega\bar{\epsilon}^{1/2} = p$ ($|\alpha| \ll 1$ usually) is just the condition for Bragg reflection from the helical structure. If we transform Eq. (5) back into the lab frame, we have

$$\vec{E}_\pm = \overleftrightarrow{R}^{-1} \cdot \vec{E}_{T\pm}$$

$$= \left(\frac{\hat{x} + i\hat{y}}{\sqrt{2}} A_{R\pm} + \frac{\hat{x} - i\hat{y}}{\sqrt{2}} A_{L\pm} e^{i4\pi z/p}\right) \mathcal{E}_{t\pm} \times \exp\left[i\left(\frac{m_\pm \omega \bar{\epsilon}^{1/2}}{c} - \frac{2\pi}{p}\right) z - i\omega t\right] , \quad (8)$$

where

$$A_{L\pm} = (1 + f_\pm)/\sqrt{2}, \quad A_{R\pm} = (1 - f_\pm)/\sqrt{2} .$$

This equation shows that in the lab frame, each mode is a superposition of two components, a left circularly polarized one with an effective refractive index $n_{L\pm} = (m_\pm - \lambda')\bar{\epsilon}^{1/2}$ and a right circularly polarized one with $n_{R\pm} = (m_\pm + \lambda')\bar{\epsilon}^{1/2}$. The ratios of A_R to A_L for the two modes are plotted in Fig. 1. We notice that the plus mode is always nearly circularly polarized as long as $\lambda'^2 \gg \alpha^2$, and the minus mode is nearly circularly polarized everywhere except for $\lambda'^2 < \alpha^2$ and for $1 - |\alpha| < \lambda'^2 < 1 + |\alpha|$. For a

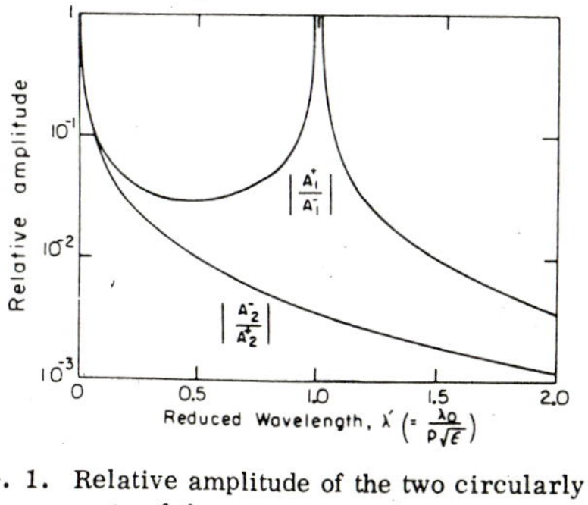

FIG. 1. Relative amplitude of the two circularly polarized components of the two propagating modes in the lab frame as a function of the reduced wavelength. Here, $\alpha = 0.03$ and the helical structure is assumed right handed. For a left-handed structure, the ratios are inverted.

right-handed cholesteric liquid crystal, the dominant components for the two modes are A_{L+} and A_{R-}, respectively. Since these two dominant circular components have different phase velocities ($n_{L+} \neq n_{R-}$), the medium possess an optical activity

$$R(\text{rad/unit length}) = (\omega/2c)\,(n_{R-} - n_{L+}) \simeq -\pi\alpha^2/4p\lambda'^2\,(1 - \lambda'^2) . \quad (9)$$

It is well known that the optical activity of a cholesteric liquid crystal can be as high as 300 rad/cm.[5] The above equation is of course not valid for $\lambda' \sim 0$ and 1, since then our approximation breaks down.

We can also understand the problem by recognizing the analog between this problem and the problem of electron wave propagation in a one-dimensional periodic lattice. The latter problem is well presented in elementary textbooks on solid-state physics.[13] We can use the same approach to solve our problem directly in the lab frame.

According to Bloch's theorem,[13] the wave solution can be written in the form

$$\vec{E} = \vec{u}(\kappa, z)\, e^{i\kappa z - i\omega t} , \quad (10)$$

where $\vec{u}(\kappa, z)$ is a periodic function of z with a period of $\frac{1}{2}p$, and κ is limited to the first Brillouin zone between $2\pi/p$ and $-2\pi/p$. Substitution of Eq. (10) into Eq. (2) yields

$$\left(\frac{\partial^2}{\partial z^2} + 2i\kappa\frac{\partial}{\partial z} - \kappa^2 + \frac{\omega^2}{c^2}\overleftrightarrow{\epsilon}(z)\right) \cdot \vec{u}(\kappa, z) = 0 . \quad (11)$$

We can expand both $\vec{u}(\kappa, z)$ and $\overleftrightarrow{\epsilon}(z)$ into Fourier series

$$\vec{u}(\kappa, z) = \sum_G \vec{C}_G(\kappa)\, e^{iGz} ,$$
$$\overleftrightarrow{\epsilon}(z) = \sum_G \overleftrightarrow{\epsilon}_G\, e^{iGz} , \quad (12)$$

where $G = N(4\pi/p)$, with N being any integer, is a reciprocal-lattice vector. Equation (11) then becomes

$$(\kappa + G)^2 \vec{C}_G - (\omega^2/c^2) \sum_{G'} \overleftrightarrow{\epsilon}_{G'} \cdot \vec{C}_{G-G'} = 0 . \quad (13)$$

This set of equations with different G's is most easily solved by expressing the vectors and tensors in the circular coordinates $(\hat{x} + i\hat{y})/\sqrt{2}$, $(\hat{x} - i\hat{y})/\sqrt{2}$, and $\hat{z}$. There, we have

$$\overleftrightarrow{\epsilon}(z) = \bar{\epsilon} \begin{pmatrix} 1 & \alpha e^{-i4\pi z/p} & 0 \\ \alpha e^{i4\pi z/p} & 1 & 0 \\ 0 & 0 & 1 \end{pmatrix} , \quad (14)$$

which has only three Fourier components $\overleftrightarrow{\epsilon}_G$ with $G = 0$ and $\pm 4\pi/p$. It is then simple to solve Eq. (13). We obtain

$$(\kappa_\pm + G + 2\pi/p)^2 = (\omega^2 \bar{\epsilon}/c^2)\,[1 + \lambda'^2 \pm (4\lambda'^2 + \alpha^2)^{1/2}] ,$$

$$\vec{E} = [(1 + f_\pm)(\hat{x} + i\hat{y}) + (1 - f_\pm)(\hat{x} - i\hat{y})\, e^{i4\pi z/p}]$$

$$\times \tfrac{1}{2}\mathcal{E}_{\xi\pm}\, e^{i(\kappa_\pm + G)z - i\omega t} \ . \quad (15)$$

These solutions are identical to those we obtained before in Eqs. (6) and (8), with $\kappa_\pm + G + 2\pi/p = m_\pm \omega \overline{\epsilon}^{1/2}/c$, remembering that $|\kappa_\pm| \leq 2\pi/p$.

We can find the dispersion curves for the two modes from Eq. (15). Either extended or reduced zone schemes can be used. Analogous to the band gap in the electronic band structure, the dispersion curve for the negative mode has a gap between $\lambda'^2 = 1 - |\alpha|$ and $1 + |\alpha|$ at $\kappa = \pm 2\pi/p$. This is just the reflection band of the cholesteric liquid crystal. However, in contrast to the case of electrons in a periodic lattice,[13] we have here only one gap for only one of the two normal modes.

Usually, one prepares a sample of cholesteric liquid crystal by holding the material between two glass substrates. In order to know how each mode is being excited, we have to solve the problem with the proper boundary conditions. This has been done by de Vries.[11] Using his result, we plot in Fig. 2, as a function of λ', the polarizations of the incident fields which should feed exclusively into each of the two modes. For $\lambda'^2 \gg |\alpha|$, both of them are close to circular polarization. The polarization of the incident field which feeds into the minus mode should have the same handedness as the helicity of the cholesteric structure.

Knowing the characteristics of linear wave propagation, we can then discuss the nonlinear optical effects in a cholesteric liquid crystal. In Sec. III, we shall consider the problem of third-harmonic generation along the helical axis in such a medium. Emphasis is on the derivation of collinear phase-matching conditions.

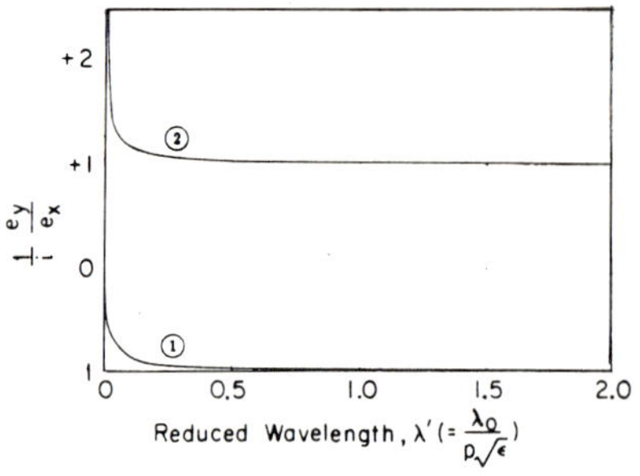

FIG. 2. Polarization of the normally incident light which feeds exclusively into the two propagating modes in a left-handed cholesteric medium with $\alpha = 0.03$ and $\overline{\epsilon} = 2.25$. Curves 1 and 2 are for plus and minus modes, respectively. e_y/e_x indicates the ratio of the field components along $\hat{y}$ and $\hat{x}$.

III. THEORY OF THIRD-HARMONIC GENERATION IN CHOLESTERIC LIQUID CRYSTALS

A. Third-Harmonic Generation

Third-harmonic generation along the helical axis in a cholesteric medium is governed by the nonlinear wave equation

$$\left[\frac{\partial^2}{\partial z^2} + \left(\frac{3\omega}{c}\right)^2 \overleftrightarrow{\epsilon}(z, 3\omega)\right] \cdot \vec{E}(z, 3\omega) = -4\pi\left(\frac{3\omega}{c}\right)^2 \vec{P}^{(3)}(z, 3\omega) , \quad (16)$$

where the nonlinear polarization $\vec{P}^{(3)}(z, 3\omega)$ has the form

$$\vec{P}^{(3)}(z, 3\omega) = \overleftrightarrow{\chi}^{(3)}(z, 3\omega) \vdots \vec{E}(z, \omega)\, \vec{E}(z, \omega)\, \vec{E}(z, \omega) \ . \quad (17)$$

The nonlinear susceptibility tensor $\overleftrightarrow{\chi}^{(3)}(z, 3\omega)$ is also a periodic function of z.

In the rotating frame, Eq. (16) becomes

$$\left[\frac{\partial^2}{\partial z^2} + \frac{4\pi}{p}\overleftrightarrow{\sigma}\frac{\partial}{\partial z} - \left(\frac{2\pi}{p}\right)^2 + \left(\frac{3\omega}{c}\right)^2 \overleftrightarrow{\epsilon}_T\right] \cdot \vec{E}_T(z, 3\omega) = -4\pi\left(\frac{3\omega}{c}\right)^2 \vec{P}_T^{(3)} \quad (18)$$

with

$$\vec{P}_T^{(3)} = -4\pi(3\omega/c)^2 \overleftrightarrow{\chi}_T^{(3)} \cdot \vec{E}_T(z, \omega)\, \vec{E}_T(z, \omega)\, \vec{E}_T(z, \omega) \ .$$

The transformed $\overleftrightarrow{\chi}_T^{(3)}$ is now independent of z and has the form for a birefringent material with four independent elements.[16] Then, the components of $\vec{P}_T^{(3)}$ along $\hat{\xi}$ and $\hat{\eta}$ are

$$\begin{aligned} P_{T\xi}^{(3)}(3\omega) &= C_{11} E_\xi(\omega)\, E_\xi(\omega) E_\xi(\omega) + C_{12} E_\xi(\omega)\, E_\eta(\omega)\, E_\eta(\omega) , \\ P_{T\eta}^{(3)}(3\omega) &= C_{21} E_\eta(\omega)\, E_\xi(\omega) E_\xi(\omega) + C_{22} E_\eta(\omega)\, E_\eta(\omega)\, E_\eta(\omega) . \end{aligned} \quad (19)$$

To solve Eq. (18), we use the usual slowly varying-amplitude approximation.[17] Let

$$\vec{E}_T(z, 3\omega) = [\hat{e}_+ \mathcal{E}_+(z, 3\omega)\, e^{ik_+ z} + \hat{e}_- \mathcal{E}_-(z, 3\omega)\, e^{ik_- z}] e^{-i\omega t} , \quad (20)$$

where

$$\left|\frac{\partial^2 \mathcal{E}_+}{\partial z^2}\right| \ll \left|2k_+ \frac{\partial \mathcal{E}_+}{\partial z}\right| , \qquad \left|\frac{\partial^2 \mathcal{E}_-}{\partial z^2}\right| \ll \left|2k_- \frac{\partial \mathcal{E}_-}{\partial z}\right| .$$

The expressions for the unit vectors $\hat{e}_+$ and $\hat{e}_-$, obtained from Eqs. (5) and (6), are

$$\hat{e}_\pm = [1/(1 + |f_\pm|^2)^{1/2}]\, (\hat{\xi} + i f_\pm \hat{\eta}) \ . \quad (21)$$

Then, substitution of Eq. (20) into Eq. (18) yields

$$\sum_j \left(2ik_j + \frac{4\pi}{p}\overleftrightarrow{\sigma}\right)\hat{e}_j\, e^{ik_j z - i3\omega t}\, \frac{\partial}{\partial z}\mathcal{E}_j(z, 3\omega) = -4\pi\left(\frac{3\omega}{c}\right)^2 \vec{P}_T^{(3)} . \quad (22)$$

The scalar products of $\hat{e}_1^\dagger$ and $\hat{e}_2^\dagger$ with the above equation give

$$A\, e^{ik_+z} \frac{\partial}{\partial z}\mathcal{E}_+ + B\, e^{ik_-z}\frac{\partial}{\partial z}\mathcal{E}_- = -4\pi\left(\frac{3\omega}{c}\right)^2 \hat{e}_+^\dagger\cdot \vec{P}_T^{(3)} ,$$

$$C\, e^{ik_+z} \frac{\partial}{\partial z}\mathcal{E}_+ + D\, e^{ik_-z}\frac{\partial}{\partial z}\mathcal{E}_- = -4\pi\left(\frac{3\omega}{c}\right)^2 \hat{e}_-^\dagger\cdot \vec{P}_T^{(3)} , \quad (23)$$

where

$A = 2ik_+ + (4\pi/p)\hat{e}_+^\dagger\cdot\overleftrightarrow{\sigma}\cdot\hat{e}_+$, $B = \hat{e}_+^\dagger\cdot[2ik_- + (4\pi/p)\overleftrightarrow{\sigma}]\cdot\hat{e}_-$,

$C = \hat{e}_-^\dagger\cdot[2ik_+ + (4\pi/p)\overleftrightarrow{\sigma}]\cdot\hat{e}_+$, $D = 2ik_- + (4\pi/p)\hat{e}_-^\dagger\cdot\overleftrightarrow{\sigma}\cdot\hat{e}_-$.

From Eq. (23), we can easily obtain expressions for $\partial\mathcal{E}_+/\partial z$ and $\partial\mathcal{E}_-/\partial z$. Then, in the low depletion limit using the parametric approximations,[17] we find

$$\mathcal{E}_+(z=l, 3\omega) = 4\pi\left(\frac{3\omega}{c}\right)^2\left(\frac{D\hat{e}_+^\dagger - B\hat{e}_-^\dagger}{AD-BC}\right)\cdot\overleftrightarrow{\chi}_T^{(3)} : \sum_{l,m,n=\pm} \hat{e}_l\hat{e}_m\hat{e}_n\mathcal{E}_l(\omega)\mathcal{E}_m(\omega)\mathcal{E}_n(\omega)\,\frac{\sin(\frac{1}{2}\Delta k_{+lmn}l)}{\frac{1}{2}\Delta k_{+lmn}}\, e^{i\Delta k_{+lmn}l/2} ,$$

$$\mathcal{E}_-(z=l, 3\omega) = 4\pi\left(\frac{3\omega}{c}\right)\left(\frac{A\hat{e}_-^\dagger - C\hat{e}_+^\dagger}{AD-BC}\right)\cdot\overleftrightarrow{\chi}_T^{(3)} : \sum_{l,m,n=\pm} \hat{e}_l\hat{e}_m\hat{e}_n\mathcal{E}_l(\omega)\mathcal{E}_m(\omega)\mathcal{E}_n(\omega)\,\frac{\sin(\frac{1}{2}\Delta k_{-lmn}l)}{\frac{1}{2}\Delta k_{-lmn}}\, e^{i\Delta k_{-lmn}l/2} , \quad (24)$$

where

$$\Delta k_{\pm lmn} = k_l^{(\omega)} + k_m^{(\omega)} + k_n^{(\omega)} - k_\pm^{(3\omega)} .$$

Note that the waves can propagate along the helical axis in either direction, and, correspondingly, the wave vectors k can be positive or negative. The generated third-harmonic intensity is given by Eq. (7), which is of course independent of the coordinate system we choose. If the beam has a finite cross section, then the total third-harmonic power is[18]

$$I_\pm(3\omega) = \int S_\pm(3\omega, x, y)\,dx\,dy . \quad (25)$$

B. Phase Matching or Conservation of Linear Momentum

Third-harmonic generation is most efficient when a certain phase-matching condition $\Delta k_{\pm lmn} = 0$ is satisfied. When this happens, the corresponding phase-matched term dominates over the other terms in the summation in Eq. (24). Since we allow both the fundamental and the third harmonics to be in either mode and to propagate along the z axis in either direction, we can have many different phase-matching conditions in a cholesteric liquid crystal by taking all possible sign combinations in the following equation:

$$\pm|k_\pm^{(3\omega)}| = \pm|k_\pm^{(\omega)}| \pm |k_\pm^{(\omega)}| \pm |k_\pm^{(\omega)}| . \quad (26)$$

Through the dependence of k on the pitch p, the phase-matching conditions can be achieved by adjusting p. However, Eq. (26) cannot be satisfied for all the 40 different sign combinations. If we assume $\bar{\epsilon}^{(3\omega)} > \bar{\epsilon}^{(\omega)}$, then only 12 of them are possible, and another three can be conditionally satisfied. They are listed in Table I, where a bar over k indicates a backward propagating mode ($\bar{k} = -k$).

Physically, phase matching is achieved here through compensation of momentum mismatch by the lattice momentum. This can be seen as follows. We realize from Eq. (6) that for the plus mode in the region $\lambda'^2 \gg \alpha^2$ and for the minus mode in the region $\alpha^2 \ll \lambda'^2$ and $(1-\lambda')^2 \gg \alpha^2/4\lambda'$, the wave vectors $|k_\pm|$ can be well approximated by

$$|k_\pm| = |k_0 \pm 2\pi/p| , \quad (27)$$

where $k_0^{(\omega)} = \omega\bar{\epsilon}^{1/2}/c$ is the average wave vector. This approximation is excellent for the first 11 mode combinations in Table I, and is marginal for the 12th, since α is usually about 0.03. Consequently, using Eq. (27), we can rewrite these phase-matching conditions in terms of $k_0^{(\omega)}$, $k_0^{(3\omega)}$, and the unit lattice momentum $Q = 4\pi/p$ as shown in the second column of Table I. We notice that in all cases the average momentum mismatch between the fundamental and the third harmonic is compensated by the lattice momentum Q or $2Q$. From these expressions we can easily calculate the values of p satisfying the phase-matching conditions if $\bar{\epsilon}^{(3\omega)}$ and $\bar{\epsilon}^{(\omega)}$ for the material are known. For the materials we have used in our experimental investigation, $\bar{\epsilon}^{(3\omega)}$ and $\bar{\epsilon}^{(\omega)}$ are typically 2.30 and 2.18, respectively, and $\alpha \approx 0.03$. The corresponding pitches for phase matching of different mode combinations are given in the last column of Table I for illustration.

The approximation of Eq. (27) is not valid for the last three mode combinations in Table I. It is clear that since the momentum mismatches in these cases are small, the pitches for phase matching must be long. Then, as seen from Eq. (6), the birefringence factor α can no longer be neglected in the expression for $k_\pm$. In fact, the phase-matching conditions for these last three mode combinations may not always be satisfied. The conditions for phase matching in these cases are given in Table I.

The more rigorous interpretation of phase matching or momentum matching here is to use the concept of waves propagating in a periodic medium.

TABLE I. Mode combinations for possible phase matching of harmonic generation assuming $\bar{\epsilon}^{(3\omega)} > \bar{\epsilon}^{(\omega)}$. The phase-matching conditions are expressed in terms of the average wave vectors using the approximation of Eq. (27) in the second column, and in terms of the wave vectors for Bloch wave functions in the third column. The superbars indicate backward propagating modes. For typical cholesteric materials with $\bar{\epsilon}^{(\omega)} = 2.18$, $\bar{\epsilon}^{(3\omega)} = 2.30$, and $\alpha = 0.03$, the approximate pitches for the various phase-matching cases are given in the last column.

Mode combination for phase matching			Pitch for phase matching (μm)
1. $k_+^{(\omega)} + 2k_-^{(\omega)} = \bar{k}_-^{(3\omega)}$	$3k_0^{(\omega)} = \bar{k}_0^{(3\omega)} + Q$	$\kappa_+^{(\omega)} + 2\kappa_-^{(\omega)} = \bar{\kappa}_-^{(3\omega)}$	0.24
2. $3k_-^{(\omega)} = \bar{\kappa}_+^{(3\omega)}$		$3\kappa_-^{(\omega)} = \bar{\kappa}_+^{(3\omega)} + Q$	0.24
3. $\bar{k}_+^{(\omega)} + 2k_-^{(\omega)} = \bar{k}_+^{(3\omega)}$		$\bar{\kappa}_+^{(\omega)} + 2\kappa_-^{(\omega)} = \bar{\kappa}_+^{(3\omega)}$	0.35
4. $k_+^{(\omega)} + \bar{k}_+^{(\omega)} + k_-^{(\omega)} = \bar{k}_-^{(3\omega)}$	$2k_0^{(\omega)} + \bar{k}_0^{(\omega)} = \bar{k}_0^{(3\omega)} + Q$	$\kappa_+^{(\omega)} + \kappa_-^{(\omega)} + \bar{\kappa}_+^{(\omega)} = \bar{\kappa}_-^{(3\omega)}$	0.35
5. $2k_-^{(\omega)} + \bar{k}_-^{(\omega)} = \bar{k}_-^{(3\omega)}$		$2\kappa_-^{(\omega)} + \bar{\kappa}_-^{(\omega)} = \bar{\kappa}_-^{(3\omega)}$	0.35
6. $3k_-^{(\omega)} = \bar{k}_-^{(3\omega)}$	$3k_0^{(\omega)} = \bar{k}_0^{(3\omega)} + 2Q$	$3\kappa_-^{(\omega)} = \bar{\kappa}_-^{(3\omega)} + Q$	0.47
7. $2k_+^{(\omega)} + \bar{k}_-^{(\omega)} = k_+^{(3\omega)}$		$2\kappa_+^{(\omega)} + \bar{\kappa}_-^{(\omega)} = \kappa_+^{(3\omega)} + Q$	0.69
8. $k_+^{(\omega)} + k_-^{(\omega)} + \bar{k}_-^{(\omega)} = k_-^{(3\omega)}$	$2k_0^{(\omega)} + \bar{k}_0^{(\omega)} = k_0^{(3\omega)} - Q$	$\kappa_+^{(\omega)} + \kappa_-^{(\omega)} + \bar{\kappa}_-^{(\omega)} = \kappa_-^{(3\omega)} + Q$	0.69
9. $2k_+^{(\omega)} + \bar{k}_+^{(\omega)} = k_-^{(3\omega)}$		$2\kappa_+^{(\omega)} + \bar{\kappa}_+^{(\omega)} = \kappa_-^{(3\omega)} + Q$	0.69
10. $\bar{k}_+^{(\omega)} + 2k_-^{(\omega)} = \bar{k}_-^{(3\omega)}$	$2k_0^{(\omega)} + \bar{k}_0^{(\omega)} = \bar{k}_0^{(3\omega)} + 2Q$	$\bar{\kappa}_+^{(\omega)} + 2\kappa_-^{(\omega)} = \bar{\kappa}_-^{(3\omega)}$	0.70
11. $2k_+^{(\omega)} + \bar{k}_-^{(\omega)} = k_-^{(3\omega)}$	$2k_0^{(\omega)} + \bar{k}_0^{(\omega)} = k_0^{(3\omega)} - 2Q$	$2\kappa_+^{(\omega)} + \bar{\kappa}_-^{(\omega)} = \kappa_-^{(3\omega)}$	1.4
12. $3k_+^{(\omega)} = k_+^{(3\omega)}$	$3k_0^{(\omega)} = k_0^{(3\omega)} - Q$	$3\kappa_+^{(\omega)} = \kappa_+^{(3\omega)} - Q$	17
13. $3k_+^{(\omega)} = k_-^{(3\omega)}$		$3\kappa_+^{(\omega)} = \kappa_-^{(3\omega)} + NQ$	$\epsilon_\eta^{1/2}(3\omega) \geq \epsilon_\xi^{1/2}(\omega)$[a]
14. $2k_+^{(\omega)} + k_-^{(\omega)} = k_-^{(3\omega)}$		$2\kappa_+^{(\omega)} + \kappa_-^{(\omega)} - \kappa_-^{(3\omega)} + NQ$	$3\epsilon_\eta^{1/2}(3\omega) \geq 2\epsilon_\xi^{1/2}(\omega) + \epsilon_\eta^{1/2}(\omega)$[a]
15. $k_+^{(\omega)} + 2k_-^{(\omega)} = k_-^{(3\omega)}$		$\kappa_+^{(\omega)} + 2\kappa_-^{(\omega)} = \kappa_-^{(3\omega)} + NQ$	$3\epsilon_\eta^{1/2}(3\omega) \leq \epsilon_\eta^{1/2}(\omega) + 2\epsilon_\eta^{1/2}(\omega)$[a]

[a]Condition for phase matching.

As we showed in Sec. III A, the em waves propagating in such a medium should have the form of Bloch functions, Eq. (10), with κ being the wave vectors. It is really κ's which we should use in discussing conservation of linear momentum or phase matching. In the third column of Table I, we write the phase-matching conditions for the various mode combinations in terms of κ's. Again, we notice that in many cases, the mismatches between κ's are compensated by the lattice momentum $Q = 4\pi/p$. In analogy to electron-electron interaction in a periodic lattice, these phase-matched processes can then be called the coherent optical umklapp processes. They are coherent since third-harmonic generation is a coherent process. For the other cases where phase matching can be achieved with no lattice momentum involved, the processes are correspondingly the normal processes.[13] It is clear that we can also express the phase-matching conditions for the last three mode combinations in terms of κ's and $G = N(4\pi/p)$, but the integer N in these cases depend on the actual values of $\bar{\epsilon}$ and α.

Finally, we realize that as $p \to \infty$, the cholesteric medium becomes a simple birefringent medium, and the waves propagating along the z axis have two linearly polarized modes. The last three combinations in Table I can then be written as

$$3[\epsilon_\eta^{1/2}(3\omega) - \epsilon_\eta^{1/2}(\omega)] = n[\epsilon_\xi^{1/2}(\omega) - \epsilon_\eta^{1/2}(\omega)] \quad (28)$$

with $n = 3$, 2, and 1, respectively. Equation (28) shows that the color dispersion is balanced by birefringence. This is how phase matching is usually achieved in a birefringent medium.

It has been suggested that phase matching can probably be achieved in cholesteric liquid crystals through compensation of color dispersion by circular birefringence because of their large optical rotary power.[2] This would be the case if, in the mode combinations 12–15 of Table I with all waves propagating in the same direction, the modes were circularly polarized. However, in all these cases, because of the large pitches for possible phase matching, the modes are far from being circularly polarized.

C. Conservation of Angular Momentum

The optical fields can also exchange angular momentum with the cholesteric medium in the third-harmonic generation process. The medium has a local twofold symmetry about the helical axis, and therefore in the process of converting three fundamental photons into one third-harmonic photon can exchange an angular momentum in units of $2\hbar$ with the optical fields.[19] Since each photon of circular polarization carries an angular momentum of $\pm\hbar$, any polarization combination in the third-harmonic generation satisfies the requirement of conservation of angular momentum. Thus, for example, for the sixth mode combination in Table I, the modes involved are nearly circularly polarized, and creation of a third-harmonic photon leaves an angular momentum of $2(2\hbar)$ to the medium. Similarly, one can show that conservation of angular momentum is satisfied for all other mode combinations.

In Sec. IV, we show our experimental results, which verify the above theoretical predictions.

IV. EXPERIMENTS

A. Sample Preparation and Related Measurements

In our experiments we have used a number of different cholesteric liquid-crystalline materials in order to observe the various predicted phase-matching peaks in third-harmonic generation at a convenient temperature. Most of the phase-matching conditions (1–11) in Table I require a cholesteric liquid crystal with a pitch less than 1.5 μm. To observe them, we used mixtures of cholesteryl oleyl carbonate, cholesteryl nonanoate, and cholesteryl chloride.[20] The ratio of cholesteryl oleyl carbonate to cholesteryl nonanoate was unity by weight in all our mixtures, but the concentration of cholesteryl chloride varied for different phase-matching conditions as listed in Table II. Figure 3 shows, from our measurements, the pitch as a function of the concentration of cholesteryl chloride at 20 and 40 °C. The temperature variation of the pitch between 20 and 40 °C can be crudely estimated from the two curves in Fig. 3.[21] The sample used to observe the 12th phase-matching condition was a mixture of cholesteryl chloride and cholesteryl myristate (1.75 to 1 by weight).[20] This mixture has a pitch which is variable from -1.7 μm to $\pm\infty$ to $+2\mu m$ by varying the temperature from 20 to 68 °C. (A negative pitch denotes a left-handed helical structure.) This allowed us to observe the phase-matching peaks at both -17.3 and $+17.4$ μm.

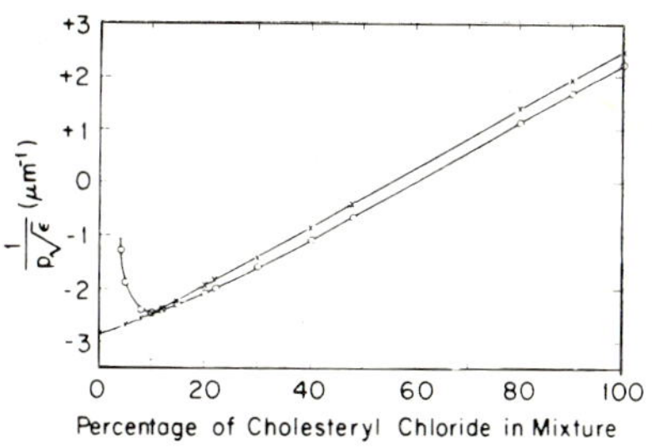

FIG. 3. $(p\bar{\epsilon}^{1/2})^{-1}$ as a function of the concentration of cholesteryl chloride in a mixture containing equal amounts of cholesteryl oleyl carbonate and cholesteryl nonanoate at 20 and 40 °C. A negative p indicates a left-handed cholesteric structure. Mixtures containing $\gtrsim 80\%$ cholesteryl chloride are "supercooled" liquid-crystal mixtures at these temperatures and are unstable. When freshly prepared, they last only for a few minutes before transforming to crystals. [(O): 20 °C, (×): 40 °C.]

Samples were prepared by placing a few drops of the mixture on a clean, rubbed glass window. A second window was then pressed upon the first with a ring of Teflon or Mylar as the spacer. Clear and uniform samples up to ~ 250 μm in thickness were obtained with the helical axis more or less perpen-

TABLE II. Empirical data on the various cholesteric mixtures used to observe phase-matched third-harmonic generation. Predicted phase-matching pitches are calculated using Eqs. (6) and (26).

	Mode combination for phase matching	Concentration of cholesteryl chloride (%)	$\bar{\epsilon}^{(\omega)}$	$\bar{\epsilon}^{(3\omega)}$	α	Predicted pitch for phase matching
1.	$k_+^{(\omega)} + 2k_-^{(\omega)} = \bar{k}_-^{(3\omega)}$	0	2.17 ± 0.01	2.27 ± 0.01	0.027 ± 0.002	− 237 ± 2 nm
2.	$3k_-^{(\omega)} = \bar{k}_+^{(3\omega)}$					
3.	$\bar{k}_+^{(\omega)} + 2k_-^{(\omega)} = \bar{k}_+^{(3\omega)}$					
4.	$k_+^{(\omega)} + \bar{k}_+^{(\omega)} + k_-^{(\omega)} = \bar{k}_-^{(3\omega)}$	22	2.18	2.29	0.028	− 352 ± 2 nm
5.	$2k_-^{(\omega)} + \bar{k}_-^{(\omega)} = \bar{k}_-^{(3\omega)}$					
6.	$3k_-^{(\omega)} = \bar{k}_-^{(3\omega)}$	30	2.18	2.30	0.027	− 472 ± 3 nm
7.	$2k_+^{(\omega)} + \bar{k}_-^{(\omega)} = k_+^{(3\omega)}$					
8.	$k_+^{(\omega)} + k_-^{(\omega)} + \bar{k}_-^{(\omega)} = k_-^{(3\omega)}$	40	2.19	2.31	0.030	− 689 ± 5 nm
9.	$2k_+^{(\omega)} + \bar{k}_+^{(\omega)} = k_-^{(3\omega)}$					
10.	$\bar{k}_+^{(\omega)} + 2k_-^{(\omega)} = \bar{k}_-^{(3\omega)}$					
11.	$2k_+^{(\omega)} + \bar{k}_-^{(\omega)} = k_-^{(3\omega)}$	48	2.20	2.32	0.030	− 1377 ± 10 nm
12.	$3k_+^{(\omega)} = k_+^{(3\omega)}$	a	2.19	2.30	0.029 0.027	− 17.3 ± 1 μm + 17.4 ± 1 μm

[a]Mixture of 1.75 parts of cholesteryl chloride and 1 part of cholesteryl myristate by weight.

dicular to the normal of the surface. Under a polarizing microscope, however, small regions of >10 μm in size were discernible with slightly different optical properties, indicating domains with somewhat different orientations of the helical axis. Bragg reflection of a collimated beam from such a sample spread over a cone of about 12°, suggesting that the helical axis had a distribution of around 6° about the normal of the surface. Each domain, however, seemed to extend over the entire thickness of the sample. The effect of multidomains on phase-matched third-harmonic generation will be discussed in Sec. V. Most of our experiments were done with 130-μm-thick samples. Samples with randomly oriented domains were easily obtained by increasing the sample thickness or by subjecting samples to thermal shocks.

Prediction of the precise pitches for phase matching requires the measurement of $\bar{\epsilon}$ and α [see Eqs. (26) and (6)]. Since $\bar{\epsilon}$ is an average over the two principal dielectric constants in the ordered phase, we measured $\bar{\epsilon}(\omega)$ and $\bar{\epsilon}(3\omega)$ with $\lambda_\omega = 1.06$ μm in the isotropic liquid phase where the random orientations of the molecules gives the desired average. It was assumed that the small variation of $\bar{\epsilon}$ due to temperature and phases could be neglected. We used the prism method with a mercury-arc lamp as the light source and a filtered photomultiplier tube as the detector.[22] The results are given in Table II.

The birefringence factor α is most easily obtained by measuring the optical activity of the sample [see Eq. (9)]. We measured the optical activity at the He-Ne laser frequency (6328 Å) and assumed a negligible dispersion for α. For all samples, α decreased with increasing temperature.[4,23] An example is shown in Fig. 4. In Table II, we list our measured values of α for the various mixtures at the temperatures where phase matching occurs. We realize, from our discussion in Sec. III, that

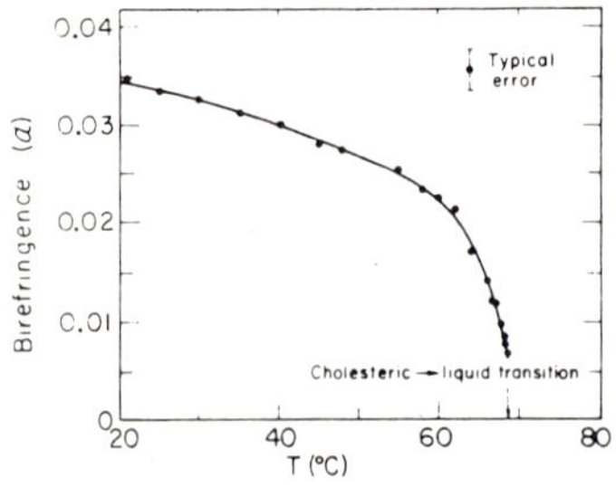

FIG. 4. Variation of the birefringence factor α as a function of temperature at 6328 Å for the mixture of cholesteryl chloride and cholesteryl myristate (1.75:1 by wieght). The solid curve is a smooth fit to the data points.

all these phase-matching conditions in Table II do not depend critically on the values of α. Therefore, the measurements of α here need not be very accurate.

From the measured values of $\bar{\epsilon}$ and α, we can calculate the pitches for phase matching for the various mode combinations using Eqs. (6) and (26). These predicted pitches are also given in Table II.

We used three different techniques to measure the pitches of our samples. For pitches significantly larger than optical wavelengths, two techniques are convenient. Direct observation of the samples under a polarizing microscope can reveal the periodic structure as alternating light and dark bands,[24] whose periodicity is $\frac{1}{2}p$. This method requires knowing the orientation of the helical axis in the region being examined, and the axis must have a component perpendicular to the microscope axis. A generally more convenient technique is the diffraction method.[25] In a sample with uniformly oriented helical axes, a laser beam propagating normal and polarized normal to the helical axis will generate diffraction spots at angles θ with respect to the helical axis satisfying $\cos\theta = 2N\lambda_0/p$, where N is an integer and $\lambda_0 = 2\pi\omega/c$. If the sample has many domains of different orientations, then diffraction arcs or rings will appear. From the diffraction pattern, one can find the pitch. For pitches in the vicinity of the visible spectrum, the pitch may be deduced from measurements of reflectivity or transmissivity as a function of wavelength. The position of the center of the reflection band corresponds to $p = 2\pi c/\omega\bar{\epsilon}^{1/2}$. This is most easily measured in thin ($\lesssim 10$ μm) samples, for which the band has a true peak; in thicker samples, there is a broad band of total reflectivity for the minus mode, making the center of the band more difficult to determine.

In the regions where the above three techniques of pitch measurements overlap, we found good agreement. Temperature dependences of the pitch for the various mixtures were also measured. An example is given in Fig. 5 for the mixture used to observe the 12th phase-matching condition. In Table III, we list the temperatures at which the predicted phase matching would occur in the various mixtures. Most of the uncertainty in the predicted temperatures arose from the measurements of $\bar{\epsilon}$, which entered the theoretical expressions explicitly and also implicitly through the experimentally determined α and p. Table III also gives the rate of change of pitch with temperature in the phase-matching regions.

B. Experimental Arrangements for Third-Harmonic Generation

We used a mode-locked Nd:glass laser as the fundamental pump source. A typical pulse train lasted about 200 nsec with the individual pulses separated by about 7 nsec. The total energy in each

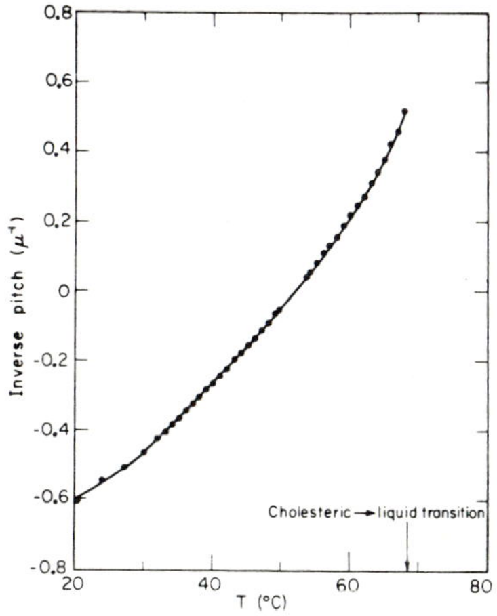

FIG. 5. Variation of the inverse pitch $1/p$ with temperature for the mixture of cholesteryl chloride and cholesteryl myristate (1.75 : 1 by weight). The experimental uncertainty is $\pm 0.005\ \mu m^{-1}$. The solid curve is a smooth fit to the data points.

train was about 0.03 J and the pulsewidth of individual pulses was about 7 psec as measured by two-photon fluorescence technique.[26] The beam diameter was about 2 mm. Mode-locked pulses were used in the experiments because their high peak intensities greatly enhanced the total third-harmonic power generated.

The experimental arrangement for measuring phase-matched third-harmonic generation with the fundamental propagating only in one direction is shown in Fig. 6. The third harmonic generated either forward or backward was detected by a photomultiplier after the appropriate filters. The setup with the fundamental propagating in both directions is shown in Fig. 7. A movable mirror was used to obtain laser light propagating in the backward direction. A stressed fused silica plate was used as a variable retardation plate in front of the mirror to control the polarization of the backward propagating laser beam. The laser beam was always circularly polarized for two reasons: First, as we discussed in Sec. II, circular polarizations are nearly the optimum polarizations for the incoming beam to excite the normal modes in all the samples of interest to us; and second, since no third harmonic can be generated in an isotropic medium by a circularly polarized beam,[27] this eliminates the background third-harmonic radiation generated from various components except the liquid crystal along the optical path. The third-harmonic blocking filters in front of the quarter-wave plate are necessary to eliminate the third harmonic generated before the quarter-wave plate. The samples were immersed in a water bath with temperature controlled to within ± 0.02 °C. Typically the temperature was swept very slowly (0.1–0.01 °C/min) through the predicted phase-matching region. In order to minimize fluctuations,[28] the third-harmonic signal from the liquid-crystalline sample was always

TABLE III. Predicted and observed phase-matching temperatures and relative third-harmonic peak intensities in various cases. dp/dT is rate of change of pitch with temperature at the phase-matching temperatures. A negative pitch indicates a left-handed helical structure.

	Mode combination for phase matching	Predicted phase-matching temperature (°C)	$\frac{dp}{dT}$ (nm/°C)	Observed phase-matching temperature (°C)	Relative phase-matched third-harmonic intensity
1.	$k_+^{(\omega)} + 2k_-^{(\omega)} = \bar{k}_-^{(3\omega)}$	38 ± 3	0.7		
2.	$3k_-^{(\omega)} = \bar{k}_+^{(3\omega)}$			39.5	2×10^{-4}
3.	$\bar{k}_+^{(\omega)} + 2k_-^{(\omega)} = \bar{k}_+^{(3\omega)}$			30.5	1×10^{-2}
4.	$k_+^{(\omega)} + \bar{k}_+^{(\omega)} + k_-^{(\omega)} = \bar{k}_-^{(3\omega)}$	32 ± 2	1.3		
5.	$2k_-^{(\omega)} + \bar{k}_-^{(\omega)} = \bar{k}_-^{(3\omega)}$			30.5	3×10^{-3}
6.	$3k_-^{(\omega)} = \bar{k}_-^{(3\omega)}$	38.2 ± 1	3.6	38.1	1×10^{-4}
7.	$2k_+^{(\omega)} + \bar{k}_-^{(\omega)} = k_+^{(3\omega)}$			29.9	3×10^{-1}
8.	$k_+^{(\omega)} + k_-^{(\omega)} + \bar{k}_-^{(\omega)} = k_-^{(3\omega)}$	30.1 ± 0.6	8.5		
9.	$2k_+^{(\omega)} + \bar{k}_+^{(\omega)} = k_-^{(3\omega)}$			29.9	1×10^{-2}
10.	$\bar{k}_+^{(\omega)} + 2k_-^{(\omega)} = \bar{k}_-^{(3\omega)}$	31.1 ± 0.6	8.5	31.2	3×10^{-2}
11.	$2k_+^{(\omega)} + \bar{k}_-^{(\omega)} = k_-^{(3\omega)}$	33.3 ± 0.3	39	33.6	4×10^{-4}
12.	$3k_+^{(\omega)} = k_+^{(3\omega)}$	49.4 ± 0.2 54.2 ± 0.2	6400	49.3 54.1	1 1

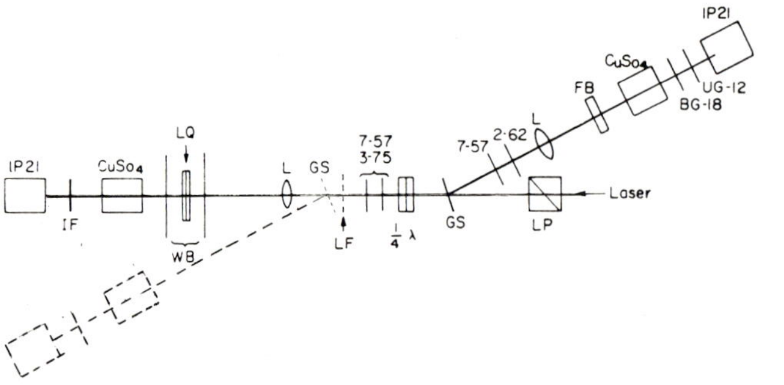

FIG. 6. Experimental arrangement for observing third-harmonic generation in cholesteric liquid crystals: LP (linear polarizer); GS (glass slide); LF (laser attenuation filter); L (lens); LQ (liquid crystalline sample); WB (water bath); IF (interference filter at 0.353 μm); IP21 (photomultiplier); $\frac{1}{4}\lambda$ (quarter-wave plate at 1.06 μm); FB (fuchsin dye cell). Corning and Schött glass filters are labeled by their catalog numbers.

normalized against the reference third-harmonic signal generated in a phase-matched solution of fuchsin basic dye dissolved in hexafluoracetone sesquihydrate.[3]

C. Experimental Results on Phase-Matched Third-Harmonic Generation

Consider first the phase-matching condition $3k_+^{(\omega)} = k_+^{(3\omega)}$ (12 in Table II). The setup in Fig. 6 was used for the experiments. Since the pitch of the sample can vary from left to right handedness, there should be two phase-matching peaks. The one at -17.3 μm (49.4 °C) should be generated by right circularly polarized fundamental waves and the other at $+17.4$ μm (54.2 °C) should be generated by left circularly polarized fundamental waves. The generated third harmonics should have opposite polarizations in the two cases. The experimental results are shown in Fig. 8. The two peaks appear at temperatures within 0.1 °C of the predicted phase-matching temperatures. They were generated, respectively, by right and left circularly polarized laser light as predicted. The theoretical phase-matching curve calculated by assuming monochromatic pump waves in a uniformly ordered liquid crystal is shown in Fig. 8. The experimental peaks are definitely broader with no clear fine structure at the wings. The difference between theory and experiment will be discussed in Sec. V. Since the molecular structures for $p < 0$ and $p > 0$ are different and $\overleftrightarrow{\chi}^{NL}$ could vary accordingly, we would not expect the two phase-matching peaks to have the same magnitude. Experimentally, we found that the two peaks were different in height, but their difference was within the 20% experimental accuracy. We also found that the two peaks were actually oppositely polarized. They had an elliptical polarization with the ratio of the two circularly polarized components being 5 ± 1. The theoretical ratio, derived from Eq. (8) or (15) is 4.8. Comparison of the phase-matched third-harmonic signals from the liquid crystal and from the fuchsin basic dye solution yields $I_{LQ}^{(3\omega)}/I_{Dye}^{(3\omega)} \sim 0.1$. We also measured the phase-matched third-harmonic generation from samples with different thickness. The third-harmonic intensities were indeed proportional to the square of the sample thickness.

Consider next the phase-matching conditons $3k_-^{(\omega)} = \bar{k}_+^{(3\omega)}$ and $3k_-^{(\omega)} = \bar{k}_-^{(3\omega)}$ (2 and 6 in Table II). The setup in Fig. 6 was used for the measurements. The samples had left-handed helical structure, and therefore, the incoming laser beam was left circularly polarized to feed efficiently into the minus mode. The third-harmonic signals generated in the backward direction were detected. The experimental results are shown in Figs. 9 and 10. There in-

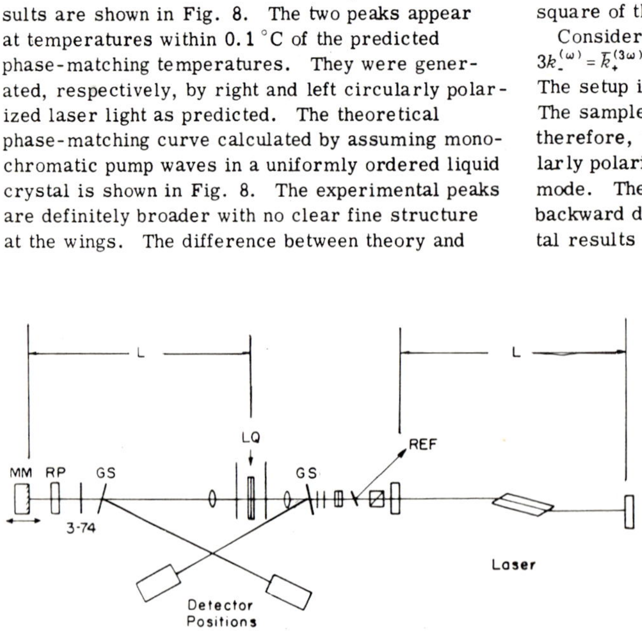

FIG. 7. Experimental arrangement for observing phase-matched third-harmonic generation in cholesteric liquid crystals requiring the simultaneous presence of laser light propagating both forward and backward. The two optical paths marked L are equal to ensure overlapping of the picosecond mode-locked pulses in the sample. MM (movable mirror); RP (retardation plate); REF (reference arm for creating third-harmonic reference signals). The unlabeled optical components are the same as in Fig. 6.

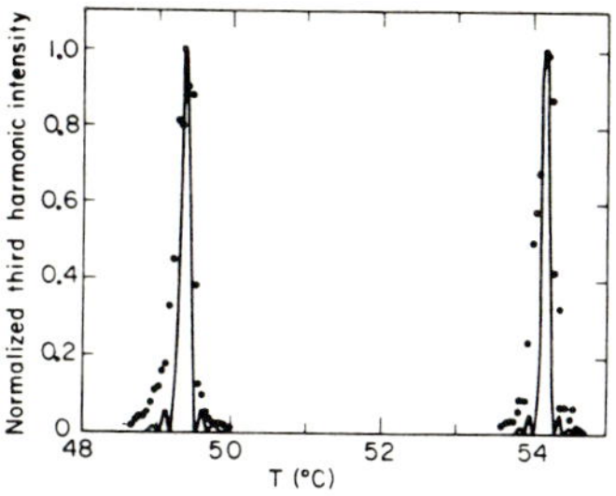

FIG. 8. Phase-matching peaks for mode combination 12 observed with a sample 130 μm thick. The peak at the lower temperature is generated by right circularly polarized fundamental waves and the one at the higher temperature by left circularly polarized fundamental waves. The solid line is the theoretical phase-matching curve and the dots are experimental data points. The uncertainty in the experimental third-harmonic intensity is about 20%.

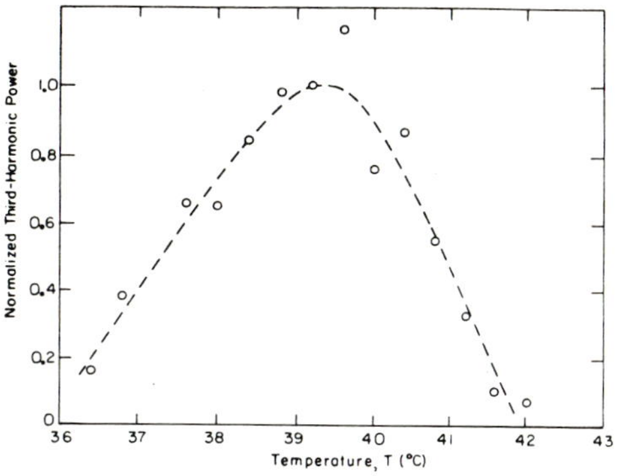

FIG. 9. Phase-matching peak for mode combination 2 observed with a sample 130 μm thick. The circles are experimental points with an uncertainty of 20%. The curve is a smooth fit to the data.

deed exist peaks at the predicted phase-matching temperatures. However, the widths of the peaks are several times broader than those of the theoretical phase-matching curves for a monochromatic input laser beam. This broadening is due mainly to the large spectral content of the laser, as will be discussed later. As a further confirmation of the theoretical predictions, we found, in the same temperature ranges, no phase-matching peak for third-harmonic generation in the forward direction, and also no peak for backward third-harmonic generation if the fundamental was right circularly polarized. We also found that the phase-matched third-harmonic outputs were nearly left and right circularly polarized (a ratio larger than 10 between two circularly polarized components) for the two cases, as predicted by the theory. The intensities of the phase-matched third harmonic generated by these mode combinations relative to that of mode combination 12 are given in Table III.

To observe the phase-matching conditions 1, 4, and 8 in the Tables, it would require the forward propagating fundamental waves to be in both plus and minus modes. Then, the incoming laser beam would have to be nearly linearly polarized. Consequently, large third-harmonic background signals would be created by the various components along the optical path. Our attempts to observe the phase-matching peaks on top of the background in these cases were not successful.

The remaining phase-matching conditions in the Tables except 13–15 require the simultaneous presence of fundamental waves propagating in opposite directions along the helical axis. This was accomplished in two ways. One method was to construct the sample cell with a front-surface mirror in contact with the back of the sample. Each mode-locked pulse reflected back from the mirror overlaps with itself in the sample. Alternatively, we put a movable mirror at a distance beyond the sample equal to the optical length of the laser cavity (see Fig. 7). Then, each mode-locked pulse reflected back by the mirror meets the next pulse in the train in the sample cell. Both methods yielded the same results. The samples all had left-handed helical structure. Right and left circularly polarized light fed efficiently into plus and minus modes, respectively. Note that the sense of circular polarization is reversed upon reflection from a mirror. If a $\frac{1}{4}\lambda$ retardation plate is inserted in front of the mirror, then the backward propagating light has the same circular polarization as the incoming beam. According to theory, the third-harmonic output in all these cases should be nearly circularly polarized.

Figure 11 shows the observed phase-matching

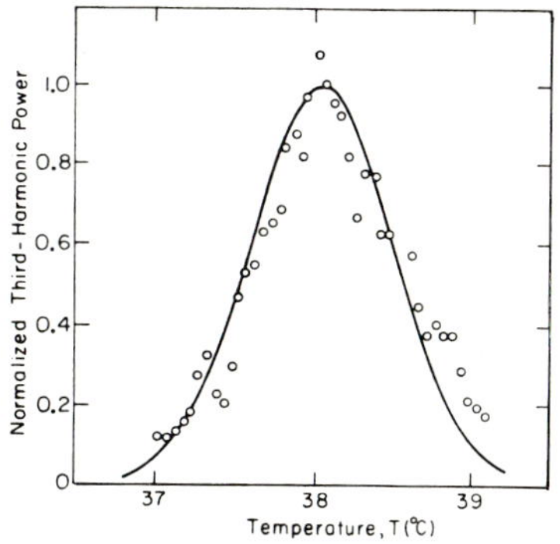

FIG. 10. Phase-matching peak for mode combination 6 observed with a sample 130 μm thick. The circles are the experimental data and have about a 20% uncertainty. The solid line is a theoretical phase-matching curve, assuming a spectral content of 150 Å for the laser pulses.

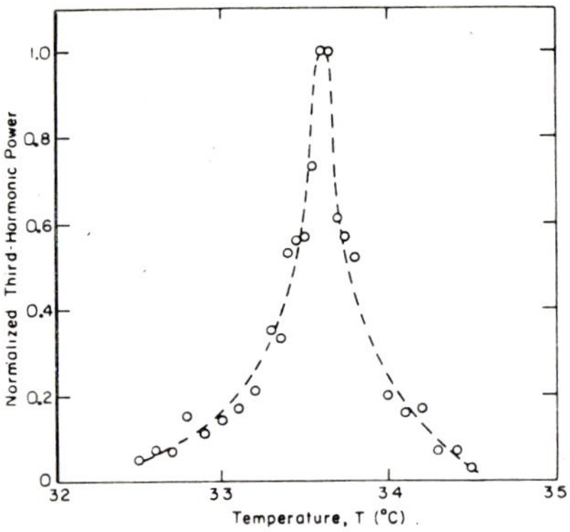

FIG. 11. Phase-matching peaks for mode combination 11 observed with a sample 130 μm thick. The circles are experimental points with an uncertainty of 20%. The curve is a smooth fit to the data.

peak at the predicted temperature for the 11th phase-matching condition. This peak had a relatively weak intensity and was observed with a reflecting back on the sample cell. No polarization measurement was attempted.

In Fig. 12, the observed small peak at 31.2 °C came from 10 and the large peak at 29.9 °C from 7 as predicted. Although the phase-matching peaks of 8 and 9 should also appear at 29.9 °C, they were not excited with the given laser polarizations. The large peak had a 30:1 ratio of right-to-left circular components as expected. When the retardation plate was inserted with the incoming beam right circularly polarized, 9 was excited and the same phase-matching peak was observed with a much smaller amplitude. The peak due to 8 was not observed as mentioned earlier.

The phase-matching peaks of 3–5 should appear at the same temperature. With the incoming beam circularly polarized and without the $\frac{1}{4}\lambda$ retardation plate in position, only 3 was excited. The observed phase matching peak shown in Fig. 13 appeared at the predicted temperature. Also as predicted, the third harmonic was generated only in the backward direction and had a 10:1 polarization ratio between the right and left circular components. With the retardation plate inserted, 5 was excited and the same phase-matching peak was observed with a smaller amplitude. The third-harmonic output was nearly left circularly polarized as predicted. For reasons mentioned earlier, the peak due to 4 was not observed.

For the above cases, we also verified the necessity for the simultaneous presence of laser light propagating both forward and backward in order to generate phase-matched third-harmonic radiation. By translating the movable mirror over a distance larger than 1 mm forward or backward from the position of maximum output, we essentially reduced the third-harmonic signal to zero. From such measurements, we could also deduce the pulsewidth of the mode-locked pulses (see Sec. V). In Table III, the peak intensities of the various phase-matching peaks relative to that of 12 are listed. In all these cases, the observed phase-matching peaks were much broader than the theoretical predictions assuming a monochromatic pump beam. The broadening was due to the broad spectral content of the mode-locked pulses, as we shall discuss in Sec. V.

The phase-matching conditions 13 and 14 cannot be satisfied in the liquid-crystalline materials we used, but 15 can probably be satisfied at $p \sim 50$ μm in the mixture used to observe phase matching of 12. Our experimental effort to detect such a phase-

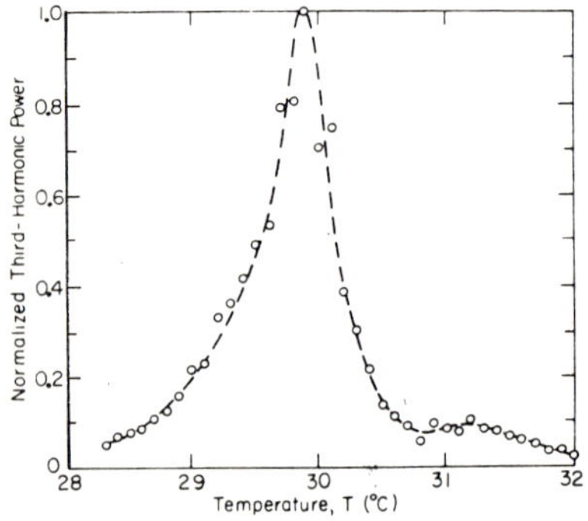

FIG. 12. Phase-matching peak for mode combinations 7 and 10 observed with a sample 130 μm thick. The circles are experimental points with an uncertainty of 20%. The curve is a smooth fit to the data.

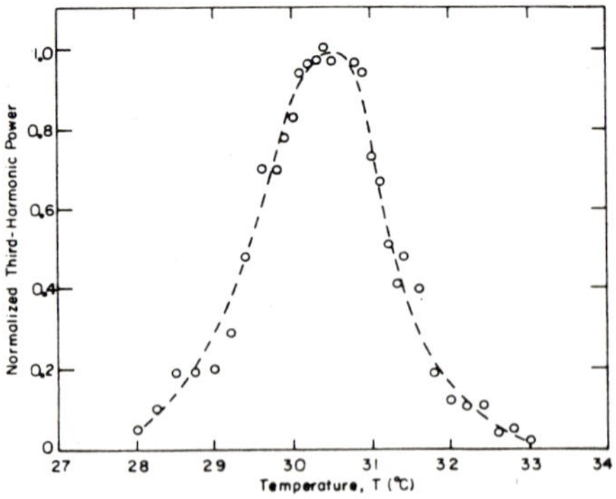

FIG. 13. Phase-matching peak for mode combination 3 observed with a sample 130 μm thick. The circles are experimental points with an uncertainty of about 20%. The curve is a smooth fit to the data.

matching peak was largely frustrated because the sample quality deteriorated when the pitch became large,[23,29] and because the large third-harmonic background generated from other components could not be eliminated in this case.

V. DISCUSSION

We have seen in Sec. IV that the experimental results agree very well with the theoretical predictions in Sec. III. Therefore, this confirms the model of Oseen and de Vries[10,11] which assumes that optically, a cholesteric liquid crystal can be treated as a twisted birefringent medium.

If the over-all molecular arrangement in planes perpendicular to the helical axis has no inversion symmetry, then second-harmonic generation would be possible in a cholesteric liquid crystal, and would also be phase matchable. The phase-matching condition $2k_+^{(\omega)} = k_+^{(2\omega)}$ would be satisfied at $p = 28$ μm in the mixture used to observe the phase-matched third-harmonic generation 12. Experimentally, we observed no second-harmonic phase-matching peak, suggesting that the contrary is true. Durand and Lee[7] and Goldberg and Schnur[8,9] have also found no second-harmonic generation in liquid crystals and have come to the same conclusion.

We notice that all the observed phase-matching peaks had widths greater than those predicted theoretically for a monochromatic laser and a perfect liquid-crystal sample. In the latter case, the full width at half-maximum of a phase-matching peak in terms of pitch variation would be given by

$$\delta p_1 \simeq \pi \Big/ L\left(\frac{d(\Delta k)}{dp}\right), \tag{29}$$

where L is the sample thickness and $d(\Delta k)/dp$ is the derivative of the phase mismatch Δk evaluated at $\Delta k = 0$. The value of $(\delta p)_1/p$ for each phase-matching peak is given in Table IV.

Surface effects on a liquid-crystal sample are usually important. The molecular orientation on the boundary layers appears to remain unchanged despite changes in external parameters.[5,30] This constraint would not allow a perfect helical structure of arbitrary pitch to fit between the two interfaces. There would in general be a distortion of the helical structure amounting to $\sim \frac{1}{2}p$ over the sample thickness. This implies a pitch distribution with a width $\delta p_2/p \sim p/2L$, the values of which at phase matching are given in Table IV. For the same sample thickness, this effect clearly increases with a pitch, and appears to be a strong contribution to the observed width of the 12th phase-matching peak occurring at $|p| \simeq 17$ μm. There, in terms of temperature, the two widths corresponding to δp_1 and δp_2 are 0.2 and 0.1 °C, respectively. The observed full width at half-maximum is 0.3 °C.

That the mode-locked pulses are far from being monochromatic would also introduce broadening in the phase-matching peaks. The different frequency components of the laser pulses yield phase-matched third-harmonic (or more generally, sum-frequency) generation at different pitches. By assuming a spectral content of the mode-locked laser pulses of about 150 Å (full width at half-maximum of a Gaussian pulse), presumably due mainly to frequency chirping, we obtain a convolution width $\delta p_3/p = 7.5\times10^{-3}$ for all the phase-matching peaks. This value agrees quite well with the observed widths for all cases except 12 as indicated in Table IV, and therefore, the spectral content of the laser pulses is believed to be the dominant contribution for these cases.

As we mentioned earlier, our samples were actually composed of many domains with different orientations of the helical axis spread over a cone of about 12°. This might lead to additional broadening of the observed phase-matching peaks, since different domains would be at phase matching at different pitches if we allow the third-harmonic to be noncollinear with the fundamental. However, this effect would be small if the collection angle of the third-

TABLE IV. Various possible contributions to the widths of the phase-matching peaks together with the predicted and the observed widths of the peaks. $d(\Delta k)/dp$ is the rate of change of the phase mismatch with pitch at phase matching. δp_1, δp_2, δp_3, and δp_4 for a sample 130 μm thick due, respectively, to the inherent width, the surface effect on helical structure, the spectral content of laser pulses, and the domain structure in the samples.

Mode combination for phase matching	$\frac{d(\Delta k)}{dp}$ (μm^{-2})	$\frac{\delta p_1}{p}$	$\frac{\delta p_2}{p}$	$\frac{\delta p_3}{p}$	$\frac{\delta p_4}{p}$	Predicted width (°C)	Observed width (°C)
2. $3k_-^{(\omega)} = \bar{k}_+^{(3\omega)}$	220	4.0×10^{-4}	9.1×10^{-4}	7.5×10^{-3}	8×10^{-5}	3.0	2.8
3. $\bar{k}_+^{(\omega)} + 2k_-^{(\omega)} = \bar{k}_+^{(3\omega)}$	100	9.5×10^{-4}	1.4×10^{-3}	7.5×10^{-3}	7×10^{-5}	2.2	1.7
6. $3k_-^{(\omega)} = \bar{k}_-^{(3\omega)}$	110	4.6×10^{-4}	1.8×10^{-3}	7.5×10^{-3}	8×10^{-5}	1.3	1.0
7. $2k_+^{(\omega)} + \bar{k}_-^{(\omega)} = k_+^{(3\omega)}$	27	1.3×10^{-3}	2.6×10^{-3}	7.5×10^{-3}	7×10^{-5}	0.7	0.6
11. $2k_+^{(\omega)} + \bar{k}_-^{(\omega)} = k_-^{(3\omega)}$	13	5.8×10^{-4}	5.3×10^{-4}	7.5×10^{-3}	7×10^{-5}	0.4	0.4
12. $3k_+^{(\omega)} = k_+^{(3\omega)}$	0.04	3.3×10^{-2}	6.7×10^{-2}	7.5×10^{-3}	5×10^{-3}	0.3	0.3

harmonic detector is small. In our experiments, the third-harmonic detector aligned with the beam had an acceptance angle of ~ 3°. The laser beam at the sample, even when it was focused, had a convergence angle less than 1°. This would contribute a width $\delta p_4/p$ to the phase-matching peak. The estimated values of $\delta p_4/p$ for the various cases are given in Table IV. They appear to be always less than the observed widths. In all cases except 12, there is a loss of third-harmonic power due to the fact that only domains with helical axes tipped at an angle less than 0.8° to 2.3° contribute to the observed third harmonic. Compared to a perfect sample, this loss should be roughly a factor of 3–6.

From our measurements of the relative intensities of the various phase-matching peaks, we can deduce through Eq. (24) in the phase-matching approximation the nonlinear susceptibility elements C_{11}, C_{12}, C_{21}, and C_{22}, assuming that they are the same for different mixtures. Using the measured intensities for the phase-matching peaks 2, 3, 6, 7, 11, and 12 (corrected for their different widths), we found from computer calculations that all four C's are nearly equal. Uncertainties in their values were large and came mainly from our relative intensity measurements. Because of the necessarily different experimental arrangement for the various phase-matching cases, correction for the differences introduced uncertainties. Different experimental runs also yielded slightly different results, presumably due to differences in sample quality. Taking all these factors into consideration, we can only conclude that C_{11}, C_{12}, C_{21}, and C_{22} are equal to within 10% and their values are uncertain within a factor of 2. Comparison with the third-harmonic intensity generated in a phase-matched solution (45 gm/liter) of fuchsin dye in hexafluoracetone sesquihydrate[2] gave $|C/\chi^{NL}_{\text{dye}}| \sim 0.2$. This value could increase by a factor of 2 if the liquid-crystal sample were of a single domain.

We realize that if C's are equal, then $\overleftrightarrow{\chi}^{(3)}$ would be independent of z even in the lab frame. We can then find the solution of third-harmonic generation directly in the lab frame using Eq. (16). The optical fields in the equation can be written from Eq. (8) as

$$\vec{E}_\pm(z,\omega) = [(\hat{x}+i\hat{y})A_{R\pm}\, e^{i\omega n_{R\pm} z/C} + (\hat{x}-i\hat{y})A_{L\pm}\, e^{i\omega n_{L\pm} z/C}] \times (\mathcal{E}_{\ell\pm}/\sqrt{2})\, e^{-i\omega t} \quad . \quad (30)$$

Phase-matched third-harmonic generations should occur when individual components of $\vec{E}_\pm$ are phase matched, and the corresponding intensities can be calculated knowing $A^{(\omega)}_{R\pm}$ and $A^{(\omega)}_{L\pm}$. For example, we have phase matching when

$$2n^{(\omega)}_{R+} + n^{(\omega)}_{L+} = 3n^{(\omega)}_{R+}$$

or

$$2n^{(\omega)}_{L+} + n^{(\omega)}_{R+} = 3n^{(3\omega)}_{L+} ,$$

which can be easily shown to be equivalent to the phase-matching condition 12. The corresponding third-harmonic field has $A^{(3\omega)}_{R+} \propto A^{(\omega)}_{R+}A^{(\omega)}_{R+}A^{(\omega)}_{L+}$ and $A^{(\omega)}_{L+}A^{(\omega)}_{L+}A^{(\omega)}_{R+}$. The relative intensities of the various phase-matching peaks calculated this way agree moderately well with the experimental results except for 2, 6, and 11, where the predicted intensities depend more critically on the inequality of the C's.

With C's being equal, the expected intensity of the phase-matching peak 10 should be four times greater than that of 7. However, our experimental data show that 10 is only $\frac{1}{10}$ as intense as 7, or about 40 times weaker than predicted. The discrepancy is due to strong reflection of left circularly polarized laser light from the sample, since the laser frequency is close to the reflection band of the sample. Measurements of transmission through the sample showed that only $\frac{1}{10}$ of the left circularly polarized light was transmitted. Since this phase-matching case involves two left circularly polarized fundamental fields (and one right), the strong reflection would decrease the observed third-harmonic intensity by a factor of 10^{-2}, assuming that the 90% reflection happens at the interface.

Phase-matched third-harmonic generation would of course occur in any cholesteric liquid crystal as long as the helical pitch can be adjusted to the correct value. We tried the experiment on an entirely different cholesteric material. Poly-γ-benzyl-L-glutamate (PBLG), a synthetic α-helix protein, dissolved in dioxane is cholesteric for concentrations from ~ 0.1 to ~ 0.5 g of PBLG per g of solvent.[31] The dielectric constants are comparable to those of the cholesterol-derived materials and hence the pitches for phase matching are approximately the same. A pitch of ~ 17 μm is realizable by adjusting the concentration of PBLG, and hence phase-matched third-harmonic generation 12 should be possible. We did not use temperature to tune the pitch because of the slow response of PBLG samples. Instead, we used many samples with different concentrations to yield different pitches. We did observe a peak in the third-harmonic intensity around $p = 17$ μm with the laser beam polarized to feed efficiently into the plus mode, and no peak for the opposite laser polarization. There was a moderate amount of scattering in the data, presumably due to the use of many different samples.

Recently, measurements of ultrashort pulses have attracted much attention.[26,32–34] In all cases, except the case of Treacy using the compression technique,[33] the experiments measure only the pulsewidth. Both the second-harmonic-generation technique and the two-phonon-fluorescence technique are inherently symmetric. They measure the auto correlation function $G^{(2)}(\tau)=\int|E(t)|^2|E(t+\tau)|^2dt$, which is characterized by $G^{(2)}(\tau)=G^{(2)}(-\tau)$, and therefore cannot yield any information about the pulse shape. Phase-matched third-harmonic generation has also been used for pulsewidth measurements,[34] but only under circumstances where a symmetrized form of the third-order autocorrelation function was observed. Here, in principle, the technique can provide information about the pulsewidth as well as pulse asymmetry. If phase matching requires two fundamental photons in one mode and one in the other mode, then we can measure the correlation function $G^{(3)}(\tau)=\int_{-\infty}^{\infty}|E(t)|^4|E(t+\tau)|^2dt \neq G^{(3)}(-\tau)$. This correlation function is symmetric $[G(\tau)=G(-\tau)]$ only if the pulse is symmetric. Therefore, from the symmetry of $G(\tau)$, we can deduce information about the pulse asymmetry.[4,35] Most of the phase-matching conditions in cholesteric liquid crystals satisfy this requirement.

To demonstrate the technique, we split the laser beam into two beams with proper polarizations to excite the phase-matching peak 7. The two beams, after traveling about the same optical path, met each other at the sample from opposite sides. A variable optical delay in one arm allowed continuous variation of the relative arrival time τ of the two pulses. Our results are shown in Fig. 14. The curve shows an average pulsewidth of about 7.5 psec and a pulse asymmetry in the sense that the trailing edge of the pulse was steeper than the leading edge. This agrees with the result of Treacy.[33] Assuming an asymmetric pulse constructed from two half-Gaussian curves joined at their maxima, with their widths differing by a factor of 5.5, we obtained a correlation curve which fits well with the data, as shown in Fig. 14. However, this curve is not very sensitive to the pulse asymmetry.[35] Consequently, unless the third-harmonic generation can be measured very accurately, the technique cannot yield very good quantitative measure of the pulse asymmetry. Prior knowledge of the pulse shape would also be necessary for more accurate conclusion. Since better signal-to-noise ratio would help the resolution, we should probably use crystals such as calcite as the nonlinear medium in such measurements, where phase-matched third-harmonic generation can be achieved with two fundamental photons in the ordinary mode and one in the extraordinary mode.[36] There, with the fundamental and the third harmonic propagating in the same direction, the signals can be much stronger. Ideally, one would like to obtain information about the pulse asymmetry for individual pulses instead of averages over all the pulses in a train.

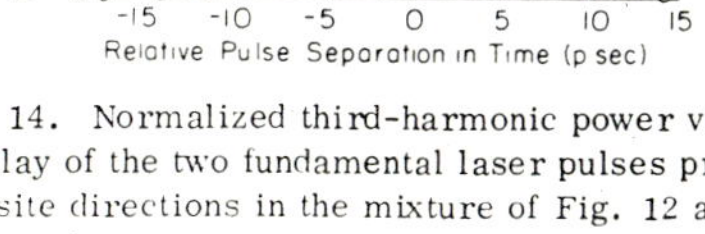

FIG. 14. Normalized third-harmonic power vs relative time delay of the two fundamental laser pulses propagating in opposite directions in the mixture of Fig. 12 at the phase-matching temperature 29.9 °C. The circles are experimental points with an uncertainty of 20%. The solid curve is obtained from theoretical calculation by assuming that the mode-locked pulses are made of two half-Gaussian curves joined at their maxima, with the leading half 5.5 times broader than the lagging half.

Our maximum observed energy conversion from the laser frequency to the third harmonic was small. The most efficient case was 12. Without focusing, the 130-μm-thick samples converted about 10^{-14} of the laser energy to the third harmonic. By focusing the laser beam down to 0.1 mm diameter, we could increase the conversion by a factor of $\sim 4\times 10^2$. Although thicker samples of good quality could probably be made with the help of external fields, the conversion efficiency would still be too small for such samples to be useful as practical third-harmonic generators.

We can of course also have other types of nonlinear interaction phase matched in a cholesteric liquid crystal. We realize that it is the periodicity of the medium which makes phase matching easily achievable here. It is clear that other media with periodicity in the optical range can also be used. In fact, our theoretical analysis in Secs. II and III, using the Bloch picture should be applicable to all such materials. Bloembergen and Sievers have considered possibilities for phase matching of a number of nonlinear interactions in periodic layers of GaP and GaAs.[37] This layer medium lacks inversion symmetry, thus permitting the observation of second-order nonlinear effects. However, it is difficult to fabricate and not so directly tunable. The great advantage of cholesteric liquid crystals is their inherent and variable periodicity.

VI. CONCLUSIONS

We have shown theoretically and experimentally that third-harmonic generation can be collinearly phase matched in cholesteric liquid crystals. Phase matching is achieved since the momentum mismatch between the fundamental and the third harmonic is compensated by the lattice momentum which is present due to the periodicity of the helical structure of the cholesteric medium. Many different phase-matching conditions exist. Analogous to electron-electron interaction in a periodic lattice, they can be identified as normal and umklapp processes, respectively. Most of the predicted phase-matching conditions were confirmed experimentally. In several cases, phase-matched third-harmonic generation occurs only when the fundamental waves appear simultaneously propagating in opposite directions. These processes can be used to measure the width and asymmetry of the mode-locked pulses. We also attempted to observe second-harmonic generation in cholesteric liquid crystals. The negative results indicate that the molecular arrangement of the liquid crystals in a plane perpendicular to the helical axis has an over-all inversion symmetry. The theoretical discussion in this paper can be extended to other types of media with periodicity in the optical range.

---

*This work was performed under the auspices of the U.S. Atomic Energy Commission.

†Permanent address: Physics Department, Williams College, Williamstown, Mass.

[1]J. Giordmaine, Phys. Rev. Letters 8, 19 (1962); P. D. Maker, R. W. Terhune, M. Nisenoff, and C. M. Savage, *ibid.* 8, 21 (1962).

[2]H. Rabin and P. P. Bey, Phys. Rev. 156, 1010 (1967); C. K. N. Patel and N. Van Tran, Appl. Phys. Letters 15, 189 (1969).

[3]J. P. Bey, J. F. Guiliani, and H. Rabin, Phys. Rev. Letters 19, 819 (1967); R. K. Chang and L. K. Galbraith, Phys. Rev. 171, 993 (1968).

[4]J. W. Shelton and Y. R. Shen, Phys. Rev. Letters 25, 23 (1970); 26, 538 (1971).

[5]See, for example, G. W. Gray, *Molecular Structure and the Properties of Liquid Crystals* (Academic, New York, 1962); I. G. Chistyakov, Usp. Fiz. Nauk 89, 563 (1966) [Sov. Phys. Usp. 9, 551 (1967)]; G. H. Brown, J. W. Doane, and V. D. Neff, CRC Critical Rev. Solid State Sci. 1, 303 (1970).

[6]I. Freund and P. M. Rentzepis, Phys. Rev. Letters 18, 393 (1967).

[7]G. Durand and C. H. Lee, Mol. Cryst. 5, 171 (1968).

[8]L. S. Goldberg and J. S. Schnur, Appl. Phys. Letters 14, 306 (1969).

[9]L. S. Goldberg and J. S. Schnur, Radio Electron. Eng. (GB) 39, 279 (1970).

[10]C. W. Oseen, Trans. Faraday Soc. 29, 883 (1933).

[11]H. de Vries, Acta Cryst. 4, 219 (1951).

[12]I. Freund, Phys. Rev. Letters 21, 1404 (1968).

[13]See, for example, C. Kittel, *Introduction to Solid State Physics*, 3rd ed. (Wiley, New York, 1966).

[14]D. W. Berreman and T. J. Scheffer, Mol. Cryst. Liquid Cryst. 11, 395 (1970); D. Dreher, G. Meier, and A. Saupe, *ibid.* 13, 17 (1971).

[15]See, for example, C. P. Slichter, *Principles of Magnetic Resonance* (Harper & Row, New York, 1963), p. 11.

[16]P. N. Butcher, *Nonlinear Optical Phenomena* (Ohio State University Engineering Publications, Columbus, Ohio, 1965).

[17]See, for example, N. Bloembergen, *Nonlinear Optics* (Benjamin, New York, 1964).

[18]As long as the field variation across the beam profile is not too rapid, we can use the ray approximation to find $|\mathcal{E}_{\pm}(3\omega, x, y)|^2$ at local points.

[19]H. J. Simon and N. Bloembergen, Phys. Rev. 171, 1104 (1968); C. L. Tang and H. Rabin, Phys. Rev. B 3, 4025 (1971).

[20]These chemicals were obtained from Aldrich Chemical Co., Eastman Organic Chemicals, and Varilight Corporation, and were used without further purification.

[21]For more information on this three-component cholesteric system see: J. L. Fergason, Appl. Opt. 7, 1729 (1968); J. L. Fergason, *ibid.* 7, 1729 (1968); J. L. Fergason, Am. J. Phys. 38, 425 (1970); L. Melamed and D. Rubin, Appl. Opt. 10, 1103 (1971).

[22]See, for example, F. W. Sears, *Optics* (Addison-Wesley, Reading, Mass., 1949), p. 47.

[23]H. Baessler, T. M. Laronge, and M. M. Labes, J. Chem. Phys. 51, 3213 (1969).

[24]C. Robinson, Tetrahedron 13, 219 (1961).

[25]E. Sackmann, S. Meiboom, L. C. Snyder, A. E. Meixner, and R. E. Dietz, J. Am. Chem. Soc. 90, 3567 (1968).

[26]J. A. Giordmaine, P. M. Rentzepis, S. L. Shapiro, and K. W. Wecht, Appl. Phys. Letters 11, 216 (1967).

[27]P. P. Bey and H. Rabin, Phys. Rev. 162, 794 (1967).

[28]J. Ducuing and N. Bloembergen, Phys. Rev. 133, 1493 (1964).

[29]P. Kassubek and G. Meier, Mol. Cryst. Liquid Cryst. 8, 305 (1969).

[30]P. Chatelain, Bull. Soc. Franc. Mineral. Crist. 66, 105 (1943).

[31]C. Robinson, J. C. Ward, and R. B. Beevers, Discussions Faraday Soc. 25, 29 (1958).

[32]J. A. Armstrong, Appl. Phys. Letters 10, 16 (1967).

[33]E. B. Treacy, Appl. Phys. Letters 14, 112 (1969).

[34]R. C. Eckardt and C. H. Lee, Appl. Phys. Letters 15, 425 (1969).

[35]H. P. Weber and R. Dandliker, Phys. Letters 28A, 77 (1968).

[36]P. D. Maker and R. W. Terhund, Phys. Rev. 137, A801 (1965).

[37]N. Bloembergen and A. J. Sievers, Appl. Phys. Letters 17, 483 (1970).

Mol. Cryst. Liq. Cryst., 1983, Vol. 98, pp. 1–12
0026-8941/83/9804–0001/$18.50/0

Optical Second Harmonic Generation in Various Liquid Crystalline Phases†

M. I. BARNIK, L. M. BLINOV‡, A. M. DOROZHKIN and N. M. SHTYKOV

Organic Intermediates and Dyes Institute, Moscow, 103787, USSR, B. Sadovaya, 1

(Received December 9, 1982)

The zero-field and field-induced optical second harmonic generation (SHG) was investigated for the nematic and smectic A phases of various liquid crystals. The components of the cubic non-linear susceptibility tensor were measured for substances with different molecular structure. The phase-matched SHG was observed for all the compounds investigated. The directions of the phase synchronisms as well as the corresponding non-linear susceptibilities were determined for the ee-o and oe-o interactions. The zero-field phase-matched SHG was observed for the oe-o interaction. It was accounted for by a multipolar mechanism.

INTRODUCTION

Fifteen years have passed since the first observations of the optical second harmonic generation (SHG) from liquid crystals;[1] however, up to now, there is no final answer to the question concerning the nature of the effect. The problem is of fundamental importance because it is closely related to the symmetry properties of liquid crystals.

The first successful observations of SHG from cholesteryl carbonate were explained in terms of a lack of the inversion center in small liquid crystalline "swarms".[1] Later,[2,3] without any success, there has been carried out extensive search for the SHG in nematic, cholesteric and smectic (non-ferroelectric) liquid crystals, and the previous result was accounted for by the presence of solid crystals in the cholesteric phase.

---

†Presented at the Ninth International Liquid Crystal Conference Bangalore, 1982.

‡At present, in the Institute of Crystallography of U.S.S.R. Acad. Sci., Leninsky prosp. 59, Moscow, 117333.

The second turn of the discussion around the nature of a SHG in nematic liquid crystals arised when the SHG was observed in oriented layers of 4-methoxybenzylidene-4′-butylaniline (MBBA).[4] The phenomenon has been explained by the lack of the symmetry center in the nematic phase. The zero-field SHG in MBBA was also investigated[5] but the nature of the effect was connected with the flexoelectric polarization of surface layers. Such a polarization has to remove the inversion center in surface liquid crystalline layers and to allow the SHG to be detectable. Another explanation of the zero-field SHG in terms of the electric quadrupolar interaction was suggested in.[6]

Recently there was also considerable activity in the investigations of the SHG in nematic and cholesteric phases induced by an external electric field.[7,8] Such experiments allow the high order molecular hyperpolarizabilities to be calculated.[9-11] The SHG was also observed in a ferroelectric (chiral smectic C*) liquid crystal.[12]

The aim of this paper is to investigate the optical SHG, both the zero-field and field induced in liquid crystals differing by their structure and molecular parameters (for example, by the value and direction of their permanent dipole moment).

In experiments with the field-induced SHG we always tried to obtain the phase-matched generation, having in mind to study the zero-field SHG in the same synchronism directions when the external field is switched off. The search for zero-field SHG in the phase-matched conditions increases markedly the sensitivity of an experimental set-up and allows the pump beam intensity to be lowered. The study of the field-induced SHG was also used for the calculation of the number of parameters of the liquid crystalline medium (a cubic non-linear susceptibility Γ_{ijkl}, a quadratic hyperpolarizability β_{ijk}[9-11] and the discussion of their connection with molecular structure. In contrast to isotropic liquids, the value of Γ_{ijkl} depends not only on molecular parameters but on the orientational order parameters $\langle P_2 \rangle$ and $\langle P_4 \rangle$ of liquid crystal as well. The latter opens the possibility to measure $\langle P_2 \rangle$ and $\langle P_4 \rangle$.

EXPERIMENTAL TECHNIQUE

Four liquid crystalline compounds were studied: 4-*n*-pentyl-4′-cyanobiphenyl (5CB), 4-*n*-octyl-4′-cyanobiphenyl (8CB), 4-*n*-hexyloxy-4′-*n*-pentyl-α-cyanostilbene (HOPCS) and MBBA. All the compounds have the nematic phase in convenient temperature range, and, in addition, 8CB has the smectic A phase. Both 5CB and 8CB have positive dielectric anisotropy ($\varepsilon_a = \varepsilon_{\|} - \varepsilon_{\perp} \approx +10$) while HOPCS and MBBA have the negative ε_a values ($\varepsilon_a = -5$ and -0.5, respectively). The difference in signs of ε_a is due to the different orientation of the dipolar cyano-group in the two cases, parallel with the long molecular axes for 5CB and 8CB and nearly perpendicular to them for HOPCS and MBBA.

The sketch of the experimental set-up is shown in Figure 1. A Q-switched Nd-YAG laser, operating at 1.06 μm and a pulse repetition 2–12.5 Hz was used to provide the fundamental (pump) beam. The peak power was 200–300 kW. The beam was focused with a 43 cm lens so that the power density on the sample placed in a thermostate was about 100–200 MW·cm.$^{-2}$ For investigation the field-induced SHG, short pulses ($t_p = 20$ μs) of high voltage (U_p = 4kV) were provided by an electrical generator. The pulse duration was chosen from the condition $\tau < t_p < T$, where τ is the relaxation time for dipolar (Debye) polarization, and T is the director reorientation time. Under such a condition, molecular dipoles are oriented by the field but the Fredericks transition does not take place. The sensitivity of our set-up was about 30 photons of the optical second harmonic per single laser pulse. The cell temperature was stabilized with an accuracy of 0.1° K.

In experiments, there were measured the intensities of the second harmonic $I_{2\omega}^{\|,\perp}$ and the coherence length of $l_{\|,\perp}$ in the liquid crystalline phases and the corresponding parameters $I_{2\omega}^{ISO}$, l_{iso} for the isotropic phase. From these

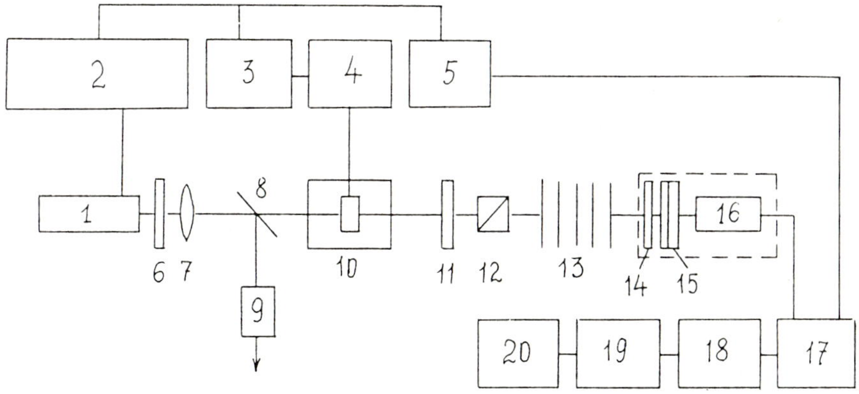

FIGURE 1 Experimental set-up: 1. Q-switch Nd-YAG laser; 2. laser controlling unit; 3. device for synchronization laser pulses with voltage pulses driving a liquid crystal cell; 4. high voltage pulse generator; 5. strobe-pulse generator (pulse duration 0.7 μs); 6. optical filter for fundamental beam; 7. long-focus lens; 8. beam splitter; 9. photocell for measuring the power of the fundamental beam; 10. thermostate with a liquid crystal cell; 11., 14 optical attenuators for fundamental beam; 12. analyzer; 13. neutral filters; 15. interference filter for the optical second harmonic (532 nm); 16. photomultiplier; 17. analog switch with a preamplifier; 18. amplifier and oscilloscope; 19. integrator; 20. digital voltmeter.

values the cubic susceptibilities $\Gamma_{\|,\perp}$ and Γ_{iso} were calculated. The coherence lengths and second harmonic intensities in maxima of Maker's oscillations were measured using wedge-form cells, Figure 2a. For the z-axis being oriented along the director of a liquid crystal, and the subscripts $\|$ and $\perp$ referring to the z-direction we have $\Gamma_{\|} \equiv \Gamma_{zzzz}$ and $\Gamma_{\perp} \equiv \Gamma_{xxxx}$. The values $l_{\|}$, $I_{2\omega}^{\|}$ and $\Gamma_{\|}$ were determined from experiments with homogeneously oriented layers when the directions of the dc field, and the optical fields of the first and second harmonics were all parallel with the director. The value $l_{\perp}$, $I_{2\omega}^{\perp}$ and $\Gamma_{\perp}$ were measured using homeotropically oriented layers when the directions of all the fields were collinear and perpendicular to the director.

To obtain the homogeneous orientation of a liquid crystal glass plates were covered by a film of polyvinyl alcohol and rubbed with cotton or other tissue. For the homeotropic orientation a film of chromium stearylchloride was used. The quality of orientation was periodically checked using a polarization microscope. The inner angle of wedge-form cells was measured by means of a goniometer. The accuracy of measurements of the coherence length and second harmonic intensities was 3% and 10%, respectively. The absolute values of susceptibilities Γ were calculated using the d_{11}-component of crystalline quartz as a reference ($d_{11} = 8 \cdot 10^{-10}$ CGSU).

The rectangular prisms, Figure 2b, were used for the investigation of SHG under the phase-matched conditions. One of the acute angles of the prisms was chosen to be equal to either 30° or 45° respectively for homeotropically and homogeneously oriented layers. The synchroneous angles Θ_s were measured and the corresponding (synchroneous) non-linear susceptibilities Γ_s and χ_s were calculated.

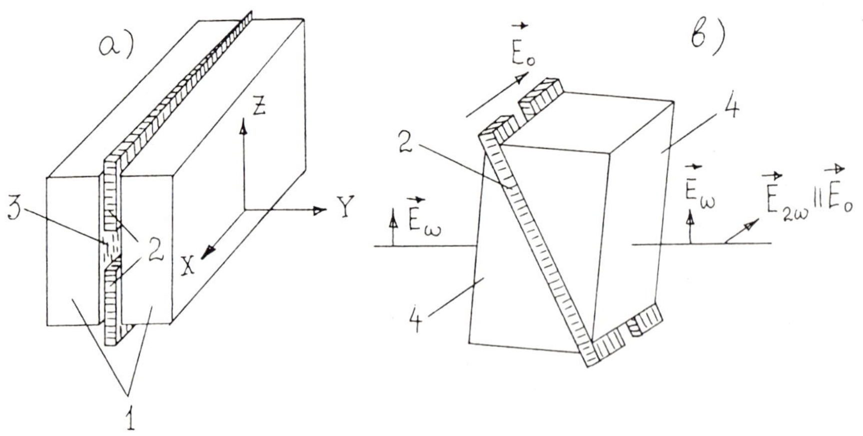

FIGURE 2 Cells for measuring coherence lengths $l_{\|,\perp}$ and non-linear susceptibilities (a): cells for observation of phase-matching (b): 1. glass plates; 2. electrodes; 3. liquid crystal; 4. glass rectangular prisms; y-axis coincides with beam direction.

It should be noted that 5CB was chosen from the methodical point of view as a reference because it has been investigated previously. Our results on 5CB agree well with those from Ref. 7.

RESULTS AND DISCUSSION

A. Field-induced SHG

The measured values of coherence lengths $l_\parallel$, $l_\perp$, l_{iso} and the corresponding cubic non-linear susceptibilities $\Gamma_\parallel$, $\Gamma_\perp$, Γ_{iso} for all the compounds investigated are shown in Table. The effective cubic polarizabilities $\gamma_\parallel^*$, $\gamma_\perp^*$, γ_s^* and γ_{iso} were calculated using the isotropic (Vuks)[13] form for the local electric field. The value of β_μ (the projection of the vectorial part of the quadratic hyperpolarizability β_{ijk} on the direction of the molecular dipole moment $\vec{\mu}$) was calculated from the formula[14]

$$\gamma_{iso} = \frac{1}{5}\gamma_{iijj} + \frac{1}{5kT}g\mu_i\beta_{ijj} = \frac{1}{5}\gamma_{iijj} + \frac{1}{5kT}\mu^*\beta_\mu, \tag{1}$$

where $\mu^* = g\mu$ is an effective molecular dipole moment, g is the Kirkwood factor.

Let us discuss data from Table I and begin with the isotropic phase. It is seen that the quadratic hyperpolarizabilities are almost equal for all the substances despite the pronounced difference in molecular structure. By the way, the values are approximately two times more than that for nitrobenzene. Probably, the main contribution to the β_μ value results from the benzene rings of molecules containing conjugated chains of π-bonds. The value of the cubic polarizability γ_{iso} depends linearly on the permanent dipole moment of molecules (the molecular dipole moments of 5CB, 8CB and HOPCS are nearly equal and markedly higher than that for MBBA). The latter result is consistent with increasing degree of the field orientation of the molecules which is proportional to the product μE.

The situation is more complicated for liquid crystalline phases. Nonlinear parameters for different liquid crystals vary to a larger extent. For all the substances investigated $\Gamma_\parallel$-values are higher than $\Gamma_\perp$. It is quite natural since, on measuring $\Gamma_\parallel$, the electric field of the fundamental beam, thanks to the nematic order, effectively interacts with the longitudinal component of a quadratic molecular polarizability which is larger than the transverse component for all the compounds studied. When measuring $\Gamma_\perp$, on the contrary, the transverse (small) components of the polarizability are responsible for SHG. Thus, even for equal degree of the field-induced molecular

TABLE I

Parameters	Liquid crystalline phase				
	Nematic phase				Smectic A phase
	HOPCS (23°C)	MBBA (23°C)	5CB (23°C)	8CB (33°C)	8CB (23°C)
$l_{\parallel}$, μm	3.4	3.3	4.8	5.85	5.4
$l_{\perp}$, μm	8.95	8.65	9.88	11.65	12.2
θ_c(ee-o), degree	27.7	26.7	26.1	—	24.5
θ_c(oe-o), degree	38.9	38.0	34.5	—	33.0
$\Gamma_{\parallel} \cdot 10^{14}$, ESU	118	109	260	205	196
$\Gamma_{\parallel} \cdot 10^{14}$, ESU	42	22	41	31	21
$r_s \cdot 10^{14}$, ESU	24	12	23	—	21
$\chi_s \cdot 10^{12}$, ESU	1.6	1.5	0.7	—	0.95
$\gamma_{\parallel}^* \cdot 10^{36}$, ESU	119	83	158	156	149
$\gamma_{\perp}^* \cdot 10^{36}$, ESU	42.5	16.7	25	23.5	16
$\gamma_c^* \cdot 10^{36}$, ESU	24.3	9.1	14	—	16
$\Gamma_{\parallel}/\Gamma_{\perp}$	2.8	4.95	6.35	6.6	9.3
		Isotropic phase			
l_{iso}, μm		5.9	5.6	7.4	8.6
$\Gamma_{iso} \cdot 10^{14}$, ESU		69	49	106	83
$\gamma_{iso} \cdot 10^{36}$, ESU		70	37.2	64.5	63
$\beta_{\mu} \cdot 10^{31}$, ESU		42	39	40.5	39.5

orientation along and across the director, the effective quadratic susceptibility parallel to the director is higher than that for the perpendicular direction.

Moreover, it is evident, that for equal quadratic polarizabilities a higher value of $\Gamma_{\parallel}$ should correspond to the molecules with bigger dipole moments as, in this case, the degree of the field orientation (the field polarization of the medium) is higher. In addition, the orientation of the dipole relative to the molecular axis is of importance. So, the cubic susceptibility $\Gamma_{\parallel}$ for both 5CB and 8CB (dipole moment $\mu \approx$ 5D is parallel to the long molecular axis) is markedly higher than that for HOPCS and MBBA (dipole moments $\mu \approx$ 4 and 2D, respectively, are directed at an angle $\theta \approx 60°$ to the long axis). The susceptibility $\Gamma_{\perp}$ is maximum for HOPCS whose molecules have maximum transverse component of the dipole moment. Relatively high values $\Gamma_{\perp}$ for 5CB and 8CB may be accounted for by an essential contribution from big longitudinal dipole moments due to the non-ideality of the nematic order, $S < 1$. It is seen from Table I that the anisotropy of the cubic susceptibility $\Gamma_{\parallel}/\Gamma_{\perp}$ increases when one follows from liquid crystals with negative dielectric anisotropy to those with positive one. Thus the values and the directions of dipolar groups play a crucial role in the magnitude of components of the cubic susceptibility tensor.

To study the SHG peculiarities for the smectic A phase, there was investigated the temperature behavior of the coherence lengths $L_{\|}$ and $L_{\perp}$ in 8CB (Figure 3), as well as the intensities of the second harmonic $I^{\|}_{2\omega}$ and $I^{\perp}_{2\omega}$. There exist simple relationships between the coherence lengths in the nematic ($l_{\|}$, $l_{\perp}$) and isotropic (l_{iso}) phases[7]

$$\frac{1}{l_{\|}} - \frac{1}{l_{iso}} = K\langle P_2\rangle; \qquad \frac{1}{l_{iso}} - \frac{1}{l_{\perp}} = \frac{K}{2}\langle P_2\rangle, \tag{2}$$

where $\langle P_2\rangle$ is an orientational order parameter. We can calculate the order parameter as a function of temperature using experimental data on $l_{\|,\perp}$. The coefficient K was determined from the condition $\langle P_2\rangle = 0.64$ at $t - t_{iso} = -17$ °K.[15] The calculated curve is given in Figure 4, where, for comparison, the temperature dependence $\langle P_2\rangle$ from the refractive index measurements[13] is also shown. Both results qualitatively agree.

The temperature behavior of non-linear susceptibilities $\Gamma_{\|}$, $\Gamma_{\perp}$ normalized to the value Γ_{iso} is illustrated in Figure 5. The values $\Gamma_{\|}$ and $\Gamma_{\perp}$ depend upon molecular parameters and the order parameter of a mesophase.[7] If the order parameter is described only by the value $\langle P_2\rangle$ then, neglecting the temperature dependence of density, we have

$$R_{\|} = \frac{\Gamma_{\|}}{\Gamma_{iso}} = \frac{F_{\|}}{F_{iso}}\left(\frac{T_{iso}}{T} + \frac{2}{7}\frac{\alpha}{\gamma_{iso}}\langle P_2\rangle\right) \tag{3}$$

$$R_{\perp} = \frac{\Gamma_{\perp}}{\Gamma_{iso}} = \frac{F_{\perp}}{F_{iso}}\left(\frac{T_{iso}}{T} - \frac{1}{7}\frac{\alpha}{\gamma_{iso}}\langle P_2\rangle\right) \tag{4}$$

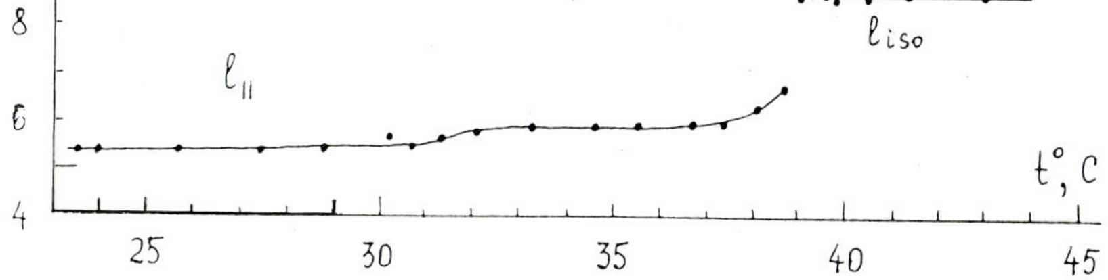

FIGURE 3 Coherence lengths of 8CB as functions of temperature.

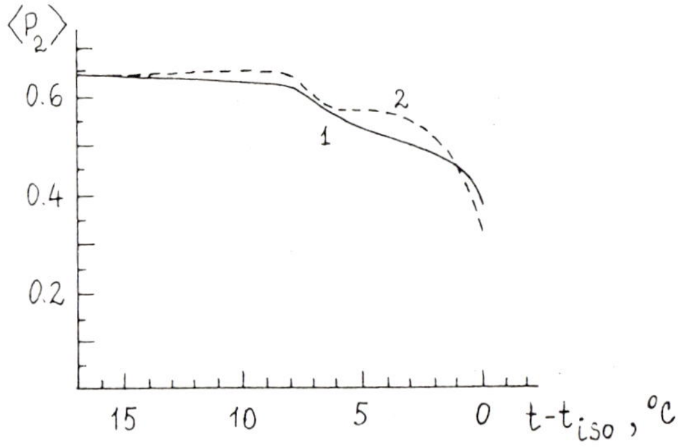

FIGURE 4 Temperature behavior of the order parameter for 8CB. 1. results of the paper;[13] 2. values calculated from formula.[1]

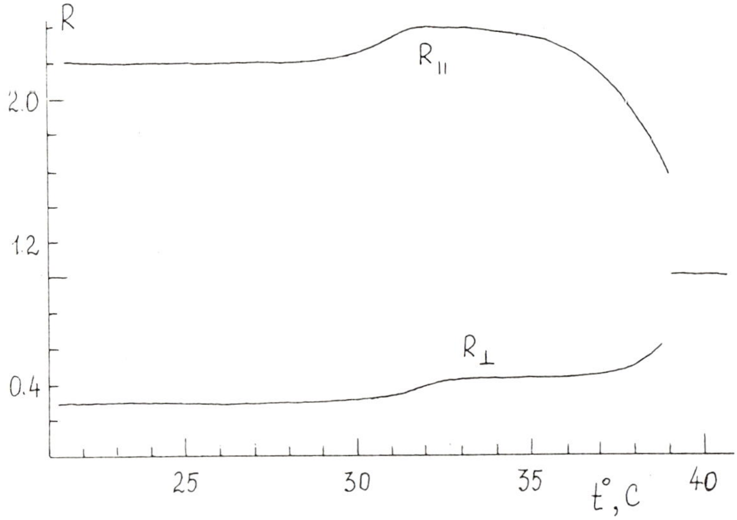

FIGURE 5 Temperature behavior of cubic non-linear susceptibilities for 8CB. Absolute value $\Gamma_{iso} = 8.3 \cdot 10^{-13}$ CGSU.

Here, $F_{\|}$, $F_{\perp}$ and F_{iso} are local field factors, α is a molecular parameter including the dipole moment, second order and third order hyperpolarizabilities. Under condition $F_{\|} = F_{\perp} = F_{iso}$ the formula[3] does not agree with the experiment for 8CB. Indeed, it follows that $\Gamma_{\|}$ should increase at the transition from the nematic to smectic A phase according to the increase in $\langle P_2 \rangle$.[3] However, in the experiment, $\Gamma_{\|}$ markedly decreases. The similar temperature behavior was earlier observed for dielectric permittivity $\varepsilon_{\|}$ of 8CB.[16] In the latter case the decrease in $\varepsilon_{\|}$ is due to the antiparallel correlation of molecular dipoles in the smectic A phase,[17] which results in a decrease in the effective dipole moment μ.* Thus, the decrease in Γ may be ascribed to the decreasing parameter α including μ^* and the hyperpolarizability β_{ijk}. It is also possible that the dipolar correlation influences the β_{ijk}-value, too. Indeed, it is difficult to ascribe quantitatively all the changes in Γ only to the decrease in μ^* calculated from the temperature dependence of $\varepsilon_{\parallel}$.

For all the substances investigated we managed to observe the phase-matched SHG of the ee-o and oe-o types. The corresponding directions, characterized by an angle Θ_s between the liquid crystal director and the wave vector of the fundamental beam, are also shown in the table. The values for the non-linear (synchroneous) susceptibility Γ_s for the ee-o interaction is presented, too. The curve of the phase synchronism for the smectic A phase of 8CB is shown in Figure 6.

B. Zero-field SHG

Figure 7 shows the intensities for the phase-matched SHG in MBBA measured for both ee-o and oe-o types of the synchronism. It shows that the ee-o synchronism is observed only for the field-induced SHG, while the oe-o type is observed in both the field-induced and zero-field cases. It

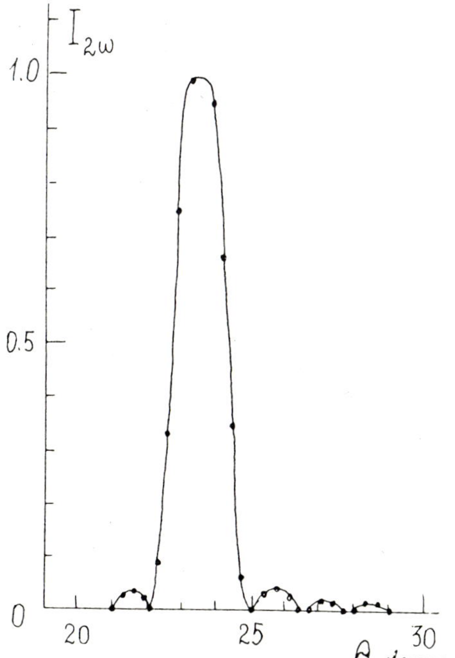

FIGURE 6 The intensity of the second harmonic $I_{2\omega}$ as a function of the angle Θ between the director and wave vector of the fundamental beam for 8CB (t = 23°C).

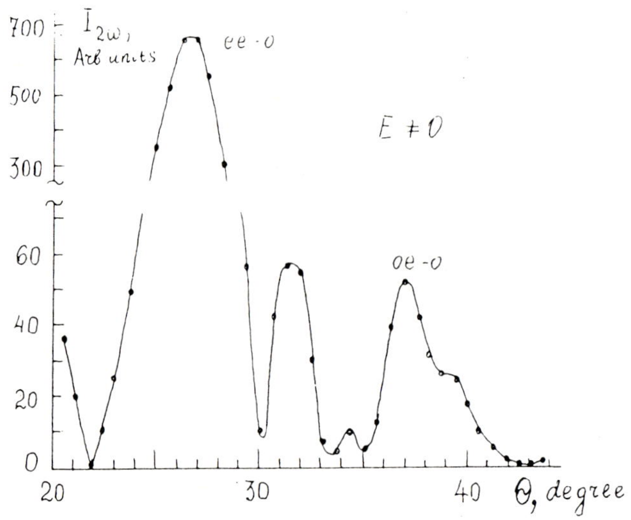

(*a*)

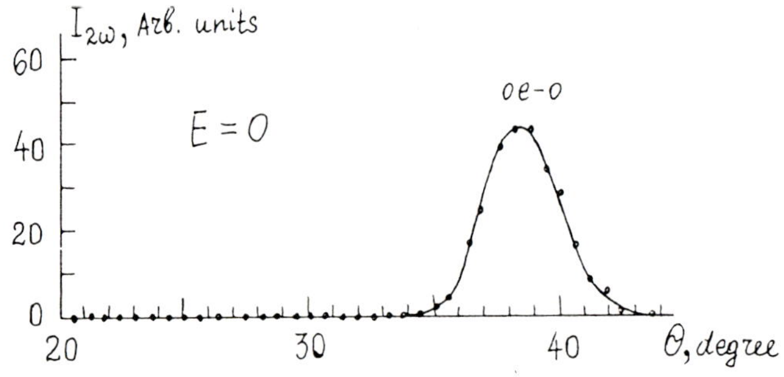

(*b*)

FIGURE 7 The intensity of the second harmonic on angle Θ for homeotropically oriented MBBA: (a) the induced field is switched on ($E = 14\,\mathrm{kV \cdot cm^{-1}}$); (b) zero-field SHG. Cell thickness 50 μm, $t = 23.5°C$.

should be noted that SHG is detectable for both homogeneously and homeotropically oriented layers. To decide whether the zero-field SHG has volume or surface nature the $I_{2\omega}$ for the oe-o synchronism was measured as a function of layer thickness (L) using a wedge-form cell and scanning a laser beam along cell. At small thicknesses, the SHG intensity $I_{2\omega}$ is proportional to L^2 (Figure 8). Such a dependence is observed up to the values $L \approx 50$ μm, and then the slope of the curve decreases. For the sake of comparison, in the same Figure curve $I_{2\omega}(L)$ is also shown for the field-induced SHG of the ee-o type. In the latter case, the dependence $I_{2\omega} \sim L^2$ is observed up to thicknesses ~100 μm. This difference in critical

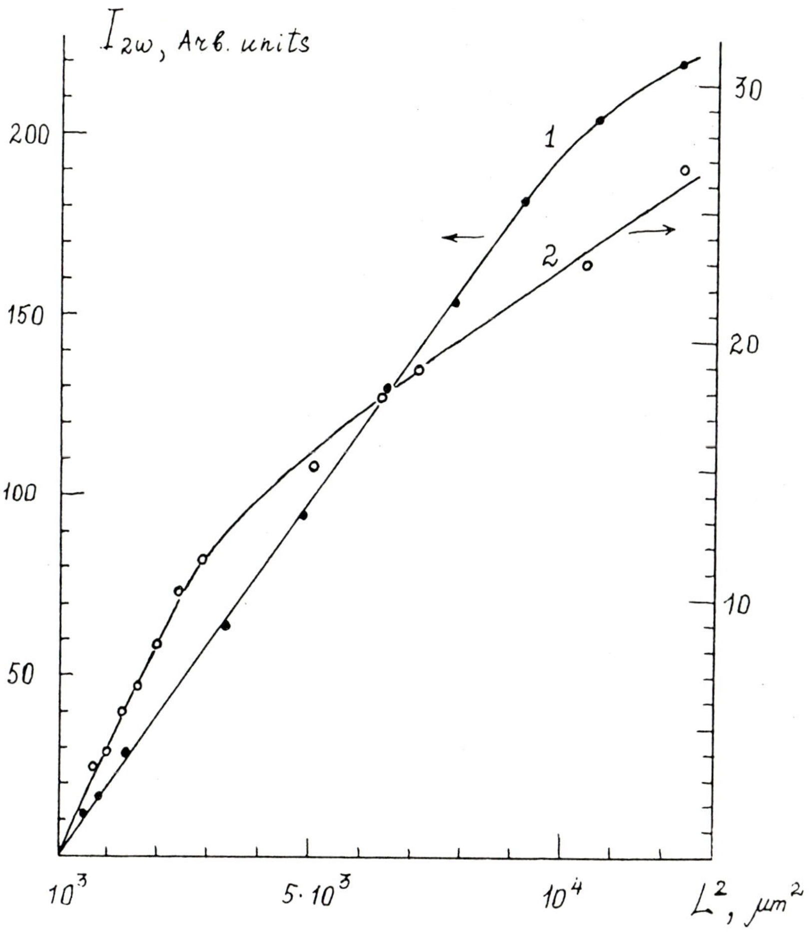

FIGURE 8 The second harmonic intensity as a function of the thickness for a homeotropically oriented layer of MBBA. 1. field-induced SHG for ee-o interaction, 2. zero-field SHG for oe-o interaction.

thicknesses appears to be due to a different character of light scattering for the two wave interaction types. Both the thickness dependence and the coincidence of observed and calculated angles for the phase-matched SHG allow us to conclude that the zero-field as well as the field-induced SHG are of volume nature and cannot be referred to the surface phenomena.

The zero-field oe-o type SHG, in our opinion, may be accounted for by a multipolar (the quadrupolar) mechanism known for the solid crystals.[18] The effective second-order susceptibilities χ_s calculated from the experimental data on the oe-o synchronism are shown in the Table. Their order of magnitude is typical of the quadrupolar SHG mechanism. So, the phenomenon of the zero-field second harmonic generation in nematic and smectic A liquid crystals can be understood from the traditional point of view on these substances as center-symmetrical media. Such an interpretation is completely consistent with the theoretical approach developed in Ref. 19.

Acknowledgments

The authors are grateful to Dr. P. Adomenas for supplying the sample of HOPCS and Dr. E. I. Kovshev for the other substances.

References

1. I. Freund, P. M. Rentzepis, *Phys. Rev. Lett.*, **18**, 393 (1967).
2. G. Durand and C. H. Lee, *Mol. Cryst.*, **5**, 171 (1968).
3. L. S. Goldberg and J. M. Schnur, *Rad. Electr. Eng.*, **39**, 279 (1970).
4. S. M. Arakelyan, G. L. Grigoryan, S. Ts. Nersisyan and Yu. S. Chilingaryan, *Zh. Eksp. Teor. Fiz.*, **80**, 1883 (1981).
5. Gu. Shi-Jie, S. K. Saha and G. K. Wong, *Mol. Cryst. Liq. Cryst.*, **69**, 287 (1981).
6. N. M. Shtykov, L. M. Blinov, A. M. Dorozhkin and M. I. Barnik, *Pis'ma Zh. Eksp. Teor. Fiz.*, **35**, 142 (1982).
7. S. K. Saha and G. K. Wong, *Appl. Phys. Lett.*, **37**, 423 (1979); *Opt. Commun.* **30**, 119 (1979); *Opt. Commun.*, **37**, 373 (1981).
8. M. I. Barnik, L. M. Blinov, A. M. Dorozhkin and N. M. Shtykov, *Zh. Eksp. Teor. Fiz.*, **81**, 1763 (1981).
9. J. F. Ward and I. J. Bigio, *Phys. Rev.*, **A11**, 60 (1975).
10. C. G. Bethea, *J. Chem. Phys.*, **69**, 1312 (1978).
11. J. L. Ouder, *J. Chem. Phys.*, **67**, 446 (1977).
12. A. N. Vtyurin, V. P. Ermakov, B. I. Ostrovski and V. F. Shabanov, *Phys. Stat. Col.*, **B107**, 397 (1981); *Kristallografiya*, **26**, 546 (1981).
13. M. F. Vuks, *Opt. Spektrosk.*, **60**, 644 (1966).
14. B. F. Levine and C. G. Bethea, *J. Chem. Phys.*, **63**, 2666 (1975).
15. R. G. Horn, *J. de Phys.*, **39**, 105 (1978).
16. C. Druon and J. M. Wacrenier, *J. de Phys.*, **38**, 47 (1977).
17. W. H. De Jeu, W. J. Goossens and P. Bordewijk, *J. Chem. Phys.*, **61**, 1985 (1974).
18. N. Bloembergen, Nonlinear Optics, W. A. Benjamin, Inc. New York, Amsterdam, 1965.
19. G. A. Lyakhov and Yu. P. Svirko, *Zh. Eksp. Teor. Fiz.*, **80**, 1307 (1981).

Phase-synchronous optical second-harmonic generation in a ferroelectric liquid crystal

M. I. Barnik, L. M. Blinov, and N. M. Shtykov

Research Institute for Organic Semifinished Products and Dyes
(Submitted 21 October 1983)
Zh. Eksp. Teor. Fiz. **86**, 1681–1683 (May 1984)

"Temperature" phase synchronism is obtained in optical second-harmonic generation in a ferroelectric smectic C^* whose helicoid is untwisted by a constant electric field. The nonlinear second-harmonic susceptibility due to spontaneous polar ordering of the molecules is estimated.

The ferroelectric smectics C^* and H^* are the only liquid-crystal phases with a polar structure that should make possible in them second-harmonic generation (SHG) via the quadratic nonlinearity $\chi^{(2)}$. Up to now, only nonsynchronous SHG was reported in a ferroelectric liquid crystal (LC).[1,2] We present here the results of an investigation of phase-synchronous SHG in the ferroelectric phase C^* with untwisted helicoid. The investigated substance was *p*-decyloxybenzylidene-*p*-amino-2-methylbutylcinnamate (DOMAMBC).

In the absence of external forces in an LC, the director and the spontaneous-polarization vector of the smectic layer form in space a helicoidal structure similar to the cholesteric, the helicoid axis being perpendicular to the plane of the smectic layers. The spontaneous polarization vector of the smectic layer lies in the plane of the layer and is perpendicular to the molecule inclination plane. Application of a dc electric field perpendicular to the helicoid axis untwists the helicoid, but this raises the problem of separating the contribution made to the SHG by the spontaneous polarization from that of the polarization induced by the electric field. This is one of the problems solved in the present paper.

The usual method of obtaining phase-synchronous SHG in a nematic phase and in a smectic-*A* is to change the angle between the pump wave vector and the LC optical axis by rotating the cell around the direction of the dc electric field [3]. To obtain phase synchronism in this case we varied the temperature to change the inclination of the molecules in layers of the smectic C^* phase. A calculation of the expected synchronism direction θ_c from the refractive indices of DOBAMBC yields about 23.5° and 33.8° for interactions of the *ee–o* and *oe–o* type, respectively. Thus, in DOBAMBC, in which the angle between the molecule inclination and the normal to the smectic layer ranges from 0 to 29°,[4] it is possible to obtain temperature-controlled phase synchronism for *ee–o* interaction.

The SHG was investigated with a previously described[3] setup. We used cells with uniform LC layer thicknesses 18.4 and 90 μm, similar in configuration to the planar cell of Ref. 3, the helicoid axis being perpendicular to the walls. The dc electric field was perpendicular to the helicoid axis.

The temperature dependence of the second-harmonic intensity $I_{2\omega}$ is shown in Fig. 1. Synchronism is observed at the temperature $T_{C^*-A} - T \approx 8$ °C, corresponding[4] to an average angle $\sim 23°$ between molecule inclination and the normal to the smectic layers, and in good agreement with the calculations. The half-width of the synchronism curve and the degree of polarization $\rho = 0.95$ of the second-harmonic radiation attest to good homogeneity of the LC layer orientation.

The dependence of $I_{2\omega}$ on the electric field intensity E_0 is shown in Fig. 2. The curve has three distinct sections: *a*—the helicoid is not yet untwisted and the SHG signal is weak; *b*—the helicoid becomes untwisted and the second-harmon-

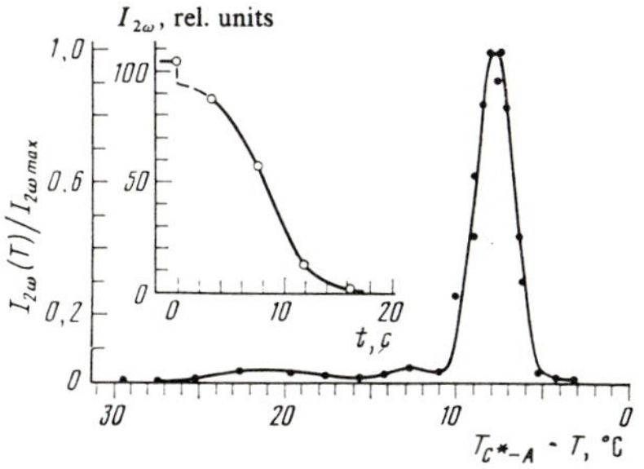

FIG. 1. Temperature dependence of the second harmonic of light. LC layer thickness μm, dc electric field intensity $E_0 = 3$ kV/cm. The inset shows the dependence of the second-harmonic intensity on the time elapsed from the instant of turning off the helicoid-untwisting field. LC layer thickness 18.4 μm, $E_0 = 4.1$ kV/cm, $T_{C^*-A} - T = 16.1$ °C.

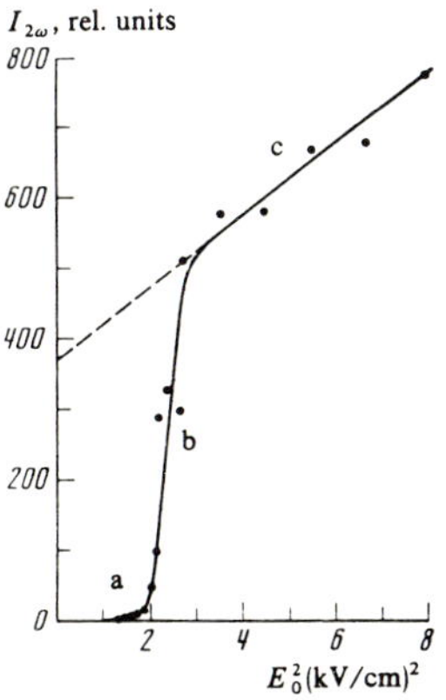

FIG. 2. Second-harmonic intensity at the synchronism maximum vs dc electric field intensity. LC layer thickness 90 μm.

 0038-5646/84/050980-02$04.00

ic intensity grows rapidly because the synchronization condition sets in; c—the helicoid is completely untwisted. The growth of $I_{2\omega}$ on the last section of the curve is obviously governed by the known mechanism of optical SHG induced by the electric field. Extrapolation of this section to zero field yields a nonzero second-harmonic intensity, thus confirming the presence of spontaneous polar ordering of the molecules in the smectic C^* phase. The appreciable slope of the plot of $I_{2\omega}$ vs E_0 on section c of Fig. 2 is evidence that the contributions made to the SHG by the spontaneous and field-induced polarizations are comparable at the given temperature, and that it is impossible to estimate their individual shares in this experiment. Such an estimate, however can be obtained by investigating the kinetics of the falloff of $I_{2\omega}$ if the helicoid relaxation time after turning off the untwisting field is long enough. This condition is realized in the low-temperature region of the C^* phase. The decrease of $I_{2\omega}$ with time after turning off the electric field at the temperature $T_{C^*-A} - T = 16.1$ °C is shown in the inset of Fig. 1; the synchronism was achieved in this case by rotating the cell. At the instant $t = 0$ the electric field is turned off, and since the field-induced ordering of the dipole relaxes within a time on the order of 10^{-7} sec, the SHG after a time $t = 3.3$ sec is certainly due only to the spontaneous polar ordering of the molecules. Comparing the values of $I_{2\omega}$ at these two instants of time we see that at the given lowered temperature, in an untwisting field of the order of 4 kV/cm, the electric-field-induced SHG mechanism makes no significant contribution to the resultant value of $I_{2\omega}$. This may possibly be due to an increased contribution to $I_{2\omega}$ from the spontaneous polar ordering of the molecules with decreasing temperature, as manifest in particular by the increase of the spontaneous polarization P_s.[4] The nonlinear second-order susceptibility $\chi^{(2)}$ calculated from the value of $I_{2\omega}$ at $t = 3.3$ sec is 3.5×10^{-12} cgs esu. An estimate of $\chi^{(2)}$ from the value of P_s for DOBAMBC, according to the known relation between them,[5] agrees with the experimentally obtained value.

In conclusion, the authors thank L. A. Beresnev for help with the preparation of the experiment.

[1]V. F. Shabanov, V. P. Ermakov, E. M. Aver'yanov and B. I. Ostrovskiĭ, Abstracts, All-Union Scientific-Technical Conf. on the Interaction of Laser Radiation with Liquid Crystals, Erevan, Erevan State Univ., 1978, p. 58.

[2]A. I. Vtyurin, V. P. Ermakov, B. I. Ostrovskiĭ, and V. F. Shabanov, Kristallografiya **26**, 546 (1981) [Sov. Phys. Crystallography **26**, 309 (1981)].

[3]M. I. Barnik, L. M. Blinov, A. M. Dorozhkin, and I. M. Shtykov, Zh. Eksp. Teor. Fiz. **81**, 1763 (1981) [Sov. Phys. JETP **54**, 935 (1981)].

[4]B. I. Ostrovskiĭ, A. Z. Rabinovich, and A. S. Sonin, Pis'ma Zh. Eksp. Teor. Fiz. **25**, 80 (1977) [JETP Lett. **25**, 70 (1977)].

[5]J. Jerphagnon, Phys. Rev. **B2**, 1091 (1970).

Translated by J. G. Adashko

THE ORIENTATIONAL MECHANISM OF NONLINEARITY AND THE SELF-FOCUSING OF He–Ne LASER RADIATION IN NEMATIC LIQUID CRYSTAL MESOPHASE (THEORY AND EXPERIMENT)

N.F. PILIPETSKI, A.V. SUKHOV, N.V. TABIRYAN and B.Ya. ZEL'DOVICH
Institute of Problems of Mechanics, Academy of Science USSR, 117526 Moscow, USSR

Received 11 September 1980
Revised manuscript received 30 December 1980

A giant optical nonlinearity of self-focusing type in the oriented mesophase of nematic liquid crystals (NLC) due to the director reorientation under the action of a light wave field is predicted. Self-focusing of He-Ne laser radiation with power $\sim 10^{-2}$ W and power density ~ 50 W/cm^2 in a planar oriented 60 μm thick NLC layer has been carried out experimentally. The measured value of the nonlinearity effective constant $\epsilon_2 = 0.07$ cm^3/erg corresponds to theoretical predictions, and turns out to be larger than the CS_2 nonlinearity by $\approx 10^9$ times.

1. Introduction

In the last years the nonlinear optics of liquid crystals attracts much attention, see survey [1]. A lot of investigations have been done concerning the LC isotropic phase near the transition point. Nonlinear optical effects have been essentially less investigated for LC mesophase. This might be connected with the tentative idea that for a noticeable value of self-action and interaction through a medium great volumes of interaction are necessary, while thicknesses of transparent NLC cells are practically restricted in size to ~ 100 μm [2]. From the papers dealing with NLC oriented mesophase let us note the paper [3,4], where the second harmonic generation has been investigated, and the paper [5], where about the thermal self-focusing and self-defocusing of the laser radiation is reported.

In papers [6,7] it is reported that the orientational mechanism of nonlinearity of the LC oriented mesophase can cause very large values of nonlinear optical constants: the constants of light amplification in the stimulated scattering in NLC [6] and in smectic LC [7]. The point is, that variations of the NLC director direction lead to very large changes of its optical properties, because $\epsilon_a = n_\parallel^2 - n_\perp^2 \sim 0.5$–$1$ ($n_\parallel$ and $n_\perp$ are the refractive indexes for the waves polarized along and perpendicular to the director, respectively). On the other hand, an initial homogeneous orientation of the NLC director can be achieved due to rather weak actions, e.g. rubbing, external electric or magnetic fields. Therefore, even rather weak light fields can noticeably change the director orientation and, thereby, give very strong nonlinear optical effects.

In the present paper very strong effects of optical nonlinearity of self-focusing type are considered theoretically and the results of the experiment, where He-Ne laser beam external self-focusing has been observed, are reported.

2. Theory

The equation for the evolution of the unit vector $\boldsymbol{n}(\boldsymbol{r}, t) = \boldsymbol{n}^0 + \delta\boldsymbol{n}(\boldsymbol{r}, t)$ of the LC director under the influence of the light field

$$\boldsymbol{E}_{\text{real}} = \tfrac{1}{2}[\boldsymbol{E}_0 \exp(\mathrm{i}\omega t) + \boldsymbol{E}_0^* \exp(-\mathrm{i}\omega t)] \tag{1}$$

in linear approximation by $\delta\boldsymbol{n}$ are presented in the paper [12]. Their stationary solution for a planarly oriented NLC layer $0 \leqslant z \leqslant L$ (see fig. 1) is the following

$$\delta n_z(z) = \frac{\epsilon_a (\boldsymbol{E}\cdot\boldsymbol{e}_x)^2 (\boldsymbol{E}^*\cdot\boldsymbol{e}_z)^2}{8\pi |E|^2} \frac{z(L-z)}{K_{11}}. \tag{2}$$

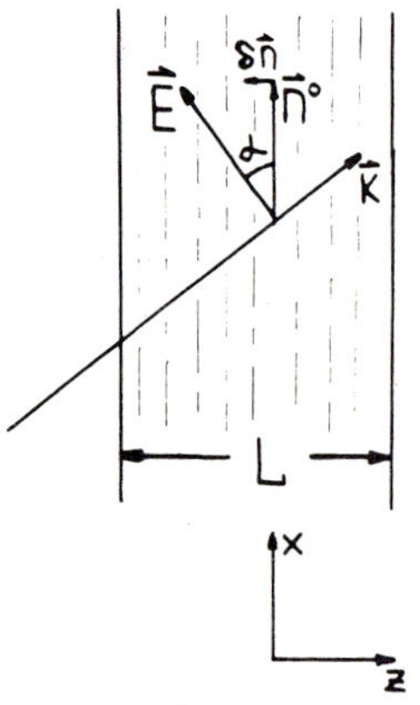

Fig. 1.

Here $\boldsymbol{e}_x$ is equal to $\boldsymbol{n}^0$ and K_{11} is Frank's constant for the given type of nematic.

Thus, the director variation is nearly equal to the ratio of the anisotropic part of the energy density of interaction between the field and nematic, $\epsilon_a|(\boldsymbol{n}\cdot\boldsymbol{E})|^2/8\pi$ to the elastic energy density K_{11}/L^2 of the nematic with characteristical deformation in uniformity size L.

The relaxation time of the stationary state, if neglecting the hydrodynamic motion, is [12]

$$\tau = \gamma L^2 \pi^{-2} K_{11} = \Gamma_1^{-1} \tag{3}$$

where γ [poise] is the relaxation constant of the director orientation in NLC mesophase.

While deducing eq. (2) it was assumed that the transverse size of the light beam $a_\perp$ is much greater than the cell's thickness $a_\perp \gtrsim L$.

From (2) it appears that an ordinary type wave, for which $(\boldsymbol{E}\cdot\boldsymbol{n}_0) \equiv (\boldsymbol{E}\cdot\boldsymbol{e}_x) = 0$ does not cause a director reorientation. For an extraordinary wave $\boldsymbol{E} = E(\boldsymbol{e}_x \cos\alpha + \boldsymbol{e}_z \sin\alpha)$ and δn_z is not zero. The phase of the extraordinary wave field is changed after passing through the cell because of the reorientation (2). The magnitude of this change is

$$\delta\varphi(0) \approx \frac{\epsilon_a L^3 \sin^2\alpha \cos\alpha}{48\pi K_{11}\lambda} |E_0|^2. \tag{4}$$

If the incident wave is not strictly plane but has a smooth dependence

$$|\boldsymbol{E}(\boldsymbol{r}_\perp)|^2 = |E_0|^2(1 - 2|r_\perp|^2/a^2 + ...)$$

on the coordinates transverse to the beam, then it is possible to use eq. (4) for $\delta\varphi(\boldsymbol{r}_\perp)$, inserting in it the local value of intensity $|\boldsymbol{E}(\boldsymbol{r}_\perp)|^2$. As a result light self-focusing appears (see [8,9,10]), and the inverse focal distance f^{-1} of the nonlinear lens is equal to

$$f^{-1} = \frac{4c\varphi(r_\perp = 0)}{\omega a^2}$$

$$= \frac{4L\epsilon_a^2}{\pi^3 a^2} \frac{\cos\alpha \sin^2\alpha}{\sqrt{\epsilon_0}\,\Gamma_1\gamma} |E_0|^2[1 - \exp(-\Gamma_1 t)] \tag{5}$$

where only the term $m = 1$ has been retained. This expression can be compared with the nonlinear lens focal distance for the case when the medium dielectric susceptibility depends on the field in the form $\epsilon = \epsilon_0 + \frac{1}{2}\epsilon_2|E|^2$, for the same geometry

$$f^{-1} = \frac{\epsilon_2|E_0|^2}{\sqrt{\epsilon_0}} \frac{L}{a^2\cos\alpha}. \tag{6}$$

3. Experiment.

The experimental setup is shown in fig. 2. The radiation of the He-Ne laser L (LG-38) generating in the lowest transverse mode, passes through the light filters F_1, and is focused by the lens LS of the focal distance $f = 25$ cm to a thin glass cell C with NLC. The angular divergence of the passing beam in the far field zone is registered by the camera PH without the objective and with the system of light filters F_2. The nematic used in the experiment was the solution of 6% mixture A and 4% of substance

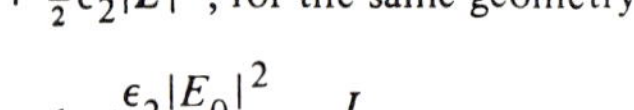

The crystal parameters are: $n_o = 1.51$, $n_e = 1.71$, $K_{11} = 8.5 \times 10^{-7}$ dyne. The crystal was at room temperature 15–20°C. The magnitude of f^{-1}, in the absence of a magnetic field is practically independent of tem-

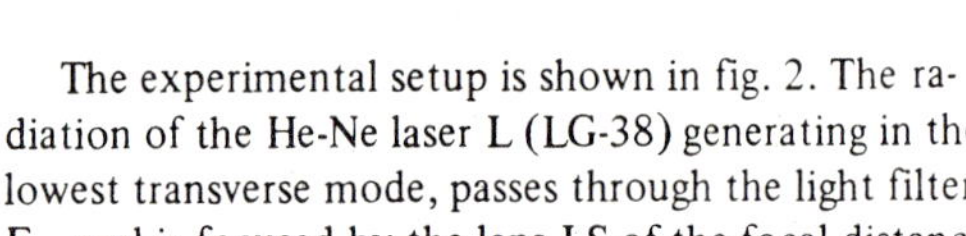

Fig. 2.

perature, because both ϵ_a^2 and K_{11} depend on temperature as the square of the order parameter. The cell thickness is 6 μm, and it has been made of 4 mm thick glass with a teflon separator. The planar orientation of the nematic is achieved by rubbing of glasses in a certain direction by diamond paste with grain size 0.5 μm. The cell plane is perpendicular to the figure plane, the vector $\boldsymbol{e}_L$ of the beam polarization and the director of NLC are in the figure plane (the polarization was verified by a Glan prism). The beam power was being changed either by filters or by a slight maladjusting of the laser resonator. The following results have been obtained: a) Noticeable change of the beam divergence is observed when the beam power is increased from 4 to 35 mW, for a cell placed in the focal waist (fig. 3a). This occurs when the wave falls obliquely to the cell. b) In the case $\alpha = 0$ a change of the beam divergence when the power is changed, is not observed (fig. 3c). c) In the case, when $\boldsymbol{n}$ is perpendicular to $\boldsymbol{e}_L$, and $\boldsymbol{e}_L$ is in the plane of the cell plate, a change of the angular structure is not observed for any incident angle. This indicates a negligible role of thermal effects in our experiment; d) In order to determine the nonlinearity sign, the cell was replaced from the caustics to the diverging beam. By this, the angular divergence decreased when the power is increased (fig. 3b). Thus we can insist that we observe self-focusing, and the orientational-polarizational dependences of the effect qualitatively coincide with the theoretical predictions. e) The ring-structure appearance in the far field zone is observed (fig. 3d). This can be explained, as is known [11], by the laser radiation phase self-modulation. The slow amplitude of laser radiation propagating along the z-direction $E_\perp(\boldsymbol{r}_\perp, z) = E_0(\boldsymbol{r}_\perp) \times \exp[\mathrm{i}\varphi_{NL}(\boldsymbol{r}_\perp)]$, can be expanded by plane waves

$$E(\boldsymbol{r}_\perp) = \int E(\boldsymbol{\theta}) \exp(\mathrm{i}\boldsymbol{\theta}\cdot\boldsymbol{r}_\perp \omega/c)\, \mathrm{d}\boldsymbol{\theta},$$

$$E(\boldsymbol{\theta}) = \int E_0(\boldsymbol{r}_\perp) \exp[\mathrm{i}\varphi_{NL}(\boldsymbol{r}_\perp) - \mathrm{i}(\omega/c)\boldsymbol{\theta}\cdot\boldsymbol{r}_\perp]\, \mathrm{d}^2\boldsymbol{r}_\perp .$$

The estimation of this integral by the method of stationary phase gives

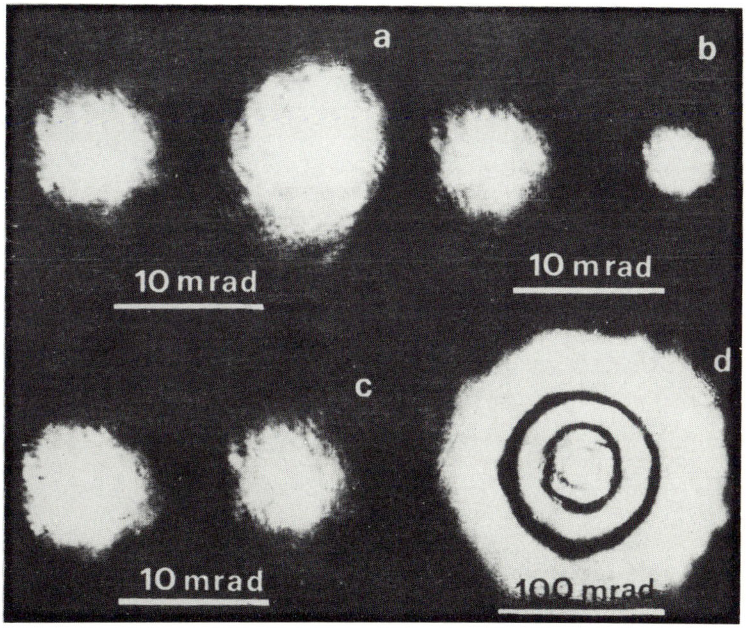

Fig. 3.

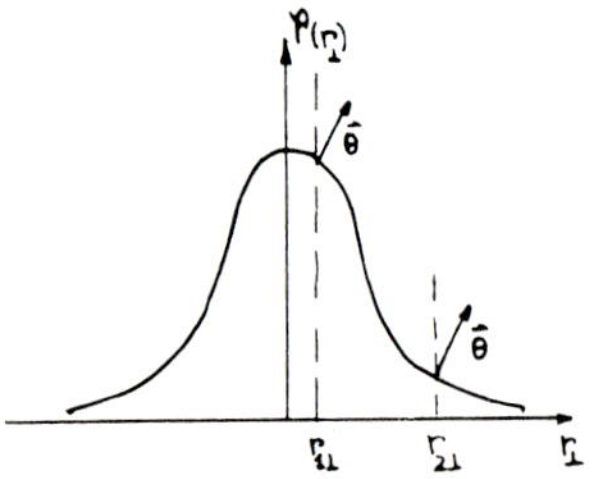

Fig. 4.

$$E(\boldsymbol{\theta}) = C_1 E_0(r_{11}) \exp[i\varphi_{NL}(r_{11}) - i(\omega/c)\boldsymbol{\theta}\cdot r_{1\perp}] + C_2 E_0(r_{2\perp}) \exp[i\varphi_{NL}(r_{12}) - i(\omega/c)\boldsymbol{\theta}\cdot r_{2\perp}]\,. \quad (7)$$

Here C_1 and C_2 are the known constants determined by the second derivatives from φ_{NL}, $r_{1\perp}$ and $r_{\perp 2}$, the two solutions of equation $\boldsymbol{\theta} = (c/\omega)\nabla\varphi_{NL}$ for the given $\boldsymbol{\theta}$, see fig. 4. The interference of two waves corresponding to the two terms in eq. (7) for $E(\boldsymbol{\theta})$ gives the ring-structure of the field in the far field zone. It is clear from fig. 4 that the maximal phase (phase in the center of the beam) φ_{max}, and, consequently, the strength of the nonlinear lens, can be roughly determined by the number of the rings N

$$\varphi_{max} \approx 2\pi N.$$

Three rings are seen in fig. 3. The corresponding magnitude of φ_{max} estimated by eq. (4) is $\varphi_{max} \approx 7\pi$.

In order to do a quantitative comparison of the parameters with the theory, the following measurements were made. By insertion into the focal waist a series of thin glass lenses with focal distances from 1.1 cm to 50 cm a graduating curve has been plotted, which allowed us to obtain the optical strength of the nonlinear lens induced in the medium by the angular divergence of the beam in the far field zone. Consider the beam intensity distribution in transverse coordinates to be gaussian. Let us define the waist radius $\boldsymbol{a}$ by the divergence θ in the far field zone, i.e. the half-width by the level e^{-2} from the maximum for the distribution $I(r_\perp) \sim \exp(-2r_\perp^2/a^2)$. Namely,

$$a = HWe^{-2}M = 2c\omega^{-1}[\theta(HWe^{-2}M)]^{-1}$$

$$= 1.18 \times 1^{-2} \text{ cm}.$$

The intensity in the beam center is

$$|E_0|^2 = 8W/cna^2 \qquad (\text{erg/cm}^3)$$

The cell was placed in the waist with an accuracy of ±0.5 mm. This has been achieved by a consecutive replacement of the lens along the beam and by finding the position, when the divergence change of the passing beam is maximal for the given W and α. The series of measurements of the dependence of the beam divergence in the far field zone on the power at the fixed external angle $\alpha_{ext} = 50°$, to which the reflection angle inside the crystal $\alpha = 32°$ corresponds, have been carried out. By the use of the graduated curve, the curve of the dependence of the nonlinear lens strength f^{-1} on the beam power has been plotted (see fig. 5). In the same way, at the fixed power $W = 30$ mW, the dependence of the nonlinear lens strength on the parameter $(\sin^2\alpha \cos\alpha)$ has been obtained, see fig. 5. Both depen-

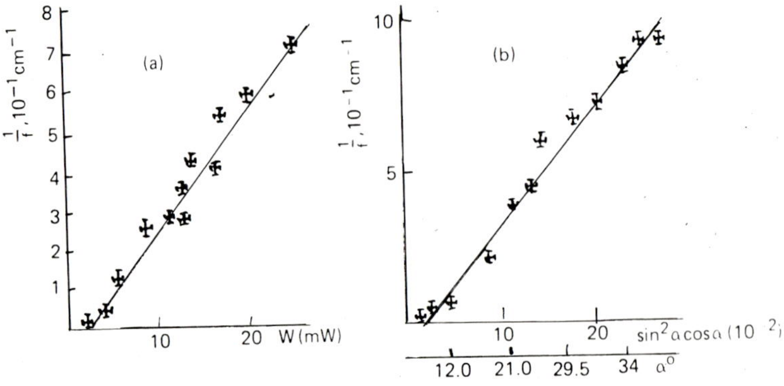

Fig. 5.

dences turn out to be linear with a good enough exactness in full accordance with the theoretical predictions. The experimental value of ϵ_2 exactly coincides with its theoretical estimation. Such an accordance can also be accidental because our inexact knowing of NLC optical and mechanical properties. The measurements of the time dependence of the nonlinear lens strength when the radiation is switched on abruptly have been also carried out. The characteristic time of establishment was $\tau \sim 10$ s. The theoretical estimation by eq. (3) gives $\tau \sim 5$ s for $\gamma \sim 1$ poise, which is typical of such a type of NLC (the exact value of γ is unknown to us). If we determine the effective value of the optical nonlinearity constant ϵ_2 from eq. (6), then, for $\alpha = 32°$ the experiment gives $\epsilon_2 = 0.07\ \mathrm{cm}^3/\mathrm{erg}$. This is approximately *nine orders higher* than the nonlinearity of such a known self-focusing medium as CS_2, for which $\epsilon_2 = 1.2 \times 10^{-10}\ \mathrm{cm}^3/\mathrm{erg}$.

References

[1] S.M. Arakelian, G.A. Lyakhov and Yu.S. Tchilingarian, UFN (in press); preprint of Yerevan Univ. K-79-2, 1979.
[2] P.G. de Gennes, The physics of liquid crystals (Clarendon Press, 1974).
[3] S.K. Saha and G.K. Wong, Appl. Phys. Lett. 34 (1979) 423.
[4] S.M. Arakelian, G.L. Grigorian, S.C. Nersisyan, M.A. Nshanian and Yu.S. Tchilingarian, Pisma ZhETF 28 (1978) 202.
[5] V. Volterra and E. Wiener-Avnear, Optics Comm. 12 (1974) 194.
[6] B.Ya. Zel'dovich and N.V. Tabiryan, Pisma ZhETF 3 (1979) 510.
[7] B.Ya. Zel'dovich and N.V. Tabiryan, Kvant. Electronica 7 (1980) 720.
[8] G.A. Ascaryan, ZhETF 42 (1962) 1567.
[9] R.Y. Chiao, E. Garmire and C.H. Townes, Phys. Rev. Lett. 13 (1964) 479.
[10] N.F. Pilipetskii and A.R. Rustamov, Pisma ZhETF 11 (196 (1965) 88.
[11] F.W. Dabby, T.K. Gustafson, J.R. Whinnery and Y. Kohanzadel, Appl. Phys. Lett. 16 (1970) 362.
[12] B.Ya. Zel'dovich, N.F. Pilipetsky, A.V. Sukhov and N.V. Tabiryan, Pisma ZhETF 31 (1980) 287.

The effect of an optical field on the nematic phase of the liquid crystal OCBP

A. S. Zolot'ko, V. F. Kitaeva, N. Kroo, N. N. Sobolev, and L. Chillag
P. N. Levedev Physics Institute, Academy of Sciences of the USSR

(Submitted 18 June 1980)
Pis'ma Zh. Eksp. Teor. Fiz. **32**, No. 2, 170–174 (20 July 1980)

We have observed a complex structure and an extraordinarily large divergence (30–40°) of an argon-laser beam emerging from an OCBP crystal placed at a focal constriction. The experimental results can be explained by a reorientation of the director in the electric field of the light wave.

PACS numbers: 61.30.Gd, 78.20.Jq

In this letter we report the experimental investigation of the effect of the optical field of an argon laser on the oriented nematic phase of the liquid crystal OCBP (octyl cyanobiphenyl). The radiation of a single-frequency CW argon laser (Spectra-Physics model 170-03) at a wavelength of $\lambda = 5145$ Å and a power of up to 120 mW was focused by a long-distance objective ($f = 210$ mm) on an OCBP crystal. The plane of the cell was perpendicular to the plane containing the wave vector **k** and director **L**. The laser radiation was polarized vertically by a polarizer, and the plane of polarization was rotated by a double Fresnel rhomb. The OCBP samples used were 50 and 150 μm thick. They were placed in a thermostatic cell with a temperature stability of 0.1°C. OCBP is in a nematic phase in the temperature interval $33° \lesssim t \lesssim 40$°C. At $t \lesssim 33$°C it is in a smectic phase; at $t \gtrsim 40$°C, an isotropic phase. Most of the experiments reported in this letter were done at a temperature of $t \simeq 37$°C, which is close to the temperature of the nematic—isotropic-liquid transition. For taking measurements at different angles a between the director **L** and wave vector **k** the cell was placed on a turntable having a vertical axis. A screen was placed at a distance of ~40 cm behind the cell, and the pattern on the screen was photographed. The cell containing OCBP was located at a constriction of the laser beam. The radius of the constriction, as estimated from the divergence of the laser beam in the far field under the assumption that the transverse intensity distribution of the beam was Gaussian, came to 5×10^{-3} cm. The field strength at the center of the beam was then $(2.3\text{–}8.8) \times 10^2$ V/cm at an incident power of 8–120 mW. The parameters of the crystal were $n_0 = 1.53$, $n_e = 1.66$, $\epsilon_\parallel = 14$, $\epsilon_\perp = 6$ ($t = 37$°C)[1], $K_{11} = 4 \times 10^{-7}$ dyn, and $K_{33} = 7 \times 10^{-7}$ dyn ($t = 34$°C).[2]

The results are set forth below.

1. When the radiation incident on the crystal was horizontally (H) polarized (the **E** vector oscillating in the plane of **k** and **L**) and the angle a was constant, the pattern observed depended on the radiation power P, i.e. on the intensity of the electric field. At low power the transmitted beam was uniform over its transverse cross section, and its divergence was small. As the power was increased, the angular divergence θ of the beam increased sharply and the beam took on a complex structure: rings appeared in the plane of the screen, which was perpendicular to **k**. They increased in number as the

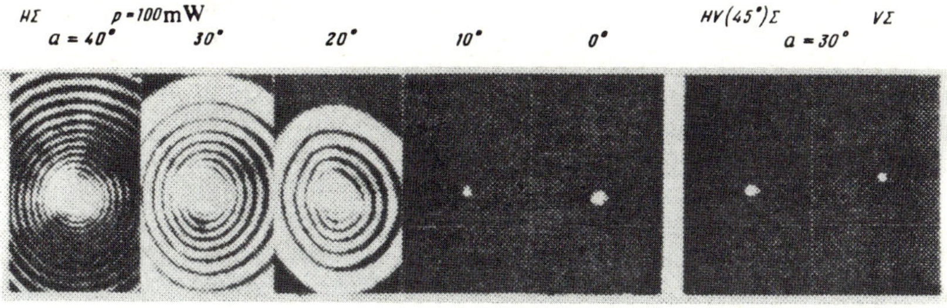

FIG. 1. Dependence of the observed pattern on the angle between the director **L** and the wave number **k** of the incident radiation (in the crystal the angles between **k** and **L** were, respectively 0, 6.5°, 13°, 19°, 24.5°).

power was increased. At powers higher than some threshold value, which was completely determined for each of the angles a, the number of rings and, hence, the angular divergence began to grow less rapidly, and the dependence of the angular divergence on the power became weak (Fig. 2).

2. For H polarization and a constant beam power P, the pattern observed in the plane of the screen depended on the angle a. It behaved with increasing angle a (0–40°) just as in the case of increasing power: the divergence angle θ increased strongly and the number of rings also increased (Fig. 1).

The rings had an irregular shape: they were larger in the vertical (V) direction (perpendicular to the plane of **k** and **L**) than in the H direction by 10–40%, depending on the angle a and power P.

3. The time interval T_d between the start of the illumination and the appearance

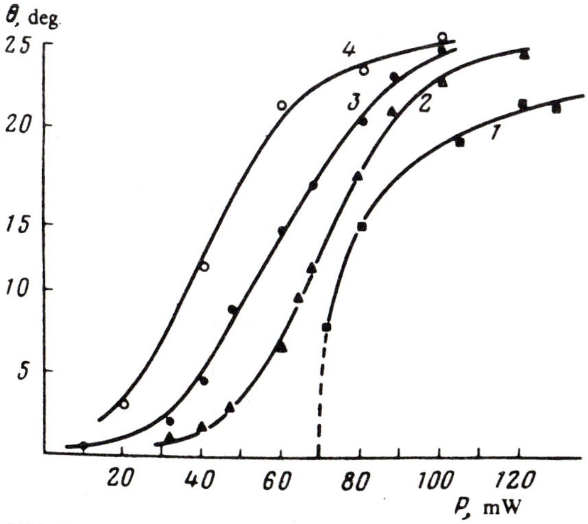

FIG. 2. Angular divergence (for the smaller dimension of the rings) of the transmitted beam as a function of the power of the laser radiation: 1—$a = 0°$, 2—$a = 10°$, 3—$a = 20°$, 4—$a = 30°$.

of the rings (the "delay time") and the time T_{st} required for establishing a stationary pattern depended on the angle a and power P. The smaller the value of a, the larger the value of T_d and the power P at which the rings appeared on the screen, and the smaller the value of T_{st}. In order of magnitude, T_d was between 10 seconds and 2 or 3 minutes, and T_{st} was from 2 to 30 seconds.

At $a = 0$, the ring structure, which was close to the onset of the region where the divergence depended weakly on the power (Fig. 2), arose almost discontinuously (over 2 or 3 sec) at $T_d \sim 30–40$ sec for $P \gtrsim 70–80$ mW. In this case $T_{st} \sim 20$ sec.

4. For V polarization of the radiation incident on the crystal and $a \gg 10°$, no rings were observed even at powers of ~ 200 mW (higher power levels were not used for fear of destroying the crystal). For $5° < a < 10°$ the rings appeared at powers of ~ 200 mW, but they were not stable: they would appear after $T_d = 30–40$ sec, and in 3 to 5 sec they would collapse; they would reappear after 30–40 sec, and in 3 to 5 sec they would again collapse, etc. At $a = 0$, as in the case of H polarization, the ring structure, which was close to the onset of the region where the divergence depended weakly on the power, arose almost discontinuously (over 2 or 3 sec) for powers $P \gtrsim 70–80$ mW. The rings were also irregular in shape, but in this case they were protracted in the H direction.

5. When the plane of polarization was rotated from H to V, the prolateness of the rings followed the plane of polarization.

6. The central spot of the ring pattern also had a complex structure. Figure 3 shows the nature of this structure and its dependence on the power at $a = 0$, as observed when the crystal was placed between crossed polarizers. The VH and HV patterns were exactly the same at low power. At a power of 104 mW the VH pattern corresponded to the HV pattern rotated by 90°; when the incident radiation was polarized at 45° to the V polarization, the VH pattern corresponded to the HV pattern rotated by 45°.

7. In crossed polarizers, the rings were extinguished only near the central spot, at

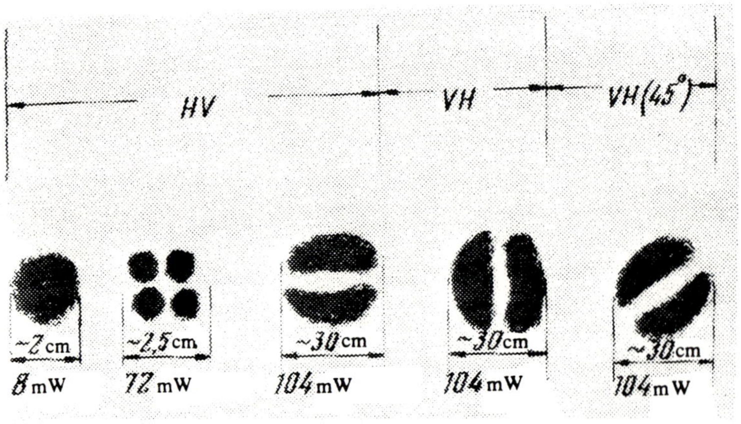

FIG. 3. Shape of the central spot when the crystal was placed between crossed polarizers.

divergence angles $\theta < 5°$. For divergence angles $\theta > 5°$ the rings were only diminished in intensity. This was true for any polarization of the incident radiation.

8. There was an extraordinarily large angular divergence of the beam on emergence from the crystal, reaching 30–40° at large powers. The total number of rings in this case was 25–35.

9. The temperature of the sample had a substantial effect. The ring structure was observed only in the nematic mesophase, and arose the more readily (the threshold power was lower and the number of rings at a given power was larger) the closer the temperature of the sample to the nematic—isotropic-liquid phase transition temperature. The ring structure was not observed in the smectic mesophase or in the isotropic liquid anywhere in the investigated interval of laser powers.

10. When the radiation incident on the crystal was circularly polarized, the ring structure was not observed. At a power of $P \sim 200$ mW there was only a small increase in the angular divergence of the beam.

In our view, all of the results of this experiment can be explained by the reorientation of the director in the electric field of the light wave, in analogy with the Fredericks transition.[3,4] Fredericks transitions are observed in uniform, constant magnetic and electric fields. Our case is more complex, but the basic features of the Fredericks effect can be clearly seen.

First of all, there was a threshold value of the field E_0 ($a = 0$) for **Lk** at which the pattern arose discontinuously for any orientation of **E** in the plane perpendicular to **k** (curve 1 in Fig. 2). Furthermore, this threshold field depended on the thickness of the sample. It was significantly larger in the 50-μm thick sample (by about a factor of three). The time required for establishing the stationary pattern was also comparable to the reorientation times of the director in external fields.[4] For angles between **L** and **k** of less than 90°, the orientational effect was observed at smaller fields.

Two more facts argue in favor of an orientational mechanism for the effects reported here: 1) the rotation of the prolateness of the rings when the plane of polarization was rotated, and 2) the periodic character of the pattern at some values of the field for small angles a.

As regards the complex structure of the beam and its extraordinarily large divergence, one can say the following. The reorientation of the director leads to a change in the refractive index and, in the general case, to a nonlinear dependence of the refractive index on the field E.[5] The nonuniformity of the field in the transverse cross section of the laser beam causes a radial dependence of the refractive index in the plane perpendicular to the wave vector **k**. The overall pattern depends on the elastic properties of the crystal and, in general, need not be radially symmetric ($K_{11} \neq K_{22} \neq K_{33}$). Thus the action of the medium on the beam passing through it is extremely complex, and this is reflected in the pattern observed. The ring structure is evidently due to nonlinear aberrations[5,6] occurring as a result of the reorientation of the director, which produces a change in the medium.

A theoretical study of the experimental results reported in this letter is in progress.

[1]D. A. Dunmur, M. R. Manterfield, W. H. Miller, and J. D. Dunleavy, Mol. Cryst. Liq. Cryst. **45**, 127 (1978).
[2]I. Janossy and L. Bata, Study of Elastic Properties near a Nematic-Smectic Transition, Preprint KFKI, Budapest, 1977.
[3]P. G. De Gennes, The Physics of Liquid Crystals, Oxford U. Press, 1974.
[4]L. M. Blinov, Electro-and Magneto-optic crystals [in Russian], Nauka, Moscow, 1978.
[5]B. Ya. Zel'dovich, N. F. Shilipetskii, A. V. Suckov, and N. V. Tabiryan, Pis'ma Zh. Eksp. Teor. Fiz. **31**, 287 (1980) [JETP Lett. **31**, 263 (1980)].
[6]S. A. Akhmanov, A. P. Sukhorukov, and R. V. Khohklov, Usp. Fiz. Nauk **93**, 19 (1967) [Sov. Phys. Usp. **10**, 609 (1968)].

Optical-Field-Induced Birefringence and Freedericksz Transition in a Nematic Liquid Crystal

S. D. Durbin, S. M. Arakelian,[a] and Y. R. Shen
Department of Physics, University of California, Berkeley, California 94720
(Received 6 April 1981)

Optical-field–induced birefringence in nematic 4-cyano-4′-pentylbiphenyl was measured with cw pump and probe beams, and the optical-field–induced Freedericksz transition was observed for the first time. The results are in quantitative agreement with the theoretical prediction.

PACS numbers: 61.30.-v, 42.65.-h

Laser-induced molecular reorientation is a common cause of optical nonlinearity in a fluid medium. In this respect, liquid crystals are often strongly nonlinear because of their large molecular anisotropy and strong correlation between molecules. The nonlinear optical properties of liquid crystals in the isotropic phase have already been studied quite extensively by a number of researchers in the past decade.[1] This is, however, not true for liquid crystals in the mesophases. Not until recently have limited theoretical and experimental studies of nonlinear optical properties of mesophases appeared in the literature.[2-4] Actually, as a result of the collective behavior of molecules, the optical nonlinearity of a mesophase can be extraordinarily large.[3] Crudely speaking, in reorienting the molecules, an optical field is equivalent to a dc field, with the optical dielectric anisotropy replacing the dc dielectric anisotropy. (A more detailed description of the optical-field–induced reorientation is outlined below.) It is known that a dc field of ~ 1 esu is sufficient to induce a significant reorientation of the molecules in a mesophase. The same is expected with an optical field of comparable strength or ~ 100 W/cm^2 in intensity. Such a field can easily be obtained by focusing a cw laser beam. In fact, a cw laser beam is best for observing the molecular reorientation effect because the response of the collective motion of the molecules is usually very slow. Two groups have recently published the observation of optical-field–induced effects in the nematic mesophase.[4] We should like to report here the first quantitative measurements of the very strong optical-field–induced reorientation of molecules in the nematic liquid crystal 4-cyano-4′-pentylbiphenyl (5CB). The results are shown to be in excellent agreement with the theory. In particular, the optical-field–induced Freedericksz transition is observed as predicted by the theory below.

In the mesophase, the average direction of molecular orientation is given by the director $\hat{n}$. For a p-polarized incident beam, the theory of field-induced molecular reorientation follows the usual derivation starting from the free-energy density[5]

$$F = \tfrac{1}{2}[K_{11}(\nabla\cdot\hat{n})^2 + K_{22}(\hat{n}\cdot\nabla\times\hat{n})^2 + K_{33}(\hat{n}\times\nabla\times\hat{n})^2 - Sn_r/c] \quad (1)$$

for a nematic liquid crystal, where K_{ii} are the Frank elastic constants. The total electromagnetic energy density $\vec{E}\cdot\vec{D}/4\pi$ is here written as Sn_r/c, where S is the magnitude of the Poynting vector, c/n_r is the ray velocity with $n_r = \{(\epsilon_\parallel + \epsilon_\perp)/\epsilon_\parallel\epsilon_\perp - [\epsilon_\perp + \Delta\epsilon(\hat{n}\cdot\hat{k})^2]^{-1}\}^{-1/2}$, $\epsilon_\perp$ and $\epsilon_\parallel$ are the optical dielectric constants parallel and perpendicular to the director $\hat{n}$, respectively, $\Delta\epsilon = \epsilon_\parallel - \epsilon_\perp$, and $\hat{k}$ is the propagation direction. Assume the sample geometry in Fig. 1 with $\hat{n}$ lying in the $\hat{x}$-$\hat{z}$ plane and being a function of z only. If $\theta(z)$ is the angle between $\hat{n}$ and $\hat{z}$ at z, then $n_x = \sin\theta(z)$ and $n_z = \cos\theta(z)$. The free-energy density becomes a function of $\theta(z)$. In the infinite-plane-wave approximation, S is a constant throughout the medium.[6] Then, minimization of the free energy $\mathfrak{F} = \int F\, d^3r$ leads to the Euler equation which can be integrated to give

$$\frac{d\theta}{dz} = \pm\left(\frac{G(\theta) - G(\theta_m)}{H(\theta)}\right)^{1/2}, \quad (2)$$

where $G(\theta) = -Sn_r(\theta)/cK_{33}$, $H(\theta) = \cos^2\theta + (K_{11}/K_{33})\sin^2\theta$, and θ_m is the value of θ at the center of the sample cell (chosen as $z = 0$). The solution of Eq. (2) with the proper boundary conditions yields $\theta(z)$ which describes the molecular reorientation by the field.

In this paper, we are interested in the case where the initial molecular alignment is homeotropic. We expect $\theta(z) = \theta(-z)$. With the boundary condition $\theta(\pm d/2) = 0$, integration of Eq. (2) yields for $-d/2 \le z \le 0$

$$z + \frac{d}{2} = \int_0^\theta \left[\frac{G(\theta') - G(\theta_m)}{H(\theta')}\right]^{1/2} d\theta'. \quad (3)$$

The value of θ_m can be obtained by setting $\theta = \theta_m$ at $z = 0$ in the above equation. A normally incident probe beam with a wavelength λ_p traversing the medium should have its extraordinary component experience an induced phase shift

$$\varphi = \frac{2\pi}{\lambda_p}\int_{-d/2}^{d/2}\left[\frac{n_0 n_e}{(n_e^2\cos^2\theta + n_0^2\sin^2\theta)^{1/2}} - n_0\right]dz, \tag{4}$$

where n_0 and n_e are respectively the ordinary and maximum extraordinary refractive indices of the nematic medium. Equations (3) and (4) have analytical solutions only in especially simple cases, but they can in general be solved numerically to give $\theta(z)$ and φ as a function of the pump field intensity $I = S = cn_0|\mathcal{E}|^2/8\pi$ and the angle α. For $\alpha = 0$, there is a threshold intensity $I_{th} = (c/n_0)(\epsilon_\parallel^2/\epsilon_\perp)(K_{33}/\Delta\epsilon)(\pi^2/d^2)$, below which no molecular reorientation can be induced. This is analogous to the dc-field-induced Freedericksz transition. Unlike the dc case, however, the optical-field-induced Freedericksz transition cannot occur for initially planar molecular alignment. This is a consequence of Mauguin's "adiabatic theorem" which states that a light beam with ordinary or extraordinary polarization rotates its polarization to follow a slow distortion of the director.[7]

The dynamic behavior of the optical-field-induced Freedericksz transition is also analogous to the dc case.[8] The initial response of the induced molecular reorientation to the laser switch-on and the long-time response to the laser switch-off are both exponential with relaxation times τ_{on} and τ_{off}, respectively.

$$\tau_{on} = \frac{c}{n_0}\frac{\epsilon_\parallel^2}{\epsilon_\perp}\frac{\gamma_1^*}{\Delta\epsilon}(I - I_{th})^{-1},$$
$$\tau_{off} = \gamma_1^* d^2/K_{33}\pi^2, \tag{5}$$

where γ_1^* is a Leslie viscosity coefficient corrected for the backflow effect. Measurements of τ_{on} and τ_{off} as a function of I allow us to determine γ_1^*, I_{th}, and K_{33} if the other parameters are known.

In our experiment, we used a 250-μm-thick sample of 5CB sandwiched between glass slides treated with dimethyl-n-octadecyl-3-aminopropyltrimethoxysilyl-chloride to give homeotropic molecular alignment. The sample was situated in an oven with a temperature stability of ± 0.05 K. A cw Ar^+ laser was used as the pump beam to induce the molecular reorientation. At high Ar^+ laser intensity ($\gtrsim 350$ W/cm^2) local heating of the sample by ~2 K was noticed. Therefore, the sample cell was placed horizontally to minimize the convection flow that may affect the measurements. A He-Ne laser was used to probe the optical birefringence φ resulting from the induced molecular reorientation. To make sure that the transverse variation of the pump beam intensity has little effect on the result, the probe beam at the sample should have a radius less than $2\pi/|\nabla\varphi|$. This was achieved by focusing the He-Ne beam tightly and using a diaphragm to limit the probe region to $\lesssim 6$ μm as compared to the Ar^+ laser spot of ~800 μm at the sample. The measured φ were collected and analyzed by a microcomputer.

To observe the optical-field-induced Freedericksz transition, the angle of incidence of the pump beam was set at ~0°. The observed birefringence versus pump intensity at $T_{NI} - T = 9.2$ K is shown in Fig. 1. It exhibits a threshold intensity at $I_{th} \simeq 155$ W/cm^2. According to Eq. (5), I_{th} can also be obtained from the measurements of τ_{on}^{-1} vs I, which exhibits a critical slowing down behavior as I approaches I_{th}. Our results are shown in Fig. 2. The linear fit to the data also yields $I_{th} = 155$ W/cm^2, in very good agreement with the direct observation. It also gives a value for the viscosity coefficient $\gamma_1^* = 0.89$ P, using $n_0 = 1.54$ and $n_e = 1.73$ for 5CB. The measured value of I_{th} allows us to find $K_{33} = 0.8\times 10^{-6}$ dyn. Then, substituting the values of γ_1^* and K_{33} in Eq. (5), we obtain $\tau_{off} \simeq 70$ sec, which is in good agreement with our directly measured value of $\tau_{off} = 63$

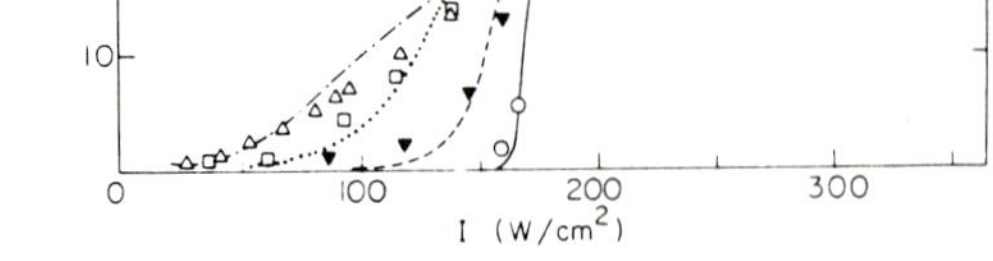

FIG. 1. Experimental points and theoretical curves for the induced birefringence at different angles α: circles and solid curve, $\alpha = 0°$; solid triangles and dashed curve, $\alpha = 3°$; squares and dotted curve, $\alpha = 11°$; open triangles and dot-dashed curve, $\alpha = 30°$. Inset shows experimental geometry.

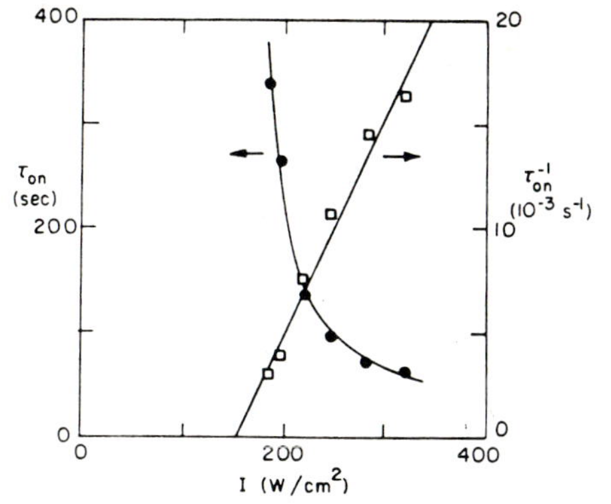

FIG. 2. Turn on time constant for $\alpha = 0°$; the straight line for τ_{on}^{-1} is least-squares fitted and determines the curve for τ_{on}.

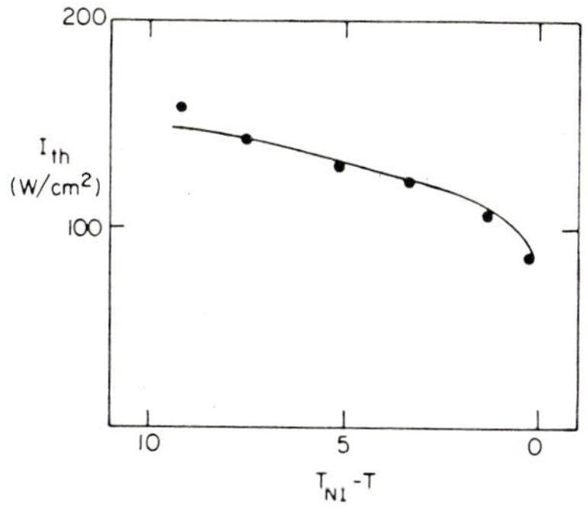

FIG. 3. Temperature dependence of threshold intensity at $\alpha = 0°$. The normalized curve shows the temperature variation of the order parameter $\langle P_2 \rangle$, computed from data in Ref. 11.

sec. We notice that our values of γ_1^* and K_{33} are fairly close to those found in the literature (Ref. 10), $\gamma_1^* = 0.67$ P and $K_{33} = 0.85 \times 10^{-6}$ dyn. The discrepancy presumably arises from the fact that the optical field had a transverse dimension of only ~800 μm.

With the pump beam incident at an angle α and the probe beam at normal incidence, the observed birefringence as a function of pump intensity, shown also in Fig. 1 for several values of α, exhibits no threshold behavior. Theoretical curves for these cases can be calculated from Eqs. (3) and (4) using our value of K_{33} and the values of K_{11}/K_{33}, n_0, and n_e from the literature.[9,10] In the actual experiment, the pump beam suffered a scattering loss in traversing the medium, and the probe beam experienced a small transverse variation of the pump intensity and hence the induced birefringence when α is finite. It is difficult to include these effects rigorously in the theoretical calculation. In obtaining the theoretical curves in Fig. 1, we simply assumed that the scattering loss would decrease the pump intensity uniformly through the sample, and the transverse variation of the induced birefringence seen by the probe beam could be accounted for by a crude averaging (which amounts to a reduction of φ by ~5% at $\alpha = 30°$). As shown in Fig. 1, the experimental data are in fair quantitative agreement with the theoretical curves, considering that no adjustable parameter was used in the calculation. The discrepancy may be due to the crude approximation used to take into account the scattering loss and the probe-beam averaging, but it is also because the theory assumes a free-energy density with no transverse variation.

We also measured, for the Freedericksz transition, I_{th} as a function of temperature, as shown in Fig. 3. We found that, more or less, I_{th} is proportional to the order parameter S. This is expected from the expression of I_{th} since $K_{33} \propto S^2$ and $\Delta\epsilon \propto S$.

From our results, we have estimated that for a pump intensity of 100 W/cm^2 at $\alpha = 30°$, the maximum reorientation angle of $\hat{n}$ is $\theta_m \sim 30°$, which corresponds to $\Delta n \sim 0.04$. This is compared to $\Delta n \sim 2 \times 10^{-11}$ in CS_2 at the same pump intensity. We therefore expect that nonlinear optical propagation effects in a nematic liquid crystal are easily observable.[4] We did, in fact, observe strong wavefront distortion of the pump-laser beam in our experiment. The nonlinearity can be further enhanced with a proper dc electric or magnetic bias field on the medium.[3,4] Novel nonlinear optical experiments arising from such large nonlinearities can be anticipated.

This work was supported by National Science Foundation Grant No. DMR-78-18826.

---

[a]On leave from Erevan State University, Erevan, Armenia, USSR.

[1]G. K. L. Wong and Y. R. Shen, Phys. Rev. Lett. 30, 895 (1973), and Phys. Rev. A 10, 1277 (1974); J. Prost and J. R. Lalanne, Phys. Rev. A 8, 2090 (1973); S. M. Arakelian, G. A. Liakhov, and Yu. S. Chilingarian, Usp. Phys. Nauk. 130, 3 (1980) [Sov. Phys. Usp. 23, 245 (1980)].

[2]J. W. Shelton and Y. R. Shen, Phys. Rev. Lett. 25, 23 (1970), and 26, 538 (1971), and Phys. Rev. A 5, 1867 (1972).

[3]R. M. Herman and R. J. Serinko, Phys. Rev. A 19, 1757 (1979); B. Ya, Zel'dovich and N. V. Tabiryan, Pis'ma Zh. Eksp. Teor. Fiz. 30, 510 (1979) [JETP Lett. 30, 478 (1979)].

[4]B. Ya. Zel'dovich, N. F. Pilipetskii, A. V. Sukhov, and N. V. Tabiryan, Pis'ma Zh. Eksp. Teor. Fiz. 30, 287 (1980) [JETP Lett. 31, 263 (1980)]; A. S. Zolot'ko, V. F. Kitaeva, N. Kroo, N. N. Sobolev, and L. Chillag, Pis'ma Zh. Eksp. Teor. Fiz. 32, 170 (1980) [JETP Lett. 32, 158 (1980)]; I. C. Khoo and S. L. Zhuang, Appl. Phys. Lett. 37, 3 (1980); I. C. Khoo, Phys. Rev. A 23, 2077 (1981).

[5]See, for example, P. Sheng, in *Introduction to Liquid Crystals*, edited by E. B. Priestley, P. J. Wojtowicz, and P. Sheng (Plenum, New York, 1975), p. 103.

[6]Because of the spatially varying anisotropy, the optical field varies in traversing the medium, but the magnitude of the Poynting vector remains constant assuming negligible loss. Not only the electricial energy, but also the magnetic energy, of the optical field participate in reorientation of the molecules. This is different from the dc case where the applied electric field is usually a constant and is solely responsible for the molecular reorientation.

[7]P. G. de Gennes, *The Physics of Liquid Crystals* (Clarendon Press, Oxford, 1974), p. 88.

[8]P. Pieranski, F. Brochard, and E. Guyon, J. Phys. (Paris) 34, 35 (1973); N. Aneva, A. G. Petrov, S. Sokerov, and S. P. Stoylov, Mol. Cryst. Liq. Cryst. 60, 1 (1980).

[9]R. G. Horn, J. Phys. (Paris) 39, 105 (1978).

[10]K. Sharp, S. Lagerwall, and B. Stebler, Mol. Cryst. Liq. Cryst. 60, 215 (1980).

[11]K. C. Chu, C. K. Chen, and Y. R. Shen, Mol. Cryst. Liq. Cryst. 59, 97 (1980).

Fredericks transitions induced by light fields

B. Ya. Zel'dovich, N. V. Tabiryan, and Yu. S. Chilingaryan

Erevan State University

(Submitted 2 December 1980)
Zh. Eksp. Teor. Fiz. **81**, 72–83 (July 1981)

The reorientation of the director of a nematic liquid crystal induced by the field of a light wave is considered. An oblique (with respect to the director) extraordinary wave of low intensity yields the predicted and previously observed giant optical nonlinearity in a nematic liquid crystal. For normal incidence of the light wave on the cuvette with a homeotropic orientation of the nematic liquid crystal, the reorientation appears only at light intensities above a certain threshold, and the process itself is similar to the Fredericks transition. The spatial distribution of the director direction is calculated for intensities above and below threshold. Hysteresis of the Fredericks transition in a light field, which has no analog in the case of static fields, is predicted.

PACS numbers: 61.30.Gd, 64.70.Ew

1. INTRODUCTION

The nonlinear optics of liquid crystals has recently received a great deal of attention (see the review[1]). In addition to the quite interesting quadratic nonlinearities (generation of the second harmonic[2]), recently there have been predictions[3-8] and experimental discoveries[4,9] of giant cubic optical nonlinearities of liquid crystals (nematic, cholesteric, and smectic), which are due to the reorientation of the director by light fields. Succeeding experimental studies[10,11] have confirmed the presence of giant optical nonlinearities for a number of specific nematic liquid crystals (NLC) and experimental geometries. In particular, a nonlinear interaction was found[11] between a normally incident light wave and a homeotropically oriented cell of the liquid crystal (OCBP), and had a characteristic threshold dependence on the intensity of the laser beam. This effect was explained qualitatively in Ref. 11 on the basis of an analogy with the Fredericks transition in the field of a light wave.

In the present study we construct a quantitative theory of the Fredricks transition in the field of a light wave. In contrast to the simplest model of the Fredericks transition in static fields (see Refs. 12 and 13), here we take into account the following two facts: 1) the amplitude for the deformation of the director above threshold must be found taking into account the distortion of the longitudinal profile of the light wave itself as it propagates in an inhomogeneous anisotropic medium; 2) if the transverse dimension of the beam is significantly smaller than the cuvette thickness, the threshold intensity itself depends on the transverse distribution of the intensity in the beam. In addition, the theory predicts that for certain liquid crystals the Fredericks transition is accompanied by hysterisis.

2. THE SYSTEM OF BASIC EQUATIONS

We shall take the free energy per unit volume of an NLC in the presence of a light field of complex amplitude **E** to be of the form

$$F\ [\mathrm{erg/cm^3}] = {}^1/_2 K_{11}(\mathrm{div}\,\mathbf{n})^2 + {}^1/_2 K_{22}(\mathbf{n}\,\mathrm{rot}\,\mathbf{n})^2 + {}^1/_2 K_{33}[\mathbf{n}\,\mathrm{rot}\,\mathbf{n}]^2 - \varepsilon_a(\mathbf{nE})(\mathbf{nE}^*)/16\pi - \varepsilon_\perp|\mathbf{E}|^2/16\pi. \quad (1)$$

Here K_{ii} are the Frank constants, **n** is a unit vector in the direction of the director (**n** and −**n** will be assumed equivalent), and the permittivity tensor of an NLC at the frequency of the light field ω is

$$\varepsilon_{ik} = \varepsilon_\perp\delta_{ik} + (\varepsilon_\parallel - \varepsilon_\perp)n_i n_k.$$

In addition, we have used in (1) the notation $\varepsilon_a = \varepsilon_\parallel - \varepsilon_\perp$.

The complex amplitude of a quasimonochromatic field $\mathbf{E}(\mathbf{r}, t)$ is determined in a self-consistent manner from the solution of Maxwell's equations with

$$\varepsilon_{ik}(\mathbf{r}, t) = \varepsilon_\perp\delta_{ik} + \varepsilon_a n_i(\mathbf{r}, t)n_k(\mathbf{r}, t).$$

However, we emphasize that to obtain the variational equations for the director $\mathbf{n}(\mathbf{r}, t)$ it is necessary to assume that the amplitude of the electric field $\mathbf{E}(\mathbf{r}, t)$ is fixed. We shall take the density of the dissipative function R (in erg/cm³ · sec) to be

$$R = {}^1/_2\eta(\dot{\mathbf{n}}\dot{\mathbf{n}}). \quad (2)$$

The variational equations for $\mathbf{n}(\mathbf{r}, t)$ have the form

$$\frac{\delta F}{\delta n_i} - \frac{\partial}{\partial x_k}\frac{\delta F}{\delta(\partial n_i/\partial x_k)} = \lambda n_i - \frac{\delta R}{\delta \dot{n}_i}, \quad (3)$$

where $\lambda(\mathbf{r})$ is an undetermined Lagrange multiplier that ensures that the condition $|\mathbf{n}| = 1$ is satisfied.

3. THE FREDERICKS TRANSITION IN BROAD LIGHT BEAMS

Let us consider a homeotropically oriented NLC cell occupying a layer $0 \le z \le L$. We shall assume that the light field has nonzero components E_x and E_z, while $E_y = 0$. The unperturbed direction of the director in $\mathbf{n}^0 = \mathbf{e}_z$. We shall also assume that the perturbation of the director does not move it out of the xz plane. Then

$$\mathbf{n}(\mathbf{r}) = \mathbf{e}_z\cos\varphi + \mathbf{e}_x\sin\varphi,$$

where $\varphi = \varphi(\mathbf{r})$. If the transverse dimensions of the beam $a_\perp$ in the xy plane are much larger than the cuvette thickness L and the beam itself is almost a plane wave, the field components E_x and E_z can be assumed to depend only on z; more precisely,

$$\mathbf{E} = [\mathbf{e}_x E_x(z) + \mathbf{e}_z E_z(z)]\exp(i\omega sx/c + ik_z z). \quad (4)$$

Here $s = \sin\alpha$ and α is the angle of incidence of the

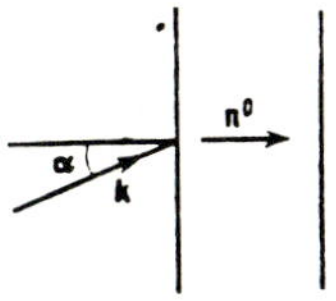

FIG. 1. Unperturbed director vector $\mathbf{n}^0$ perpendicular to the cell walls. The wave vector $\mathbf{k}$ of the light field makes an angle α with the normal to the cell walls.

light wave on the cell in air (see Fig. 1). Under these conditions the slope of the director φ also depends only on z. Now the variational equations (3) take the form

$$(K_{11}\sin^2\varphi+K_{33}\cos^2\varphi)\frac{\partial^2\varphi}{\partial z^2}-(K_{33}-K_{11})\sin\varphi\cos\varphi\left(\frac{\partial\varphi}{\partial z}\right)^2 + \frac{\varepsilon_a}{16\pi}[\sin 2\varphi(|E_x|^2-|E_z|^2)+\cos 2\varphi(E_xE_z^*+E_x^*E_z)]=\eta\frac{\partial\varphi}{\partial t}. \quad (5)$$

Let us first give a qualitative picture of the Fredericks transition. When a plane light wave falls at normal incidence on a homeotropic cell the electric field of the wave is exactly perpendicular to the director. At a positive value of ε_a it would be energetically favorable to orient the director in the direction of the field. However, this is prevented by the homeotropic orientation of the director by the curvette walls. Furthermore, in the first approximation in the light intensity the orientational effect of the field on the unperturbed director is absent, in other words, for $\mathbf{E}=\mathbf{e}_x\mathbf{E}_x$ the function $\varphi(z)\equiv 0$ is an exact solution to equation (5).

Above a certain threshold value of the intensity the solution $\varphi(z)\equiv 0$ will no longer be stable. To determine this threshold it is convenient to consider the linearized [that is, at small $\varphi(z)$] equation (5). Here we must take into account the fact that from the equation div $\mathbf{D}=0$ we have

$$k_zD_z=k_z(\varepsilon_{zx}E_x+\varepsilon_{zz}E_z)=0.$$

Therefore, to terms linear in φ we have $E_z=-\varepsilon_a\varphi E_x/\varepsilon_{\|}$ and equation (5) takes the form

$$K_{33}\frac{\partial^2\varphi}{\partial z^2}+\frac{\varepsilon_a}{4\pi}\frac{\varepsilon_\perp}{\varepsilon_\|}\frac{|E_x|^2}{2}\varphi=\eta\frac{\partial\varphi}{\partial t}. \quad (6)$$

Equation (6) is analogous to the linearized equation for the behavior of the director near the Fredericks transition in a static field $\mathbf{E}_{\text{st}}=\mathbf{e}_xE_{\text{st}}$ (cf. Ref. 13, section 4.2). There are two differences, however. First, instead of the E^2_{st} of the static case in (6) we have $|E_x|^2/2$—the mean square of the amplitude of the light field. Secondly, in a static field $\mathbf{E}_{\text{st}}$ the condition div **D** = 0 has no effect on the field vector **E** when the director is inclined. In contrast to this, in the problem with a light field, a component E_z appears because of the condition div **D** = 0.

As a result, in equation (6) we have an additional factor $\varepsilon_\perp/\varepsilon_\parallel$, the role of which reduces to a certain increase of the threshold. Expanding $\varphi(z,t)$ in a Fourier series and taking the boundary conditions $\varphi(z=0,t) = \varphi(z=L,t) = 0$ into account, we obtain from (6)

$$\varphi(z,t)=\sum_k \varphi_k \sin\frac{\pi k z}{L}\exp\Gamma_k t,$$

$$\Gamma_k=\eta^{-1}\left(\frac{\varepsilon_\perp\varepsilon_a|E|^2}{8\pi\varepsilon_\parallel}-\frac{k^2\pi^2K_{33}}{L^2}\right). \tag{7}$$

From (7) we find that for

$$|E|^2>|E|_{Fr}^2=\frac{8\pi^3K_{33}}{\varepsilon_a L^2}\frac{\varepsilon_\parallel}{\varepsilon_\perp}$$

a perturbation of the form $\sin(\pi z/L)$ begins to grow exponentially.

The steady-state value of the amplitude $\varphi=\varphi_{max}\sin(\pi z/L)$ in the regime above threshold must now be determined with allowance for the nonlinear terms in equation (5). It is very important that in this approximation we must include a solution consistent with $\varphi(z)$ to Maxwell's equations in an inhomogeneous anisotropic medium. For a wave of the form (4) these solutions can be obtained in the geometrical optics approximation (see Refs. 14 and 15):

$$E_x(z)=A(\varepsilon_{zz}-s^2)^{1/4}e^{i\psi(z)},$$

$$E_z(z)=-A\frac{\varepsilon_{xz}(\varepsilon_{zz}-s^2)^{1/4}+s(\varepsilon_\parallel\varepsilon_\perp)^{1/2}}{\varepsilon_{zz}(\varepsilon_{zz}-s^2)^{1/4}}e^{i\psi(z)}. \tag{8a}$$

The advance of the optical phase $\psi(z)$ is equal to

$$\psi(z)=\frac{\omega}{c}\int_0^L\frac{-s\varepsilon_{xz}(z')+[\varepsilon_\parallel\varepsilon_\perp(\varepsilon_{zz}(z')-s^2)]^{1/2}}{\varepsilon_{zz}(z')}dz'. \tag{8b}$$

As already noted, $s=\sin\alpha$, where α is the angle of incidence in air, and the ε_{ik} are

$$\varepsilon_{zz}=\varepsilon_\perp+\varepsilon_a\cos^2\varphi(z),\quad \varepsilon_{xz}=\varepsilon_a\sin\varphi(z)\cos\varphi(z). \tag{9}$$

The constant A can be expressed in terms of the power flux density; namely the z component of the Poynting vector is

$$P_z=c(\varepsilon_\parallel\varepsilon_\perp)^{1/4}|A|^2/8\pi$$

and is independent of z.

In the case of exactly normal incidence ($s=0$) equation (5), taking into account (8) and (9), has the following form in the stationary case:

$$(K_{11}\sin^2\varphi+K_{33}\cos^2\varphi)\frac{d^2\varphi}{dz^2}-(K_{33}-K_{11})\sin\varphi\cos\varphi\left(\frac{d\varphi}{dz}\right)^2$$
$$+\frac{\varepsilon_a\varepsilon_\parallel\varepsilon_\perp|A|^2\sin\varphi\cos\varphi}{8\pi(\varepsilon_\perp+\varepsilon_a\cos^2\varphi)^{3/2}}=0. \tag{10}$$

The exact solution of this equation will be given in Sec. 5 below. Here we shall restrict ourselves to only taking into account terms of order φ and φ^3 in this equation, a procedure valid near the Fredericks threshold. The search for a solution of the form

$$\varphi=\varphi_1\sin(\pi z/L)+\varphi_3\sin(3\pi z/L)+\ldots,$$

which automatically satisfies the boundary conditions, leads to the following expression for φ_1:

$$\varphi_1^2=2\left(1-\frac{9}{4}\frac{\varepsilon_a}{\varepsilon_\parallel}-k\right)^{-1}\frac{P-P_{Fr}}{P_{Fr}}, \tag{11}$$

$$k=(K_{33}-K_{11})/K_{33},\quad P_{Fr}\left[\frac{\text{erg}}{\text{cm}^2\cdot\text{sec}}\right]=\frac{c\varepsilon_\parallel K_{33}}{\varepsilon_a\varepsilon_\perp^{1/2}}\left(\frac{\pi}{L}\right)^2. \tag{12}$$

Here $\varphi_3\sim\varphi_1^3$. Therefore φ_1 is proportional to the square root of the excess of the intensity P above threshold P_{Fr}.

The advance of the optical phase $\psi(L)$ over the cuvette thickness L can be obtained from (8b); expanding (8b) to terms $\sim\varphi_1^2$ we obtain

$$\psi(L)=\frac{\omega}{c}\varepsilon_\perp^{1/2}L\left[1+\frac{1}{2}\frac{\varepsilon_a}{\varepsilon_\parallel}\left(1-\frac{9}{4}\frac{\varepsilon_a}{\varepsilon_\parallel}-k\right)^{-1}\frac{P-P_{Fr}}{P_{Fr}}\right]. \tag{13}$$

4. THE FREDERICKS TRANSITION IN NARROW BEAMS

In the case where the transverse dimension of the beam $a_\perp$ is smaller than or on the order of the cuvette thickness L, the Frank energy due to the transverse gradients of the director becomes dominant. Let us first estimate the order of magnitude of the threshold power of the Fredericks transition. The energy of the perturbed state is

$$W=\frac{1}{2}\varphi_1^2a_\perp^2L\left[K\left(\frac{\pi^2}{L^2}+\frac{1}{a_\perp^2}\right)-\frac{\varepsilon_a|\mathbf{E}|^2}{8\pi}\right].$$

Here $a_\perp^2L$ is the volume occupied by the disturbance and K is the Frank constant. This expression gives a stable state $\varphi_1=0$ at small $|\mathbf{E}|^2$, and instability sets in at

$$|\mathbf{E}|^2\gtrsim\frac{8\pi K}{\varepsilon_a}\left(\frac{\pi^2}{L^2}+\frac{1}{a_\perp^2}\right). \tag{14}$$

As seen from expression (14), $|\mathbf{E}|^2a_\perp^2=\text{const}$ at $a_\perp\ll L$.

In order to exactly determine the threshold of the Fredericks transition, it is necessary to solve the three-dimensional problem of the stability of the solution $\varphi=0$. We shall consider this problem for several particular cases.

a. *The single-constant approximation.* In the single-constant approximation we can assume even for a narrow beam that the vectors **E** and **n** lie in the *xz* plane. Then in the approximation linearized in $\varphi(\mathbf{r})$ we have

$$\Delta\varphi+\left(\frac{\pi}{L}\right)^2\frac{P(\mathbf{r}_\perp)}{P_{Fr}}\varphi=\frac{\eta}{K}\frac{\partial\varphi}{\partial t}. \qquad (15)$$

Here P_{Fr} is determined by formula (12) using the fact that $K_{ii}=K$.

With the substitution $\varphi(\mathbf{r},t)=\varphi'(\mathbf{r})\exp\Gamma t$, equation (15) becomes

$$\Delta\varphi'(\mathbf{r})+\left[\left(\frac{\pi}{L}\right)^2\frac{P(\mathbf{r}_\perp)}{P_{Fr}}-\frac{\Gamma\eta}{K}\right]\varphi'(\mathbf{r})=0. \qquad (16)$$

Let us first consider the case where a "ribbon" light beam falls on the medium, $P(\mathbf{r}_\perp)=P(x)$. Then after separating the variables in equation (16) we find by means of the substitution $\varphi'(\mathbf{r})=\xi(x)\chi(z)$

$$\frac{d^2\xi(x)}{dx^2}+\left[\left(\frac{\pi}{L}\right)^2\frac{P(x)}{P_{Fr}}-q\right]\xi(x)=0, \qquad (17a)$$

$$\frac{d^2\chi(z)}{dz^2}+p\chi(z)=0, \quad q-p=\Gamma\eta/K. \qquad (17b)$$

Equation (17a) is the one-dimensional Schrödinger equation for the potential $U\sim-P(x)$. It is possible to solve this equation analytically for a relation, for example, of the form

$$P(x)=P_0/\mathrm{ch}^2\beta x, \qquad (18)$$

that is, for a function which gives a good qualitative approximation of a Gaussian (Fig. 2). The eigenvalues of q for a "potential" of the form (18) will be[16]

$$q_n=\frac{\beta^2}{4}\left\{-(1+2n)+\left[1+\frac{4}{\beta^2}\left(\frac{\pi}{L}\right)^2\frac{P_0}{P_{Fr}}\right]^{1/2}\right\}^2. \qquad (19)$$

The eigenvalues of p from (17), corresponding to the eigenfunction $\chi(z)\propto\sin(\pi z/L)$ is $p=\pi^2/L^2$. The perturbation of the director field $\varphi'(\mathbf{r})$ will grow exponentially in time at $\Gamma=\eta^{-1}K(q-p)>0$. Then, defining the threshold intensity of the Fredericks transition as that value of P_0 at which a bound state first appears in the potential well $U\propto-1/\cosh^2\beta x$, we obtain

$$P_{th}=P_{Fr}(1+\beta L/\pi). \qquad (20)$$

The quantity βL can be viewed as the ratio of the cuvette thickness L to the beam diameter d: $\beta L\sim L/d$.

b. Another case which can be solved analytically and by which we can approximate the real distribution of the intensity in the beam of a light wave corresponds to a function $P(\mathbf{r})$ of the form

$$P(\rho)=\begin{cases}\text{const} & \text{for } \rho\leqslant\rho_0\\ 0 & \text{for } \rho>\rho_0\end{cases}, \qquad (21)$$

where ρ is the distance from the beam axis in the transverse direction. Writing equation (16) in cylindrical coordinates and making the substitution $\varphi'(\mathbf{r}) = \zeta(\rho)\chi(z)$, we find

$$\frac{d^2\zeta}{d\rho^2}+\frac{1}{\rho}\frac{d\zeta}{d\rho}+\left[\left(\frac{\pi}{L}\right)^2\frac{P(\rho)}{P_{\mathrm{Fr}}}-h\right]\zeta=0, \tag{22a}$$

$$\frac{d^2\chi}{dz^2}+g\chi=0, \quad h-g=\Gamma\eta/K. \tag{22b}$$

The solution of (22a) is a zeroth-order Bessel function. Let

$$P(\rho\leqslant\rho_0)>(L/\pi)^2hP_{\mathrm{Fr}}=P_{\mathrm{Fr}} \tag{23}$$

[in expression (23) we have set $h=\pi^2/L^2$, since director-field perturbations which increase with time exist only for $\Gamma=\eta^{-1}K(h-g)>0$, where g, just as p in case (a), equals π^2/L^2]. It is easy to see that the condition (23) is necessary for the existence of a nontrivial, that is, not identically equal to zero, solution to the system of Eqs. (22), compatible with the boundary conditions. Then the solution of Eq. (22a) will be

$$\zeta(\rho)=\begin{cases} c_1J_0\left(\frac{\pi\rho}{L}\left(\frac{P}{P_{\mathrm{Fr}}}-1\right)^{1/2}\right) & \text{for } \rho\leqslant\rho_0 \\ c_2K_0\left(\frac{\pi\rho}{L}\right) & \text{for } \rho>\rho_0 \end{cases}, \tag{24}$$

where c_1 and c_2 are constants and J_ν and K_ν are Bessel functions. The condition of continuity of the logarithmic derivative at the point $\rho=\rho_0$ gives

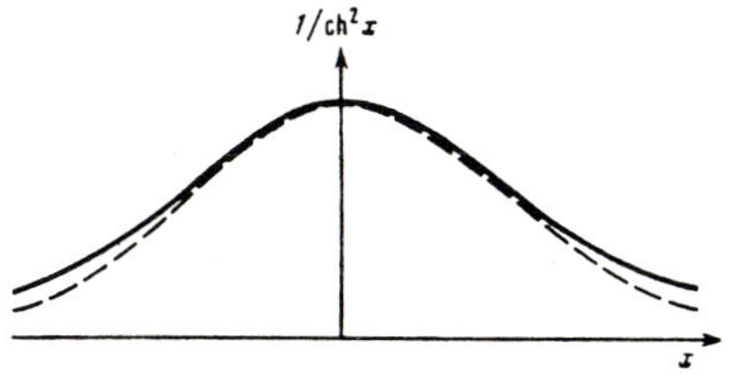

FIG. 2. Form of the function $1/\cosh^2 x$. The dashed line shows the function $\exp(-x^2)$.

$$\left(\frac{P}{P_{\mathrm{Fr}}}-1\right)^{1/2}J_1\left(\frac{\pi\rho_0}{L}\left(\frac{P}{P_{\mathrm{Fr}}}-1\right)^{1/2}\right)\Big/J_0\left(\frac{\pi\rho_0}{L}\left(\frac{P}{P_{\mathrm{Fr}}}-1\right)^{1/2}\right) =K_1\left(\frac{\pi\rho_0}{L}\right)\Big/K_0\left(\frac{\pi\rho_0}{L}\right). \tag{25}$$

In the case $\pi\rho_0/L\ll 1$, if in addition we assume that

$$\frac{\pi\rho_0}{L}\left(\frac{P}{P_{\mathrm{Fr}}}-1\right)^{1/2}\ll 1, \tag{26}$$

we have from Eq. (25)

$$P_{\mathrm{th}}=P_{\mathrm{Fr}}\left[1-2L^2/\pi^2\rho_0^2\ln(\pi\rho_0/L)\right]. \tag{27}$$

As seen from expression (27), the condition (26) is satisfied for $\ln(\pi\rho_0/L) \gg -2$, and this strengthens considerably the inequality $\pi\rho_0/L \ll 1$.

In the case where $\rho_0 \gtrsim L$, we obtain from Eq. (25)

$$P_{th} = P_{Fr}\left[1+\left(\frac{J_{01}}{\pi}\right)^2\left(\frac{L}{\rho_0}\right)^2\left(1-\frac{J_{01}LK_0(\pi\rho_0/L)}{\pi\rho_0K_1(\pi\rho_0/L)}\right)^2\right]. \tag{28}$$

Here $J_{01} \approx 2.4$ is the first zero of the Bessel function $J_0(Z)$. In the limit $\rho_0 \gg L$ formula (28) gives $P_{th} = P_{Fr}$, in agreement with the result obtained in Sec. 3 for broad beams.

In the case where the beam radius is of the order of the cuvette thickness, $\pi\rho_0/L \approx 1$, Eq. (25) can be solved graphically. As a result we find

$$P_{th} \approx 3.1 P_{Fr}. \tag{29}$$

Therefore, in all the cases studied we have $P_{th} \sim P_{Fr} \sim 1/L^2$, whereas the coefficient of proportionality is determined by the ratio ρ_0/L of the beam radius to the cuvette thickness.

In the case of more than a single constant the threshold of the Fredericks transition can be determined analytically for a ribbon beam of the form $P(\mathbf{r}_\perp) = P(y)$ polarized along the x axis. The equation linearized in $\varphi(\mathbf{r})$ in this case has the form

$$K_{22}\frac{\partial^2\varphi}{\partial y^2}+K_{33}\frac{\partial^2\varphi}{\partial z^2}+\frac{\varepsilon_a\varepsilon_\perp^{1/2}}{c\varepsilon_\parallel}P(y)\varphi=\eta\frac{\partial\varphi}{\partial t}. \tag{30}$$

Introducing the new variable $y' = (K_{33}/K_{22})^{1/2}y$, Eq. (30) can be written in the form (17) by replacing x by y and using the definition of P_{Fr} from (12). After using the solution to (17), we obtain for a light wave of the form $P(y) = P_0/\cosh^2\beta y$

$$P_{th} = P_{Fr}\left[1+\frac{\beta L}{\pi}\left(\frac{K_{22}}{K_{33}}\right)^{1/2}\right]. \tag{31}$$

Therefore, when the beam dimensions are slightly smaller than the curvette thickness the inclusion of more than one constant leads to small corrections. We can expect this to hold for other beam shapes, also.

5. HYSTERESIS OF THE FREDERICKS TRANSITIONS IN BROAD BEAMS

Formula (11) becomes invalid for the maximum angle of deviation of the director from the unperturbed direction if the parameters of the liquid-crystal medium are such that

$$B=\frac{1}{4}\left(1-\frac{9}{4}\frac{\varepsilon_a}{\varepsilon_\parallel}-k\right)\leqslant 0.$$

This occurs, for example, for the nematic crystal PAA, for which $\varepsilon_a \approx 0.9$; $K_{11} = 4.5 \times 9.5 \times 10^{-7}$ dyne [at the temperature $T = 125°C$ (Ref. 13)] and $B = -0.03$ ($B = 0.7$ for MBBA and for $B = 0.06$ for OCBP).

To study the Fredericks transition for $B \leqslant 0$ we integrate Eq. (10) with respect to z after multiplying it by $2d\varphi/dz$. After determining the integration constant in terms of the maximum angle of deviation of the director from the unperturbed direction φ_m we obtain

$$\left(\frac{d\varphi}{dz}\right)^2 = \frac{\varepsilon_\perp^{1/2} P}{cK_{33}} \frac{[1-(\varepsilon_a/\varepsilon_\parallel)\sin^2\varphi]^{1/2} - [1-(\varepsilon_a/\varepsilon_\parallel)\sin^2\varphi_m]^{1/2}}{(1-k\sin^2\varphi)[1-(\varepsilon_a/\varepsilon_\parallel)\sin^2\varphi]^{1/2}[1-(\varepsilon_a/\varepsilon_\parallel)\sin^2\varphi_m]^{1/2}}. \tag{32}$$

Taking into account the fact that the maximum angle of deviation φ_m is reached at the center of the cell, that is, for $z = L/2$, we get from (32)

$$\int_0^{\varphi_m} \left\{ \frac{(1-k\sin^2\varphi)[1-(\varepsilon_a/\varepsilon_\parallel)\sin^2\varphi]^{1/2}[1-(\varepsilon_a/\varepsilon_\parallel)\sin^2\varphi_m]^{1/2}}{[1-(\varepsilon_a/\varepsilon_\parallel)\sin^2\varphi]^{1/2} - [1-(\varepsilon_a/\varepsilon_\parallel)\sin^2\varphi_m]^{1/2}} \right\}^{1/2} d\varphi = \frac{L}{2}\left\{\frac{\varepsilon_\perp^{1/2} P}{cK_{33}}\right\}^{1/2}. \tag{33}$$

Assuming that the power density P of the radition falling on a cell of the NLC is close to the threshold value for the Fredericks transition, that is, $\varphi_m \ll 1$, let us compute the integral in (33) up to terms $\sim\varphi_m^4$. The solution of the resulting biquadratic equation for φ_m has the form

$$\varphi_m^2 = \frac{-B \pm (B^2 - 4GC)^{1/2}}{2G}, \tag{34}$$

where

$$G = \frac{1}{96}\left[\frac{11}{2} + \frac{9}{4}\frac{\varepsilon_a}{\varepsilon_\parallel} - k + \frac{63}{4}k\frac{\varepsilon_a}{\varepsilon_\parallel} - \frac{9}{2}k^2 - \frac{261}{32}\left(\frac{\varepsilon_a}{\varepsilon_\parallel}\right)^2\right],$$

$$C = 1 - (P/P_{Fr})$$

For the liquid crystals MBBA, OCBP, and PAA the values of G are almost identical at $G \approx 0.06$. Before discussing formula (34), we note that it can in principle be obtained from the condition of the minimum of the free energy after expanding it in a series in φ_m up to terms $\sim\varphi_m^6$. Apart from a coefficient that renders it dimensionless, the free energy has the form

$$\Phi = C\varphi_m^2 + {}^1/_2 B\varphi_m^4 + {}^1/_3 G\varphi_m^6. \tag{35}$$

At the point determined by formula (34) we have

$$d^2\Phi/d(\varphi_m^2)^2 = \pm(B^2 - 4GC)^{1/2}. \tag{36}$$

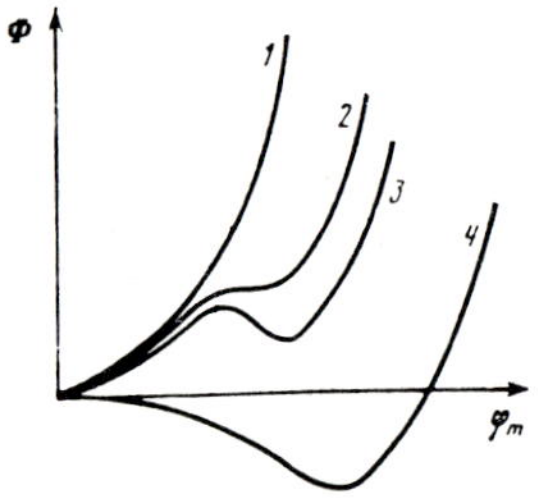

FIG. 3. Dependence of the free energy Φ on φ_m for different values of the light field power P: $1-P < P_{th}$, $2-P = P_{th}$, $3-P_{th} < P < P_{Fr}$, $-P > P_{Fr}$.

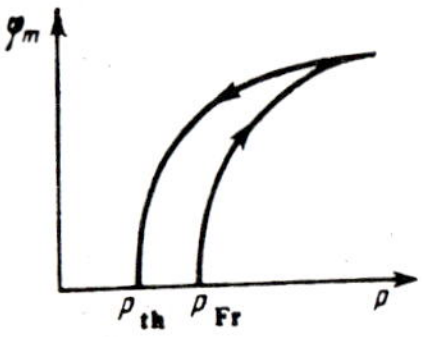

FIG. 4. Hysteresis of the Fredericks transition. The arrows indicate the direction of variation of the power of the light field P.

Therefore, we should use the plus sign in formula (34) (the condition that the free energy be a minimum).

If $B \geq 0$ the quantity φ_m will be real at $C = 1 - (P/P_{Fr})^{1/2} \leq 0$, that is, $P_{th} = P_{Fr}$. If $B < 0$, then in order that φ_m be real it is sufficient to require that the expression under the square root in (34) be positive, $B^2 - 4GC \geq 0$, from which we have

$$P_{th} = P_{Fr}(1 - B^2/4G)^2. \tag{37}$$

However, we see from (36) and (37) that when the power density of the incident radiation is $P = P_{th}$ the quantity φ_m given by (34) is a point of inflection for the free energy function (see Fig. 3). As seen from Fig. 3, for values $P_{th} < P < P_{Fr}$ the function $\Phi(\varphi_m)$ has a local minimum at $\varphi_m \neq 0$, where

$$\Phi(\varphi_m \neq 0) > \Phi(\varphi_m = 0) = 0. \tag{38}$$

The point $\varphi_m = 0$ becomes unstable only at $P > P_{Fr}$. From the above discussion it is clear that as the power density of the light wave increases from zero the Fredericks transition occurs at $P = P_{Fr}$, so that a free-energy minimum corresponding to $\varphi_m \neq 0$ and occurring at $P_{th} < P < P_{Fr}$ is unattainable because of the presence of the potential barrier. For the inverse transition, that is, as P decreases from the region of larger P_{Fr}, the free-energy minimum reached at $P < P_{Fr}$ and corresponding to $\varphi_m \neq 0$, while not an absolute minimum of the function Φ [as seen from (38)], is separated from it by the potential barrier. The barrier only disappears at $P = P_{th}$ and then we have $\varphi_m = 0$ (Fig. 4).

6. PROPAGATION OF A LIGHT WAVE AT SMALL ANGLES TO THE DIRECTOR

As already noted, an extraordinary wave propagating at an angle to the director induces strong nonlinear optical effects at very low powers $P \ll P_{Fr}$ (Refs. 3–9). When a light wave is normally incident on a cuvette with homeotropic orientation of the NLC, however, reorientation of the director is possible only above some threshold value of the power of the wave. The distortions of

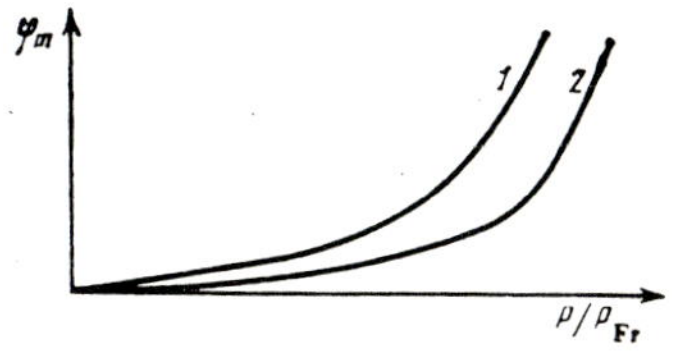

FIG. 5. Dependence of the maximum angle of deviation φ_m on P/P_{Fr} in the case of a light wave incident on a cell at small angles: *1*—$\gamma = 10°$, *2*—$\gamma = 5°$.

the field of the director in this case were discussed in Secs. 3–5. Here we shall study the nature of the reorientation of the director by the field of an extraordinary light wave in the intermediate region, that is, when the wave propagates at small angles to the director.

Substituting the expressions for the light fields (8a) into Eq. (5) and including terms up to third order in the small quantity $\varphi = \varphi_m \sin(\pi z/L)$ (that is, assuming that the power of the wave is less than or of order P_{Fr}) and of first order in $s = \sin\alpha$, we find

$$\left[\left(1-\frac{9}{4}\frac{\varepsilon_a}{\varepsilon_\parallel}\right)\frac{P}{P_{Fr}}-k\right]\varphi_m^3+2\left(1-\frac{P}{P_{Fr}}\right)\varphi_m+\frac{4\sin\gamma}{\pi}\frac{P}{P_{Fr}}=0, \qquad (39)$$

where $\sin\gamma = \varepsilon^{-1/2}\sin\alpha$ and γ is the angle of refraction of the wave in the cell of the NLC.

In the case of weak fields $P \ll P_{Fr}$ and large anisotropy of the permittivity, Eq. (39) duplicates the results obtained earlier[4]:

$$\varphi_m = \varepsilon_a |E|^2 L^2 \sin\gamma / 2\pi^4 K_{33}.$$

Since the formulas are quite awkward we shall not give the analytic form of the solution of Eq. (39). The graph of the function $\varphi_m(P/P_{Fr})$ is shown in Fig. 5 for the parameters of the liquid crystal OCBP and for different angles γ.

The behavior of the cell in the oblique-incidence case discussed here is analogous to the case of a cell in an oblique magnetic field-in both cases there is no rigorous Fredericks transition (cf. Ref. 13, §4.2.3).

7. DISCUSSION

Let us estimate numerically the value of the threshold power for the Fredericks transition for the nematic crystal OCBP used in Ref. 11. The parameters of the liquid crystal are $\varepsilon_{\parallel}^{1/2} = 1.66$, $\varepsilon_{\perp}^{1/2} = 1.53$, $K_{11} = 4 \times 10^{-7}$ dyne, $K_{33} = 7 \times 10^{-7}$ dyne, the curvette thickness is $L = 150\ \mu m$, and the beam radius is $\rho_0 = 50\ \mu m$. Substituting the Frank constant $K = 0.5\ (K_{11} + K_{33})$ into the expression for P_{Fr}, we find from formula (29)

$$P_{th} \approx 9{,}7 \cdot 10^{2}\ \text{Watt/cm}^2$$

This value is in very good agreement with the experimental value[11]

$$P_{th}^{exp} \geqslant 9 \cdot 10^{2}\ \text{Watt/cm}^2$$

For the cuvette thickness $L = 50\ \mu m$ and the same values of all the other parameters the value of the threshold power estimated from (28) is

$$P_{th} \approx 3 \cdot 10^{3}\ \text{Watt/cm}^2$$

This value is 3.1 times larger than that for the Fredericks transition in a cuvette of thickness $L = 150\ \mu m$, which is also in very good agreement with the results of Ref. 11.

The temperature dependence of the threshold power is determined by the temperature dependence of the Frank constant and of the permittivity. Bearing in mind the fact that the Frank constant K varies with temperature as the square of the order parameter $K \sim S^2$ while the anisotropy of the permittivity $\varepsilon_a \sim S$ (Ref. 12), we can separate the temperature-dependent part in expression (12) for P_{Fr} in the form $P_{Fr} \propto \varepsilon_a \varepsilon_{\parallel} \varepsilon_{\perp}^{-1/2}$. As the temperature increases, ε_a and $\varepsilon_{\parallel}$ decreases while $\varepsilon_{\perp}$ increases; from this we see that as the temperature increases the threshold power decreases. Introducing the notation

$$\varepsilon_0 = {}^1/_3 \mathrm{Sp}\,(\varepsilon_{ik}) = {}^1/_3 (\varepsilon_{\parallel} + 2\varepsilon_{\perp}),$$

we can write $P_{Fr} \propto \varepsilon_0^{1/2} \varepsilon_a$. Since $\varepsilon_0^{1/2}$ depends weakly on the temperature (for example, in the region of the MBBA mesophase $\varepsilon_0^{1/2}$ changes by only 0.01 of its value), as the point of the phase transition to an isotropic liquid is approached the threshold power decreases in proportion to ε_a, that is, to the order parameter S.

Here we note the following. At small values of $\varepsilon_a/\varepsilon_{\parallel}$ and k, as is the case near the critical point, Eq. (33) determining the maximum angle of deviation of the director φ_m as a function of the power of the light field can be simplified. For this we must expand the integrand in powers of $\varepsilon_a/\varepsilon_{\parallel}$ and k. Then the integral on the left-hand side of Eq. (33) is expressed in terms of elliptic integrals and the resulting equation implicitly determines the function $\varphi_m = \varphi_m(P)$. In the limit where $P/P_{tr} \ll 1$ (that is, $\varphi_m = \pi/2 - \delta, \delta \ll 1$) the function $\varphi_m(P)$ can be determined explicitly:

$$\delta \propto \exp[-(\pi/2)(P/P_{Fr})^{1/2}].$$

As was shown in Sec. 5, for certain types of liquid crystals the Fredericks transition can be accompanied by hysteresis. Let us calculate for the case of the well studied liquid crystal PAA the main characteristics of this phenomenon. As seen from formula (37) with $B = -0.03$ and $G = 0.06$, the powers at which the Fredericks effect in PAA is switched on and off differ little: $P_{t1} \approx 0.992 P_{tr}$. The angle corresponding to the appearance of a local minimum of the free-energy function [that is, φ_m corresponding to the inflection point of the function $\Phi(\varphi_m)$] is $\varphi_m \propto (-B/2G)^{1/2} = 0.5$.

Let us demonstrate that thermal fluctuations of the direction of the director do not lead to surmounting of the potential barrier between the minima of the free energy function. For this we note that thermal fluctuations of the direction of the director would cause the hysteresis to disappear if the mean angle of the fluctuation deviations were on the order of or larger than the difference of the angles corresponding to the local minimum and maximum of the free energy. This difference is easily calculated using formula (34) for some value of the power P such that $P_{th} < P < P_{tr}$ (that is, $0 < C < B^2/4D$). For example, for $C = B^2/8D$ we find $\Delta\varphi = 0.4$, and since $\langle(\delta\varphi)^2\rangle^{1/2} \ll 1$ for thermal fluctuations of the director the local minimum of the free energy $\Phi(\varphi_m > 0)$ is stable to thermal fluctuations.

In concluding this discussion we note that a Fredericks transition induced by light fields in a NLC with $\varepsilon_a > 0$ can occur only when the light wave is normally incident on a homeotropically oriented NLC cell. It is impossible to have a Fredericks transition induced by an ordinary light wave normally incident on a planarly oriented NLC cell. This can be understood when it is realized that the electric field of an ordinary light wave in a NLC remains perpendicular to the director as it propagates through the medium. Strong nonlinear optical effects arise if the polarization of the normally incident cell makes some angle with the director. We have considered these effects in an earlier study.[6]

The authors are grateful to E. I. Kats for a discussion of these problems.

Note added in proof. We have been kindly informed by the authors of a paper submitted to this journal that they have also studied the theory of light-induced Fredericks transitions (A. S. Zolot'ko, V. F. Kitaeva, N. N. Sobolev, and A. P. Sukhorukov, Zh. Eksp. Teor. Fiz. 81, 933 (1981) [Sov. Phys. JETP 54, No. 9 (1981)]).

[1]S. M. Arakelyan, G. A. Lyakhov, and Yu. S. Chilingaryan, Usp. Fiz. Nauk **131**, 3 (1980) [Sov. Phys. Uspekhi **23**, 245 (1980)].

[2]S. M. Arakelyan, G. L. Grigoryan, S. Ts. Nersisyan, M. A. Nshanyan, and Yu. S. Chilingaryan, ZhETF Pis. Red. **28**, 202 (1978) [JETP Lett. **28**, 186 (1979)].

[3]B. Ya. Zel'dovich and N. V. Tabiryan, ZhETF Pis. Red. **30**, 510 (1979) [JETP Lett. **30**, 478 (1980)].

[4]B. Ya. Zel'dovich, N. F. Pilipetskiĭ, A. V. Sukhov, and N. V. Tabiryan, ZhETF Pis. Red. **31**, 287 (1980) [JETP Lett. **31**, 263 (1980)].

[5]B. Ya. Zel'dovich and N. V. Tabiryan, Preprint FIAN, No. 61, 1980.

[6]B. Ya. Zel'dovich and N. V. Tabiryan, Preprint FIAN, No. 62, 1980.

[7]B. Ya. Zel'dovich and N. V. Tabiryan, Preprint FIAN, No. 63, 1980.

[8]B. Ya. Zel'dovich and N. V. Tabiryan, Kvant. Elektronika (Moscow) **7**, 770 (1980) [Sov. J. Quantum Electronics **10**, 440 (1980)].

[9]N. F. Pilipetski, A. V. Sukhov, N. V. Tabiryan, and B. Ya. Zel'dovich, Optical Communications **37**, 280 (1981).

[10]V. B. Pakhalov, A. S. Tumasyan, and Yu. S. Chilingaryan, Abstracts of Reports at the Tenth Int. Conf. on Coherent and Nonlinear Optics, Kiev, October 14–17, 1980, Part I, p. 18.

[11]A. S. Zolot'ko, V. F. Kitaeva, N. Kroo, N. N. Sobolev, and L. Chilag, ZhETF Pis. Red. **32**, 170 (1980) [JETP Lett **32**, 158 (1981)].

[12]P. G. De Gennes, The Physics of Liquid Crystals, Oxford University Press, 1974 (Russ. transl., Mir, 1977).

[13]L. M. Blinov, Elektro- i magnitooptika zhidkikh kristallov (Electro- and Magneto-optics of Liquid Crystals), Moscow, Nauka, 1978.

[14]Analiticheskie metody v teorii rasprostraneniya i difraktsii) voln (Analytic Methods in the Theory of Wave Propagation and Diffraction), ed. S. V. Butakovaya, Moscow, Sov. radio, 1970.

[15]B. Ya. Zel'dovich and N. V. Tabiryan, Zh. Eksp. Teor. Fiz. **79**, 2388 (1980) [Sov. Phys. JETP **52**, 1210 (1980)].

[16]L. D. Landau and E. M. Lifshitz, Kvantovaya mekhanika, Moscow, Nauka, 1974 (Engl. transl. Quantum Mechanics: Nonrelativistic Theory, New York, Pergamon, 1977).

Translated by Patricia Millard

 0038-5646/81/070032-06$02.40

Mol. Cryst. Liq. Cryst., 1982, Vol. 84, pp. 125–135
0026-8941/82/8401-0125 $06.50/0

The Influence of the Finite Size of the Light Spot on the Laser Induced Reorientation of Liquid Crystals

L. CSILLAG, I. JÁNOSSY, V. F. KITAEVA,* N. KROO and N. N. SOBOLEV*

Central Research Institute for Physics, Budapest 114. P.O.B. 49, Hungary
**Lebedev Physical Institute, Moscow, USSR*

(*Received July 31, 1981*)

The threshold intensity for the laser induced deformation of nematic liquid crystals is calculated as a function of the spot size of the illuminating beam. It is shown, that due to the small size of the light spot, the optical threshold is higher than that expected from the Fredericks threshold formula for static electric fields of infinite large dimension in the cell plane. Experimental data measured on a homeotrop OCB (octyl-cyano-byphenyl) sample fit well the theoretical curve. Results obtained on the temperature and wavelength dependence of the threshold can be interpreted on the basis of the given theory.

1 INTRODUCTION

As it was reported by us previously,[1] by illuminating a nematic liquid crystalline film with a focused laser beam, an increase in the beam divergence and a ring system appearing in it can be detected. This effect could be explained as an orientational deformation due to a Fredericks transition in optical fields. It has been shown, that in analogy with the static Fredericks effect, for homeotrop cell and at normal incidence this deformation starts only above a certain threshold laser power. The focused laser light, however, has a spatial intensity distribution—in the simplest case a Gaussian distribution—with a half-width of some tens of microns which is comparable with the sample thickness. Therefore it can be expected, that the actual optical threshold value differs from that of the Fredericks transition in a static electric field with practically infinite large dimensions in the cell plane.

This problem is studied in this paper both theoretically and experimentally. First we calculate the value of the laser intensity threshold by a simplified

125

model for the case of a homeotrop nematic cell and at normal incidence of the laser beam. Then experimental results on a nematic OCB (octyl-cyano-biphenyl) sample are described showing that good agreement exists between the theoretically expected and the measured values. Additional measurements are also reported on the temperature and wavelength dependence of this threshold value giving a further support to our calculations.

2 CALCULATION OF THE THRESHOLD

We investigate a homeotropic sandwich-like cell of a nematic liquid crystal (NLC), in which the L director is fixed uniformely at the boundaries, and which is illuminated with a linearly polarized and focused laser beam. It is assumed that the focal length of the focusing lens is large enough so that the laser beam can be considered as a parallel beam within the sample.

The light beam interacts with the molecules through its electric field E. The torque $\Gamma^{(E)}$ induced by E in a nematic is then

$$\Gamma^{(E)} = D \times E = \epsilon_o \epsilon_a E L (E \times L)$$

where ϵ_a is the anisotropy of the dielectric constant; $\epsilon_a = n_e^2 - n_o^2$, n_e, n_o being the extraordinary and ordinary refractive indices. We study only the cases, when $n_e > n_o$.

We choose the coordinate system so, that the cell plane coincides with the (x, y) plane (vertical plane), and the L director is parallel with the z axis, $L \parallel z$. The light beam has a polarization in the x direction (in the horizontal plane) falling perpendicularly on the cell, so the propagation vector $k \parallel z$ and L.

The stationary configuration can be determined in general from the balance between the electric and elastic torques. In our case there is always a trivial solution for the balance; the unperturbed configuration, in which $L \equiv (0, 0, 1)$. To calculate the threshold we follow the usual procedure looking for the value of the light intensity at which the linearized equation of the balance of torques has a non-trivial solution.

The exact solution of the problem is very complicated. Therefore two simplifications will be used. First, we assume that the elastic constants are equal; second, the actual Gaussian transversal intensity distribution will be replaced by a Π-shape distribution.

In the one constant approximation the elastic torque is[2]

$$\Gamma^{(e)} = KL \times \Delta L$$

where K denotes the Frank elastic constant. The director turns in this approximation only in the (x, z) plane, and both the elastic and electric torque has only a y-component.

In linear approximation the elastic torque is

$$\Gamma_y^{(e)} = K\Delta\theta$$

with $\theta = L_x$; while the electric torque can be given as

$$\Gamma_y^{(E)} = \epsilon_o\epsilon_a E_x(\theta E_x + E_z).$$

E_x is determined by the continuity of the tangential component of E; $E_x = E_o$, where E_o is the electric vector of the incoming light beam. E_z can be calculated from the condition $D_z = 0$, which leads to

$$E_z = -\frac{\epsilon_a}{n_e^2}\theta E_x$$

With the above values of E_z and E_x we get

$$\Gamma_y^{(E)} = \epsilon_o\epsilon_a \frac{n_o^2}{n_e^2} E_o^2\theta.$$

The linearized equation of torques is

$$K\Delta\theta + \epsilon_o\epsilon_a \frac{n_o^2}{n_e^2} E_o^2\theta = 0.$$

Taking into account that E_o is only a function of the distance r from the center of the laser beam, θ can be written as

$$\theta(r, z) = \theta_o(r)\sin\frac{\pi}{l} z$$

(l denotes the sample thickness), where θ_o obeys the equation

$$\frac{\partial^2\theta_o}{\partial r^2} + \frac{1}{r}\frac{\partial\theta_o}{\partial r} + \left[\frac{\epsilon_o\epsilon_a}{K}\frac{n_o^2}{n_e^2} E_o^2(r) - \left(\frac{\pi}{l}\right)^2\right]\theta_o = 0 \qquad (1)$$

In the actual experiments $E_o^2(r)$ is a Gaussian function:

$$E_o^2(r) = \epsilon^2 \exp(-r^2/r_o^2).$$

Here we take for simplicity

$$E_o^2(r) = \begin{cases} \epsilon^2 & r < r_o \\ 0 & r > r_o \end{cases}$$

where r_o is an "effective" radius of the laser beam. (The connection between r_o and the full width of the beam at half maximum (FWHM), $\delta = 2(\ln 2)^{1/2} r_o$). In this case the non-trivial solution of Eq. (1) with boundary conditions $\theta_o(0) = 1, \theta_o'(0) = 0$ and $\theta_o \to 0$ for $r \to \infty$ is

$$\theta_o(r) = \begin{cases} J_o(\lambda r) & r < r_o \\ aK_o(\mu r) & r > r_o \end{cases}$$

where J_o and K_o denote the 0-order Bessel and modified Hankel functions resp. (3) and

$$\lambda^2 = \epsilon^2 \frac{\epsilon_o \epsilon_a}{K} \frac{n_o^2}{n_e^2} - \left(\frac{\pi}{l}\right)^2; \quad \mu = \frac{\pi}{l}.$$

λ can be determined from the condition, that both θ_o and $\partial\theta_o/\partial r$ should be continuous at r_o. This condition leads to

$$\gamma \frac{J_o'(\gamma)}{J_o(\gamma)} = \kappa \frac{K_o'(\kappa)}{K_o(\kappa)} \tag{2}$$

with $\gamma = \lambda r_o$, $\kappa = \mu r_o = \pi r_o/l$. The solution of Eq. (2) is given in Figure 1.

The power of the light beam is in the cell

$$P = c r_0^2 \pi \epsilon_0 n_0 \epsilon^2 \text{ (} c \text{ light velocity)}$$

The threshold power therefore can be written as

$$P_{th} = P_o(\gamma^2(\kappa) + \kappa^2) \tag{3}$$

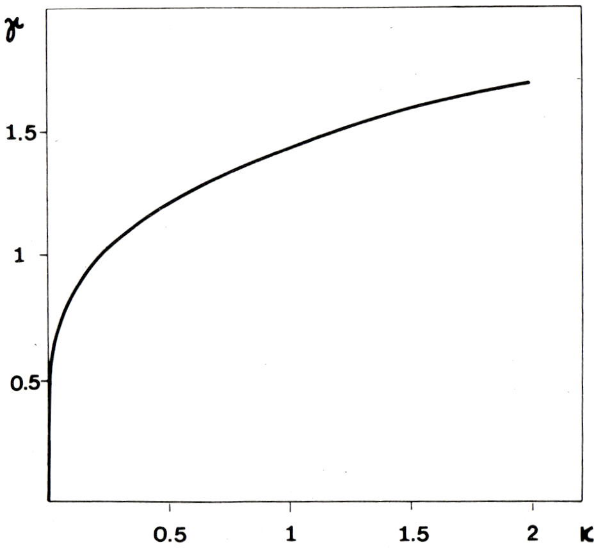

FIGURE 1 The solution of Eq. (2).

with

$$P_o = \frac{c\pi K}{n_e^2 - n_o^2} \frac{n_e^2}{n_o}. \tag{4}$$

In Figure 2, P/P_o is given as a function of r_o/l (full line). The dashed line on this figure gives the threshold, if we apply simply the Fredericks threshold formula $\epsilon_{th} = \pi/l(K/\epsilon_o\epsilon_a)^{1/2}$. The difference between the two curves is remarkable. The ratio of the two thresholds goes to infinity as $r_o/l \to 0$. Therefore we can conclude, that it is very important to take into account the finite radius of the laser beam.

It is interesting to note that even for very large r_o/l values ϵ_{th} does not reduce to the usual Fredericks formula. In this limit

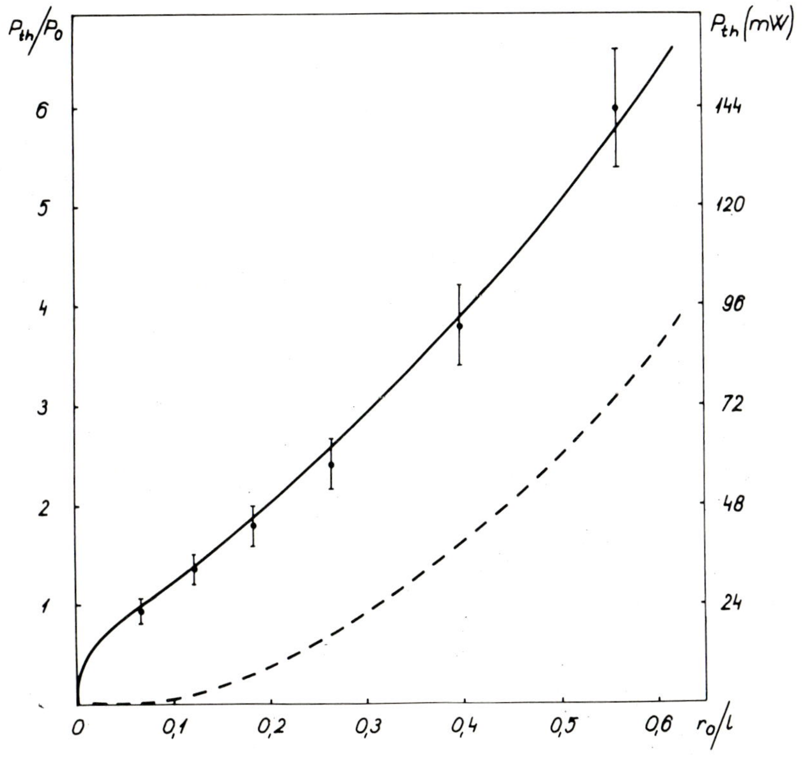

FIGURE 2 Full line: P_{th}/P_o as a function of r_o/l according to our model. Dashed line: the power corresponding to the Fredericks threshold formula. The experimental points were measured on OCB at 36.5°C; the power scale for this measurement is given at the right hand side of the figure. A good agreement between theory and experiment is obtained with $P_o = 24$ mW.

$$\epsilon_{th} = \frac{n_e}{n_o}\frac{\pi}{l}\left(\frac{K}{\epsilon_o\epsilon_a}\right)^{1/2} \qquad (r_o/l \rightarrow \infty).$$

The factor n_e/n_o comes from the coupling between the light propagation and the deformation field. The importance of this coupling has been emphasized in a previous publication.[1]

3 EXPERIMENTAL

For the experimental verification of the curve corresponding to the formula (3), two measurements were carried out. In the first exploratory experiment a homeotrop nematic MBBA cell of 150 μm thickness was illuminated with an argon ion laser beam of 65 μm diameter (FWHM).[1] At 21.5°C we got for the threshold 45 ± 10 mW, while the theoretically expected value was 35 mW, which we found to be satisfactory, taking into account the simplifying assumptions and experimental uncertainties.

In the second, more accurate and detailed investigation a homeotropic OCB sandwich-like cell of 150 μm thickness was studied. The aim of this work was to investigate the dependence of the threshold laser power first on the spot size of the focused laser beam in the middle of the nematic temperature range (36.5°C) and then its dependence on temperature in the nematic range as well as its dependence on the wavelength in the visible region.

The experimental setup was similar to that described earlier.[1] The linearly polarized beam of an argon ion laser, oscillating at 515 nm in the fundamental transversal (TEM_{00}) mode, was focused with lenses of different focal lengths to a small spot onto the NLC sample. The polarization plane of the incoming beam could be changed with the aid of a polarization rotator (Spectra Physics Model 310-21). The cell consisted of two circular glass plates separated with a teflon spacer of 150 μm thickness and sealed off with an epoxy resin. The homeotropic alignment was achieved by treating the glass surfaces with a weak solution of CTAB in chloroform.[4] The alignment was carefully tested with the conscopic method. The normal incidence of the beam was adjusted carefully with the aid of the spot reflected from the cell boundaries.

The laser power was measured with a Spectra Physics power meter (Model 404), which had a calibration accuracy of ±15%. Our power values refer always to the plane of the NLC sample i.e., the reflection losses on the lens and windows are taken into account. The intensity stability of the laser was better than 0.5%.

For focusing the laser beam, six lenses with different focal lengths (7.5 ÷ 50 cm) and of different quality were used. The lenses were always carefully adjusted for paraxial beam propagation. At a given lens, the minimal beam diameter of the focused beam could be determined with the effect itself; at a constant laser power the lens position was changed till the maximal number of fringes could be observed.

The intensity distribution in the sample plane was scanned with a pinhole of 5 μm diameter mounted on a linear stage, which could be moved across the beam with an accuracy of ± 0.2 μm. In all cases we got a nearly Gaussian intensity distribution. Some examples are given in Figure 3.

It should be mentioned here that according to the laser beam propagation laws,[5] the size of the focused beam near its minimum changes very slowly; in our case, over the 150 μm thickness of the cell, the variation of the beam diameter was less than 1% even for the lens of shortest focal length ($f = 7.5$ cm).

For determining the threshold, the diameter of the largest dark ring, observed in the transmitted beam on a screen was measured as a function of the incident laser power; this dependence proved to be nearly linear below 10 rings (Figure 4). We defined the threshold as the intersection point of this line with the power axis.

All measurements were carried out on the same OCB sample, in a time interval of about one month. During this time no ageing could be observed.

The temperature of the cell was controlled by a water thermostat, with an accuracy of ± 0.2°C.

At each measuring point, after starting the illumination we waited till the

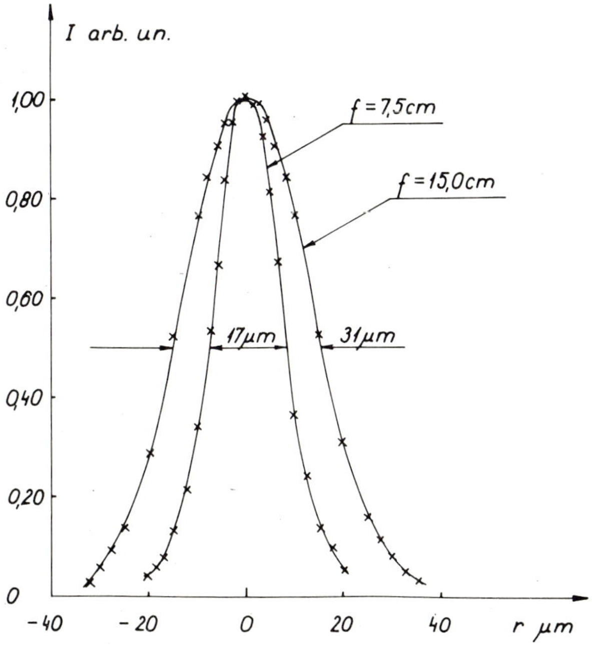

FIGURE 3 Intensity distribution of the laser beam at the sample with different lenses.

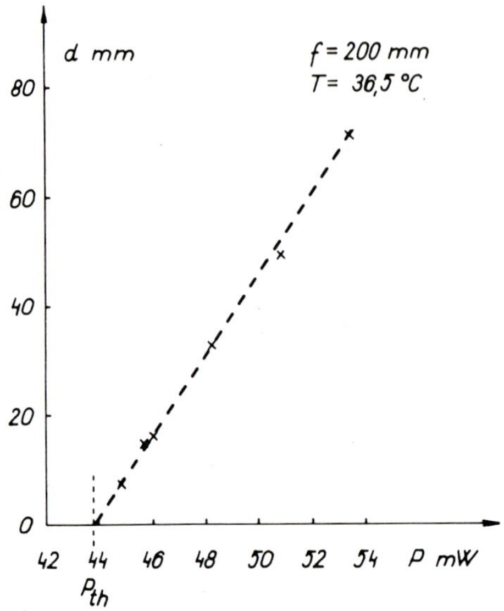

FIGURE 4 The diameter of the largest dark ring as a function of the incident laser power.

pattern had stabilized; this sometimes lasted at low power levels about 10–15 minutes. Between two measurements there was always a 1–2 minutes pause to avoid memory effects. The measurements were carried out repeatedly, at different places of the cell and with different planes of polarization of the incident beam. Places with the smallest diffuse light scattering were always chosen. The results measured under different conditions commonly did not differ more than a few mW. The spread amounted ±2 mW with the lens $f = 15$ cm and ±5 mW with that of 30 cm focus resp., i.e., the statistical error in our measurement was ±10%. It should be noted, that there may still be an additional systematic error of ±15% due to the calibration accuracy of the power meter, which could be the main source of the error in P_o and especially in the wavelength dependence of P_{th}.

4 RESULTS

The threshold intensity was measured as a function of the spot size at a fixed temperature (36.5°C). The experimental data are given in Figure 1. A good agreement between the theoretical curve and the experimental data is found with $P_o = 24$ mW.

The theoretical expression for P_o, given by Eq. (3) contains the material parameters of the nematic. In the actual case P_o can be determined however only with a certain arbitrariness, because our theoretical model assumes that the elastic constants are equal, while in reality they are different. Therefore, to compare the theoretical and experimental values, we invert Eq. (3) and evaluate an "effective" elastic constant from the threshold value. Using the experimental data of Karat *et al.*[6] $n_e = 1.678$; $n_o = 1.533$ (36.5°C, $\lambda = 515$ nm extrapolated from 546 and 589 nm) and taking $P_o = 24$ mW we get

$$K_{eff} = (0.66 \pm 0.1) \times 10^{-11} \text{ N}.$$

According to Karat *et al.*[7] the actual values of the elastic constants at this temperature are:

$$K_1 = 1.07 \times 10^{-11} \text{ N}; \quad K_2 = 0.45 \times 10^{-11} \text{ N}; \quad K_3 = 1.05 \times 10^{-11} \text{ N}.$$

Thus K_{eff} is comparable with the elastic constants measured independently.

In further experiments the temperature and wavelength dependence of the threshold was measured at a fixed spot size. The temperature dependence of the threshold was determined at $r_o = 18.6\ \mu$m. The results are given in Table I; in Figure 5 the effective elastic constant, determined from the thresholds is plotted against the temperature. For comparison the elastic data of Karat *et al.*[7] are presented in the same figure. As it can be seen, the effective elastic constant is in the whole temperature range near to K_1 and K_3 and it shows a pretransitional increase below 35°C. This increase can be evidently attributed to the sharp increase of K_2 and K_3 in the same temperature region.

The wavelength dependence of the threshold was investigated by comparing the thresholds for 515 nm (a green line of the argon laser) with that for two different wavelengths: for a blue line of the same argon ion laser (476 nm) and for the red line of an He-Ne laser (633 nm).

In the case of the blue line, the measured threshold, with a spot diameter of 30 μm ($f = 15$ cm lens) and at 36.5°C temperature:

$$P_{th}(476 \text{ nm}, 30\ \mu\text{m}, 36.5°\text{C}) = (31.2 \pm 4)\text{mW},$$

For 515 nm:

$$P_{th}(515 \text{ nm}, 30\ \mu\text{m}, 36.5°\text{C}) = (33 \pm 4)\text{mW}.$$

In the case of the red He-Ne laser, higher temperature and lens with the shortest focus ($f = 7.5$ cm) had to be used due to the smaller power available.

TABLE I

Temperature dependence of laser threshold ($f = 150$ nm, OCB homeotrop cell, $l = 150\ \mu$m, normal incidence, $\lambda = 515$ nm)

t°C	34.0	34.4	34.8	36.3	38.0	39.4
P_{th} mW	51.0	43.5	38.0	34.0	29.0	24.5

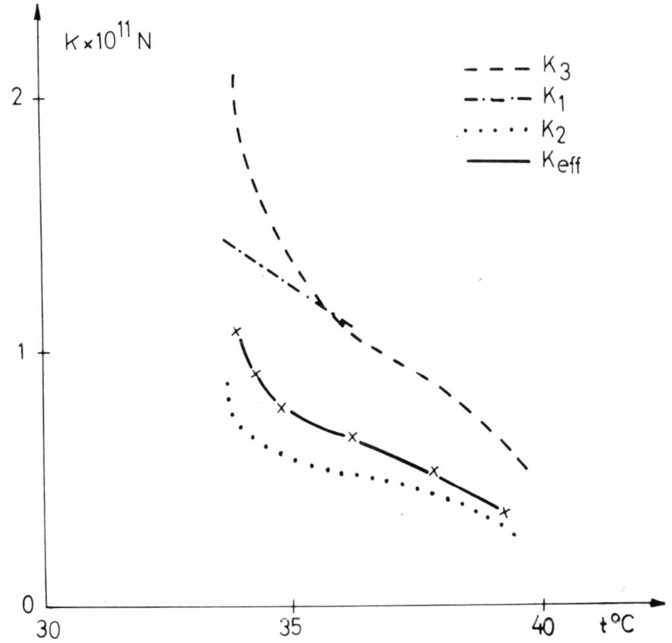

FIGURE 5 The effective elastic constant as a function of the temperature (full line). For comparison the elastic data of Karat and Madhusudana[7] are presented also.

We found with a spot diameter of 27 μm and at 38°C:

$$P_{th}(633\text{ nm}, 27\ \mu\text{m}, 38°\text{C}) = (25.6 \pm 4)\text{mW},$$

while for 515 nm:

$$P_{th}(515\text{ nm}, 27\ \mu\text{m}, 38°\text{C}) = (26 \pm 4)\text{mW},$$

In both cases the thresholds proved to be practically the same.

The theoretical formula, Eq. (3) predicts a very slow variation of P_{th} with the wavelength, i.e., as $n_e^2/n_o(n_e^2 - n_o^2)^{-1}$ changes with λ. These small changes, however are well inside the experimental error and we can conclude, that the observed very weak dependence on the wavelength gives additional support to our interpretation of the effect.

5 SUMMARY

In the paper we described a simple theoretical model of the Fredericks transition induced by optical fields. This model predicts the threshold intensity as a function of the spot size of the laser beam. The experimental data fit well the theoretical curve. From the fitting parameter, P_o, and "effective" elastic constant was determined, which turned out to be for OCB near to the actual elastic constants. The temperature dependence of K_{eff} followed well the temperature dependence of K_3; in particular the same pretransitional increase was found near to the nematic-smectic A transition. The wavelength dependence of the threshold was found to be weak, in agreement with the theory.

In conclusion, the theoretical model explains the main features of the experimental observations and provides a reliable numerical value for the threshold. This proves the underlying idea, i.e. that the observed phenomena is due to a Fredericks transition in optical field, which can be described by the continuum theory of nematics, taking into account the finite size of the deforming light beam.

The theory, used here involves two simplifications. First, it assumes a $\sqcap$ shape instead of a Gaussian. This approximation could be overcome by some numerical calculation; it is however doubtful whether this would be worthwhile if at the same time the second simplification, the one constant approximation were maintained.

The theory becomes much more complicated if the elastic constants are not assumed to be equal. As it can be shown, if $K_1 \neq K_2$ the cylindrical symmetry of the deformation is lost. Furthermore in this case the θ variable is not sufficient to describe the deformation, because the director should have an y-component too. In addition the coupling between the electromagnetic wave and the director field becomes more complicated in the presence of such deformation. All these circumstances cause considerable mathematical difficulties. On the other hand they may give rise to some original features in the cases of circularly polarized light beam or slightly oblique incidence, which cannot be described by our simplified model. Further work is planned in this direction.

References

1. L. Csillag, I. Jánossy, V. F. Kitaeva, N. Kroó, N. N. Sobolev and A. S. Zolotko, to be published in *Mol. Cryst. Liq. Cryst.*
2. P. G. de Gennes, *The Physics of Liquid Crystals,* Clarendon Press, Oxford, 1974.
3. Janke-Emde-Lösch, *Tafeln höherer funktionen,* B. G. Teubner Verlagsgesellschaft. Stuttgart, 1960.
4. J. E. Proust and L. Ter-Minassian-Saraga, *Solid State Comm.,* **11,** 1227 (1972).
5. H. Kogelnik and T. Li, *Proc. IEEE,* **54,** 1312 (1966).
6. P. P. Karat and N. V. Madhusudana, *Mol. Cryst. Liq. Cryst.,* **36,** 51 (1976).
7. P. P. Karat and N. V. Madhusudana, *Mol. Cryst. Liq. Cryst.,* **40,** 239 (1977).

Note: Since the submission of our manuscript we became aware of some new theoretical and experimental investigations closely related to our recent work:

B. Ya. Zel'dovich, N. V. Tabirian and B. S. Chilingarian, *ZhETF,* **81,** 72–83 (1981).

S. D. Durbin, S. M. Arakelian and Y. R. Shen, *Optic Letters,* **6,** 411–413 (1981).

Nonlocal radial dependence of laser-induced molecular reorientation in a nematic liquid crystal: theory and experiment

I. C. Khoo, T. H. Liu, and P. Y. Yan

Department of Electrical Engineering, The Pennsylvania State University, University Park, Pennsylvania 16802

Received June 2, 1986; accepted October 6, 1986

We present a detailed theoretical calculation, with experimental verification, of the nonlocal molecular reorientation of the nematic-liquid-crystal director axis induced by a cw Gaussian laser beam. The natures of the torque balance equations and the solutions are significantly different for normally and nonnormally incident laser beams. The nonlocal effects resulting from molecular correlation effects are particularly important for laser spot sizes that are different (smaller or larger) from the sample thickness. Experimental measurements for the transverse dependence of the molecules and the dependence of the Freedericksz threshold as a function of the laser beam sizes are in excellent agreement with theoretical results. We also comment on the effect of these nonlocal effects on transverse optical bistability.

INTRODUCTION

Studies of optical-field-induced molecular reorientations in the nematic phase of liquid crystals have revealed interesting nonlinear effects of both fundamental and applied significance.[1] Earlier studies by several groups have revealed the extraordinarily large optical nonlinearity that can be induced with relatively modest-power lasers,[2] and recent studies have shown the possibilities of utilizing these nonlinear effects in optical bistabilities and wave-mixing processes.[3] A particularly salient point of nematic response to an optical field is the strong correlation among the molecules. This correlation is manifested in the existence of molecular elastic torques within the bulk nematics and from cell boundaries. This results in a so-called nonlocal response of the nematics to a laser of finite beam size, i.e., the transverse dependence of the molecular reorientation angle exhibits a width that is in general different from the width of the laser beam. This problem has been treated briefly by several workers in various contexts.[4,5]

In these studies, however, either physically unrealistic assumptions (such as assuming that the laser transverse beam profile is a rect function) are made or the treatments are too qualitative (they apply to only one particular geometry). In some nonlinear optical processes, e.g., transverse self-phase modulations, self-focusing effects, and bistability,[6,7] a more exact description of the transverse spatial dependences of both the laser beam and the nematic reorientation are required.

In a recent study[8] we presented a calculation for the case of a linearly polarized laser incident obliquely upon a nematic film (i.e., where the laser propagation wave vector makes a finite angle with the director axis). We showed that, under physically reasonable assumptions (e.g., all the angles involved are small), the torque balance equations lend themselves to analytical solutions. Some transverse dependences of the reorientation as a function of cell geometry and optical director-axis configurations were discussed.

In this paper we present several new aspects of this problem. They include an experimental confirmation of the theoretical results for the oblique-incidence case as well as the theory for the reorientation of normally incident lasers (i.e., when the laser polarization is orthogonal to the director axis of the nematic film) and experimental results for the observed Freedericksz transition field and broadened (or narrowed) radial dependence.

THEORY

Consider the problem of a linearity polarized laser incident upon a homeotropically aligned nematic film, as shown in Fig. 1. We shall limit our discussion to the case in which $\theta < \beta < 1$, for the sake of simplicity as well as for practical considerations (because in general a small reorientation θ will contribute to a rather large change in the optical refractive index for nonlinear effects to manifest). Since the theoretical results for the $\beta \neq 0$ case have been discussed before, we will discuss here the theoretical calculation for the case $\beta = 0$.

In general, the free-energy density of the system consists of the terms from the three elastic torques (bend, splay, and twist) and the optional torque. The free-energy density term F_1 associated with the bending distortion is given by

$$\begin{aligned} F_1 &= \frac{k_1}{2}(\nabla \cdot \mathbf{n})^2 \\ &= \frac{k_1}{2}(\cos^2\theta \cos^2\phi {\theta_r}^2 - 2\sin\theta\cos\theta\cos\phi\theta_r\theta_z \\ &\quad + \sin^2\theta {\theta_z}^2), \end{aligned} \tag{1}$$

where k_1 is the elastic constant for bending, ϕ is the azimuthal angle, z the direction of propagation of the laser that coincides with the unperturbed director axis, and θ_r and θ_z are the derivatives of θ with respect to r and to z, respectively.

0740-3224/87/020115-06$02.00

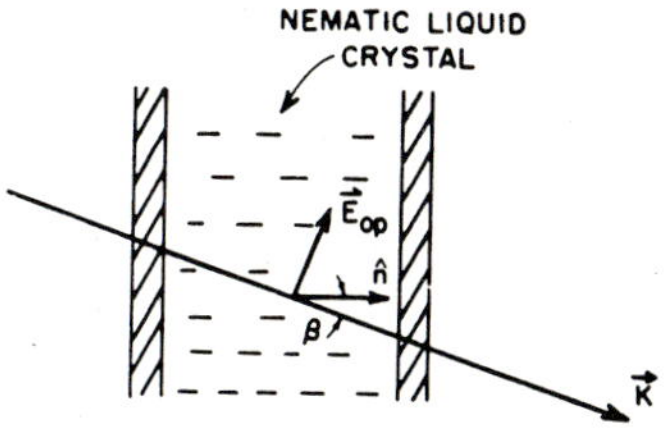

Fig. 1. A linearly polarized laser beam incident upon a homeotropically aligned nematic-liquid-crystal film.

The free-energy density term associated with twist is given by

$$F_2 = \frac{k_2}{2}(\nabla \cdot \nabla \times \mathbf{n})^2 = \frac{k_2}{2}\sin^2\phi\theta_r^{\,2}, \tag{2}$$

and that associated with splay is

$$\begin{aligned} F_3 = {} & \frac{k_3}{2}(\nabla \times \nabla \times \mathbf{n})^2 \\ = {} & (\sin\theta\sin\phi\cos\theta\cos\phi\theta_r - \cos^2\theta\cos\phi\theta_z \\ & - \sin\theta\cos\theta\theta_r)^2 + (\cos\phi\sin\theta\cos\theta\cos\phi\theta_z \\ & + \cos\phi\sin^2\theta\theta_r + \sin\theta\sin\phi\cos\theta\sin\phi\theta_Z)^2 \\ & + (\cos^2\theta\sin\phi\theta_Z + \sin\theta\cos\phi\cos\theta\sin\phi\theta_r)^2, \end{aligned} \tag{3}$$

where k_2 and k_3 are the elastic constants for twist and splay, respectively.

The total free energy of the system associated with the elastic forces is therefore given by

$$\begin{aligned} F_e = {} & \iint r\mathrm{d}r\mathrm{d}Z \int_0^{2\pi}(F_1 + F_2 + F_3)\mathrm{d}\phi \\ - {} & \iint r\mathrm{d}r\mathrm{d}Z\{[F_1(r) + F_2(r) + F_3(r)\}, \end{aligned} \tag{4}$$

where $F_1(r)$, $F_2(r)$, and $F_3(r)$ are given by

$$F_1(r) = \frac{k_1}{2}(\cos^2\theta\theta_r^{\,2} + 2\sin^2\theta\theta_Z^{\,2}), \tag{5}$$

$$F_2(r) = \frac{k_2}{2}\theta_r^{\,2}, \tag{6}$$

$$F_3(r) = \frac{k_3}{2}(2\cos^2\theta\theta_Z^{\,2} + \sin^2\theta\theta_r^{\,2}). \tag{7}$$

On the other hand, the free-energy term arising from the optical torque is given by

$$F_4(r) = \frac{-\Delta\epsilon}{8\pi}\epsilon_\perp/\epsilon_\parallel E_{op}^{\,2}(r)\sin^2\theta. \tag{8}$$

The significance of the factor $(\epsilon_\perp/\epsilon_\parallel)$ in Eq. (8) was first pointed out by Csillag *et al.*[4] and arises simply from taking into account carefully the electrodynamics of laser propagation in a birefringent medium.[4,5,9] In Eq. (8), $\epsilon_\perp$ and $\epsilon_\parallel$ are the optical dielectric constants for fields perpendicular and parallel to the director axis, respectively:

$$\Delta\epsilon = \epsilon_\parallel - \epsilon_\perp.$$

The total free energy of the system is given by

$$f = \iint r\mathrm{d}r\mathrm{d}z\{[F_1(r) + F_2(r) + F_3(r) + F_4(r)]\}. \tag{9}$$

A minimization of the free energy of the system with respect to θ gives

$$\frac{\partial}{\partial r}\left(\frac{\partial f}{\partial\theta_r}\right) + \frac{\partial}{\partial Z}\left(\frac{\partial f}{\partial\theta_Z}\right) - \frac{\partial f}{\partial\theta} = 0, \tag{10}$$

where

$$f = r[F_1(r) + F_2(r) + F_3(r) + F_4(r)]. \tag{11}$$

This gives, finally, after some lengthy and straightforward calculation, the torque balance equation

$$\begin{aligned} & [(K_1 - K_3)\sin^2\theta + K_3]\theta_{zz} + \tfrac{1}{2}(K_1 - K_3)\sin 2\theta(\theta_z)^2 \\ & + \frac{\Delta\epsilon}{8\pi}E_{op}^{\,2}(r)\sin 2\theta + \frac{1}{2}[(K_1 - K_3)\cos^2\theta + K_3 + K_2]\theta_{rr} \\ & + \tfrac{1}{4}(K_3 - K_1)\sin 2\theta(\theta_r)^2 + (K_1 - K_3)\cos^2\theta \\ & + K_2 + K_3)\theta_r/2r = 0. \end{aligned} \tag{12}$$

Even for the case of an infinite plane optical wave, Eq. (12) is extremely difficult to solve. A more meaningful and physically more insightful approach is to make a so-called one-constant approximation (i.e., $K_1 \approx K_2 \approx K_3 \approx K$). Second, in the first-order approximation, the dependence of θ on z is a simple sine wave, i.e., we assume that $\theta(r, z)$ is of the form[10]

$$\theta(r, z) = R(r)\sin\left(\frac{\pi z}{d}\right), \tag{13}$$

which obeys the hard-boundary condition $\theta(r, z) = 0$ at $z = 0$ and at $z = d$. In that case, F_1, F_2, and F_3 from Eqs. (1), (2), and (3) add up and contribute to the total elastic free energy

$$F_e^{\,1} = 2\pi k\int_0^d \mathrm{d}z\int_0^\infty r\mathrm{d}r\left[\frac{K}{2}(\theta_r^{\,2} + \theta_Z^{\,2}\right], \tag{14}$$

while the optical free-energy density term remains unchanged. The total free energy is therefore

$$F^1 = 2\pi\int_0^d \mathrm{d}z\int_0^\infty r\mathrm{d}r\left[\frac{K}{2}(\theta_r^{\,2} + \theta_Z^{\,2}) - \frac{\Delta\epsilon}{8\pi}E_{op}^{\,2}(r)\sin^2\theta\right]. \tag{15}$$

In the case of a Gaussian laser beam, $E_{op}^{\,2}(r)$ is given by

$$E_{op}^{\,2}(r) = E_{op}^{\,2}e^{-ar^2}, \tag{16}$$

where $a = 2/w_0^{\,2}$ and w_0 is the beam waist. Using Eqs. (16) and (13) in Eq. (15), and using $\sin^2\theta \sim \theta^2 - \theta^4/3$, we get, finally,

$$\begin{aligned} F^1 = \pi k\frac{d}{2}\int_0^\infty r\mathrm{d}r\Bigg[& (\mathrm{d}R/_{\mathrm{d}r})^2 + \left(\frac{\pi}{d}\right)^2 + R^2 \\ & - \frac{\Delta\epsilon}{4\pi k}E_{op}^{\,2}e^{-ar^2}R^2 + \frac{\Delta\epsilon E_{op}^{\,2}}{16\pi K}e^{-ar^2}R^3\Bigg]. \end{aligned} \tag{17}$$

The minimization of the integrand in Eq. (17), denoted as $I(R', R)$, gives

$$\frac{\mathrm{d}}{\mathrm{d}r}\left[\frac{\partial I(R', R)}{\partial R'} - \frac{\partial I(R, R')}{\partial R}\right] = 0, \tag{18}$$

$$R'' + \frac{R'}{r} + \left[be^{-ar^2} - \left(\frac{\pi}{d}\right)^2\right]R - \frac{b}{2}e^{-ar^2}R^3 = 0, \quad (19)$$

where $b = (\Delta\epsilon/4\pi k)E_{op}^2$ and the boundary conditions on R are

$$R'(0) = 0,$$
$$R(\infty) = 0. \quad (20)$$

In Eqs. (17)–(19), $R' = \partial R/\partial r$ and $R'' = \partial^2 R/\partial r^2$.

Equation (19) is an interesting nonlinear equation in $R(r)$. The occurrence of the cubic term (αR^3) is due to our expansion of sin θ to third order in θ [see the sentence just before Eq. (17)], and its inclusion is necessary for a nonvanishing solution of R. In analogy to the infinite plane-wave case, a nonvanishing value of R occurs only if the optical field exceeds a threshold value. In the infinite plane wave, approximately ($a = 0$), the threshold field is well defined by the relation

$$b_{th} = \pi^2/d^2(E_{th}^2 = 4\pi^3 K\Delta\epsilon^{-1}d^{-2}).$$

However, in the present case, because of the Gaussian function e^{-ar^2} attached to the optical-field square amplitude E_{op}^2, the equation for $R(r)$ clearly breaks up into two distinct regions: one region (region I) corresponds to the square-bracketed term in Eq. (19) being positive, while the other (region II) corresponds to the term being negative. Furthermore, since the molecules outside the laser beam exert torques on those in the central region of the beam, in competition or conjunction with the torque from the cell walls, the so-called threshold field also obviously depends on the laser beam waist.

It is instructive to compare Eq. (19), derived under the condition that $\beta = 0$, with the equation governing the radial function for $\beta \neq 0$ derived previously. We denote the reorientation angle $\theta(r, Z)$ for the case $\beta \neq 0$ as $\theta(r, Z) = R_1(r)\sin(\pi Z/d)$; then $R_1(r)$ obeys the equation

$$R_1''(r) + \frac{\dot{R}_1}{r} + \left[b\cos 2\beta e^{-ar^2} - \left(\frac{\pi}{d}\right)^2\right]R_1 + \frac{b}{2}e^{-ar^2}\sin 2\beta = 0. \quad (21)$$

Equation (21) contains a so-called "bias" term $b/2e^{-ar^2}\sin 2\beta$, i.e., there is a nonvanishing torque exerted initially by the optical field on the director axis. Optically induced reorientation is therefore possible for any finite value of E_{op}. As a result of this difference between the $\beta = 0$ case and the $\beta \neq 0$ case, the nonlocal dependences, which are manifested in the form of differences between the width of the laser beam and the width of $\theta(r)$, are quite different for the two cases. In the case of $\beta \neq 0$, the width of $\theta(r)$ is always larger than w_0, but for the $\beta = 0$ case the width can be greater or less than w_0, depending on the laser beam waist compared with the thickness of the film.

As was shown previously,[8] there is a closed-form solution to Eq. (21). Equation (20), however, does not yield any meaningful closed-form solution, but it can be readily solved numerically. In the next two sections, we discuss some of the salient points of the nonlocal dependence for the $\beta = 0$ case. We then also recall the counterpart results for the $\beta \neq 0$ case and compare these results with our experimental observations.

NUMERICAL RESULTS

Numerical results for $R(r)$ are obtained for a large range of values of w_0 (measured with respect to the thickness of the sample d).

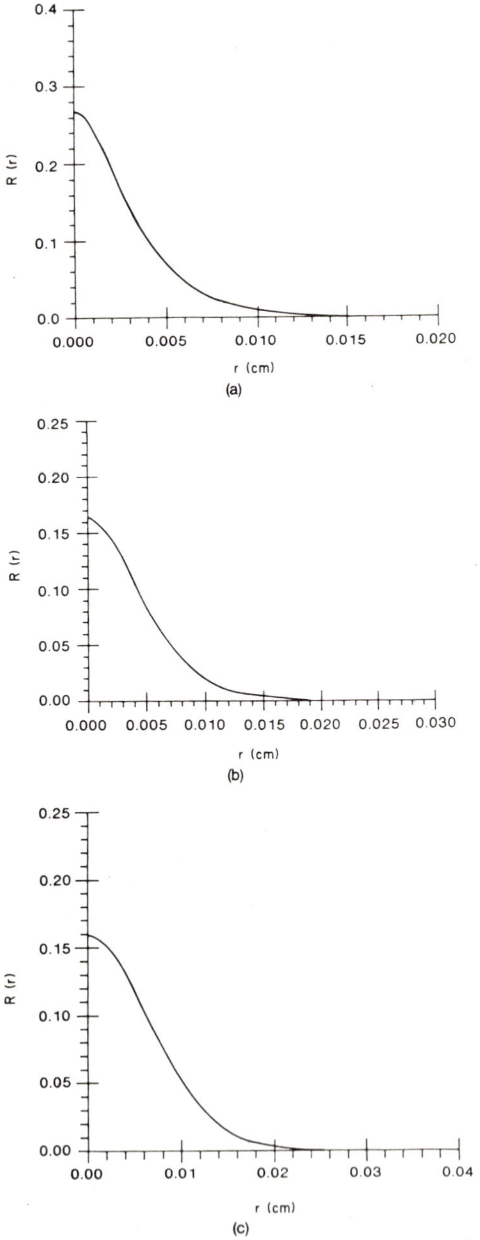

Fig. 2. (a) Solution of the reorientation transverse profile $R(r)$ for the case $w_0/d = 0.4$ [$w_0 = 40\ \mu m$, $d = 100\ \mu m$]; (b) $R(r)$ for $w_0/d = 1$; (c) $R(r)$ for $w_0/d = 2$. For all $\beta = 0$, $d = 0.01$ cm.

In the $\beta = 0$ case, director-axis reorientations do not occur until the central maximum intensity $I(0) = (nc/4\pi)E_{op}^2$ is greater than the Freedericksz threshold intensity [i.e., for $b > b_{th}^{(0)}$, where $b_{th}^{(0)} = (\Delta\epsilon/8\pi k)E_{th}^2$]. The values of b for which $\theta \neq 0$ occurs depend on w_0.

Figures 2(a), 2(b), and 2(c) show the typical dependence of $R(r)$ for $w_0/d = 0.4$, $w_0/d = 1$, and $w_0/d = 2$, respectively. In general, the director reorientations occur only after the incident optical electric field is well above the threshold field (which we shall discuss presently). The important point about Figs. 2(a)–2(c) is that the width of $R(r)$ (which is measured by the e^{-2} point; even though the curve is not a Gaussian this somewhat arbitrary approach gives us a measure of the width of the curve) with respect to the incident laser beam width is clearly dependent on the laser beam width relative to the thickness of the sample. As shown in

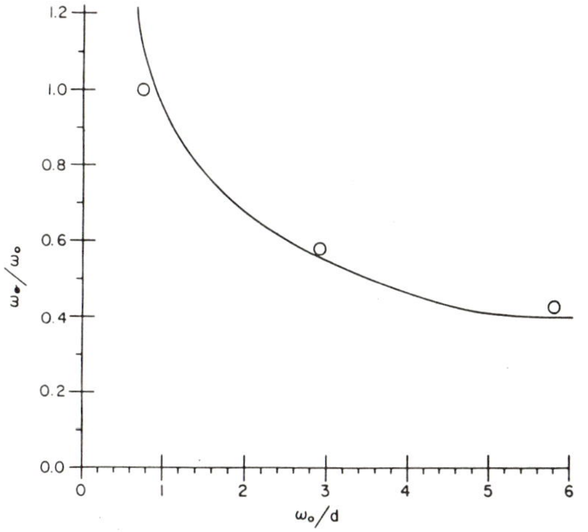

Fig. 3. Plot of the width w_0 of $R(r)$ as a function of the ratio of incident laser beam waist to the thickness of the film (w_0/d). Circles are experimentally observed points. ($\beta = 0$.)

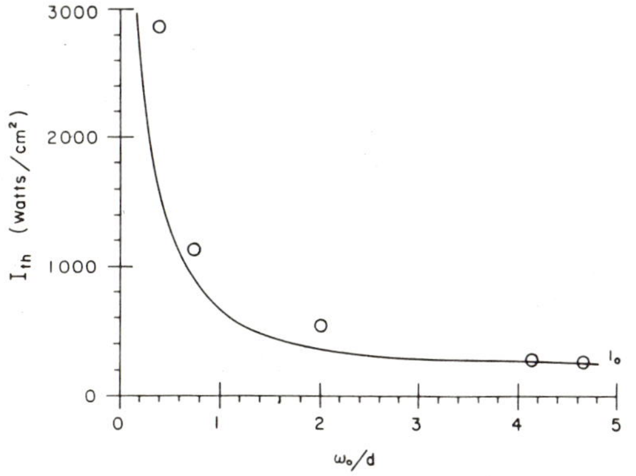

Fig. 4. Plot of the theoretically predicted threshold intensity I_{th} at which reorientation occurs versus w_0/d. The value $I^{(0)}$ corresponds to the case involving an infinite plane wave ($\beta = 0$). Circles are experimentally observed points.

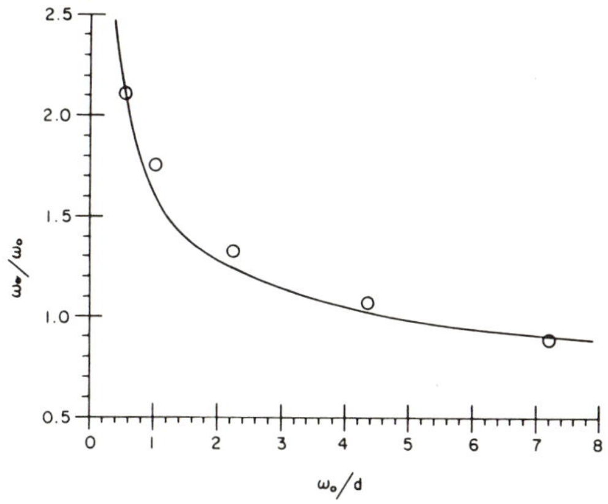

Fig. 5. Plot of the width of the reorientation transverse profile w_θ/w_0 versus w_0/d for the case $\beta \neq 0$ (from previous calculations in Ref. 6). Circles are experimentally observed points.

Fig. 2(a), where $w_0/d = 0.4$, the width of $R(r)$ (denoted as w_θ) is larger than the incident laser beam width w_0. On the other hand, Fig. 2(c) shows that for $w_0/d = 2$, one finds w_θ to be less than w_0

This variation of w_θ/w_0 with w_0/d is plotted in Fig. 3, which shows this trend, i.e., the value of w_θ/w_0 drops off monotonically from being larger than unity (for $w_0/d < 1$) to smaller than unity (for $w_0/d > 1$). At $w_0/d \sim 1$, we have $w_\theta/w_0 \sim 1$.

As we mentioned earlier, as a result of the significant elastic torque from molecules surrounding the laser beam (besides the boundary elastic torque), the field required to create finite molecular reorientation (which we shall denote b_{th}) is, in general, larger than that associated with infinite-beam-size lasers. Figure 4 shows a plot of the value of b_{th} for which nonzero reorientation $\beta(r)$ occurred. As a function of w_0/d, we note that $b_{th} \rightarrow b_{th}^{(0)}$ for $w_0 \gg d$.

For the case $\beta \neq 0$, there is no threshold field; the solution of Eq. (21) shows one other main difference, namely, that the width of the response $R_1(r)$ is *always* larger than the laser beam width. Figure 5 is a plot of w_θ/w_0 versus w_0/d from the numerical results obtained from our previous study. For $w_0/d = 2$, for example, one gets $w_\theta/w_0 \sim 1.5$ for the $\beta \neq$ case. On the other hand, for the $\beta = 0$ case, we get $w_\theta/w_0 \sim 0.7$.

EXPERIMENTS

The transverse dependence of the molecular reorientation and the associated transverse dependence of optical refractive-index change on the laser spot size play a crucial role in nonlinear processes involving the nonlinear (intensity-dependent) transverse phase shift. In self-phase modulation, and in transverse optical bistability, the observed effect depends both on the size of the laser beam and on the detailed index profile change. These concerns motivate our experimental verifications of the preceding theoretical results.

Figure 6 is a schematic of the experimental setup to measure the reorientation profile. The nematic liquid crystal used is a homeotropically aligned 100-μm-thick EM chemical E46 sample at 22°C. E46 has a $\Delta\epsilon$ of 0.7, a nematic range from −9.5 to 88°C, and little (negligible) thermal effect at the Ar laser 5145-Å line. The Ar laser is linearly polarized and focused onto the liquid crystal. The beam waist of the Ar laser on the liquid crystal is monitored with a knife edge mounted on another translator (not shown). The translator shown in the figure, with a 0.5-μm resolution, translates the He–Ne probe beam. Both the He–Ne and the Ar lasers experience a self-phase modulation effect associated with the transverse phase shift caused by the molecular reorientation (induced by the Ar^+ laser). These self-phase modulation effects are manifested in the form of an increased divergence and appearance of interference rings of the laser beams at the screen (placed 5 m away from the sample).

The liquid-crystal film is oriented such that β assumes a value of either 0° or 22° with respect to the Ar laser beam direction.

The results, as shown by the experimental data points in Figs. 3–5, show a remarkable agreement with theoretical calculation. The measured widths of the radial reorientation profile w_θ (relative to w_0) versus the waist (relative to the thickness d) are of the order of unity for $w_0/d \sim 1$ and decrease as w_0 increases (cf. Fig. 3) in the case of $\beta = 0$. On the other hand, for $\beta \neq 0$, as shown in Fig. 4, the experimental measured width w_θ is always greater than the laser beam waist w_0 for all values of w_0; w_θ approaches w_0 for large values of w_0

As is shown in Fig. 5, the experimentally observed threshold field dependence on w_0/d also follows the theoretical prediction. For E46, a 100-μm sample has a Freedericksz transition field intensity of 200 W/cm^2 (using the values $\Delta\epsilon$ = 0.3, $K \sim 10^{-7}$, d = 0.01 cm, and $n \approx n_0 \approx 1.5$). For large values of w_0/d (for example, $w_0/d \geq 5$), the observed threshold field approaches this value. However, as the incident laser beam size decreases to a value comparable with the thickness d or less, the threshold optical intensities increase dramatically. At $w_0 \sim d$, the threshold intensity increases by almost an order of magnitude. In general, the experimentally observed relative increase of the threshold field is slightly larger than the theoretical value, probably because of a systematic difference between the experimental observation of the onset of reorientation (by the appearance of the self-focusing effect on the exit Ar^+ laser beam) and also

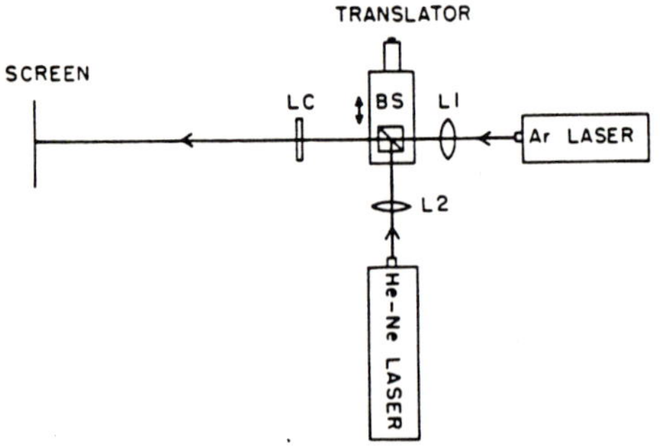

Fig. 6. Experimental setup for measuring the width of the director axis reorientation induced by a Gaussian laser beam. BS, beam splitter; L, lens; LC, liquid-crystal film.

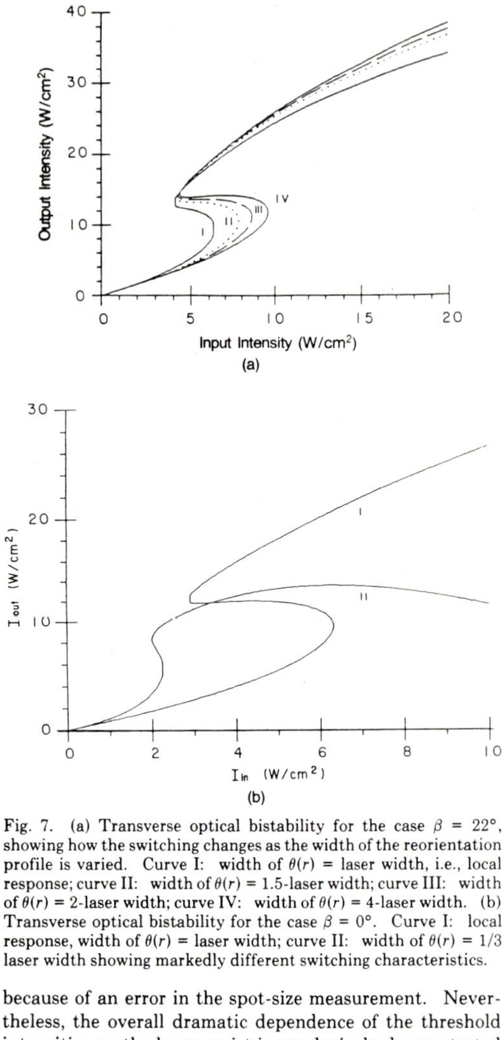

Fig. 7. (a) Transverse optical bistability for the case β = 22°, showing how the switching changes as the width of the reorientation profile is varied. Curve I: width of $\theta(r)$ = laser width, i.e., local response; curve II: width of $\theta(r)$ = 1.5-laser width; curve III: width of $\theta(r)$ = 2-laser width; curve IV: width of $\theta(r)$ = 4-laser width. (b) Transverse optical bistability for the case β = 0°. Curve I: local response, width of $\theta(r)$ = laser width; curve II: width of $\theta(r)$ = 1/3 laser width showing markedly different switching characteristics.

because of an error in the spot-size measurement. Nevertheless, the overall dramatic dependence of the threshold intensities on the beam waist is conclusively demonstrated in Fig. 5.

FURTHER REMARKS

The observed (experimentally and theoretically) broadening or narrowing of the response of the nematic reorientation could, and should, be taken into account in the study of nonlinear transverse optical effects involving a focused beam. We end this section with reference to transverse optical bistability, which recently has received considerable attention.

Figure 7(a) (calculated using the technique developed in Ref. 7) shows what would happen to the output versus input bistability (of the transmitted on-axis laser intensity) for various laser beam sizes in comparison with the thickness of the film for the case β = 22°. Although the switch-down intensities do not seem to vary much, the switch-up intensity increases by about 60% as the laser spot size is decreased from d to $d/4$. The point to note about this exercise is that in extrapolating the observed switching intensities for a large-sized beam to smaller beams (presumably for some integrated-optics consideration) these nonlocal effects should be properly accounted for. For $\beta = 0$, the fact that the observed width of $R(r)$ is in general smaller than the laser beam waist is also interesting. Figure 7(b) shows a comparison of the switching characteristics for the case when the induced index profile is smaller than the incident laser beam's waist with the case when the nonlocality is ignored. Both the switching intensities and the bistability loops are dramatically different. A detailed account of this case is clearly outside the context of this paper and is reserved for a lengthy future publication on transverse bistability.

CONCLUSION

We have presented a quantitative theoretical and experimental study of the transverse dependence of the nematic axis reorientation induced by an optical field. For most geometrical configurations, there are significant nonlocal effects, which are experimentally confirmed. Similar effects are expected in any diffusive type of optically induced nonlinearities (e.g., thermal and solid-state electronics) and are particularly pronounced when the laser beam sizes are smaller than the characteristic lengths (sample thickness, thermal diffusion length, electron–hole recombination length, etc.). In studies of transverse nonlinear optical effects, such as self-phase modulation, self-focusing, and optical switching (passive optical limiting or optical bistability), these nonlocal effects must be accounted for appropriately.

ACKNOWLEDGMENT

This research is supported by grant ECS8415387 from the National Science Foundation and by grant AFOSR-840375 from the U.S. Air Force Office of Scientific Research.

REFERENCES

1. See, for example, the review paper by I. C. Khoo and Y. R. Shen, Opt. Eng. **24,** 579 (1985).
2. N. F. Pilipetski, A. V. Sukhov, N. V. Tabiryan, and B. Ya Zel'dovich, Opt. Commun. **37,** 280 (1981); S. R. Galstyan, O. V. Garibyan, N. V. Tabiryan, and Yu. S. Chilingaryan, JETP Lett. **33,** 437 (1981); A. S. Zolot'ko, V. F. Kitaeva, N. Kroo, N. N. Sobolev, and L. Chillag, JETP Lett. **32,** 158 (1980); I. C. Khoo and S. L. Zhuang, Appl. Phys. Lett. **37,** 3 (1980); S. D. Durbin, S. M. Arakelian, and Y. R. Shen, Phys. Rev. Lett. **47,** 1411 (1981).
3. B. Ya. Zel'dovich, N. V. Tabiryan, and Yu. S. Chilingaryan, Sov. Phys. JETP **54,** 32 (1981); I. C. Khoo, Appl. Phys. Lett. **41,** 909 (1982); I. C. Khoo and S. L. Zhuang, IEEE J. Quantum Electron. **QE-18,** 246 (1981).
4. L. Csillag, J. Janossy, V. F. Kitaeva, N. Kroo, and N. N. Sobolev, Mol. Cryst. Liq. Cryst. **84,** 125 (1982).
5. E. Santamato and Y. R. Shen, Opt. Lett. **9,** 564 (1984).
6. A. E. Kaplan, Opt. Lett. **6,** 360 (1981); M. LeBerre, E. Ressayre, A. Tallet, K. Tai, and H. M. Gibbs, IEEE J. Quantum Electron. **QE-21,** 1404 (1985); J. E. Bjorkholm, P. W. Smith, W. J. Tomlinson, and A. E. Kaplan, Opt. Lett. **6,** 345 (1981); J. E. Bjorkholm, P. W. Smith, and W. J. Tomlinson, IEEE J. Quantum. Electron. **QE-18,** 2016 (1982).
7. I. C. Khoo, P. Y. Yan, T. H. Liu, S. Shepard, and J. Y. Hou, Phys. Rev. A **29,** 2756 (1984).
8. I. C. Khoo, T. H. Liu, and R. Normandin, Mol. Cryst. Liq. Cryst. **131,** 315 (1985).
9. See, for example, H. L. Ong, Phys. Rev. A **28,** 2393 (1983); B. Ya. Zel'dovich, N. V. Tabiryan, and Yu. S. Chilingaryan, Sov. Phys. JETP **54,** 32 (1981).
10. See, for example, P. G. deGennes, *The Physics of Liquid Crystals* (Clarendon, Oxford, 1984).

Mol. Cryst. Liq. Cryst., 1982, Vol. 89, pp. 287–293
0026-8941/82/8904-0287$06.50/0

Reorientation of Liquid Crystals by Superposed Optical and Quasistatic Electric Fields

L. CSILLAG, N. ÉBER, I. JÁNOSSY and N. KROÓ
Central Research Institute for Physics, H-1525 Budapest, P.O. Box 49, Hungary

and

V. F. KITAEVA and N. N. SOBOLEV
Lebedev Physics Institute, Moscow, USSR

(*Received March 1, 1982*)

The electric field of a CW laser beam can reorient a nematic liquid crystal. Experiments on the influence of a superposed quasistatic electric field are reported showing that this can reinforce or weaken the laser induced reorientation depending on the substance and geometry.

Recently some theoretical and experimental work has been published on orientational nonlinearity and optical field induced Fredericks transition in nematic liquid crystal layers.[1-5] These effects are due to the orienting action of the electric field of the light beam passing through the sample. Experimentally one observes a power dependent increase of the divergence of the light beam and a ring system appearing in it.

It is well known that the orientation of a nematic layer can be influenced by external magnetic or quasistatic fields too. Thus it can be expected that the orientational deformation caused by a light field can be controlled by external fields. The aim of the present letter is to report experimental results on such an effect.

In our experiments the external field was an electric field with $f = 10$ kHz. It was applied perpendicularly to the boundaries of the cell with the help of transparent indium oxide coatings as electrodes. The light field was produced by focusing a polarized argon ion laser beam on the sample into a spot with a radius of 40 μm. The sample thickness was in all experiments 150 μm. We confined ourselves to the case of normal incidence. The measurements consisted of determining the divergence of the laser beam as a function of the laser power and the applied voltage. As a measure of the divergence the angular diameter of the outest dark ring was measured.

The response of the nematic layer to a low frequency electric field depends crucially on the sign of the anisotropy of the static dielectric constants, $\epsilon_a^{(S)} = \epsilon_\parallel^{(S)} - \epsilon_\perp^{(S)}$, where $\epsilon_\parallel^{(S)}$ and $\epsilon_\perp^{(S)}$ denote the dielectric constants in the direction parallel and perpendicular to the nematic director. Depending on whether $\epsilon_a^{(S)}$ is positive or negative, the applied field tries to orient the molecules parallel or perpendicularly to itself, respectively. The light field always tries to orient the molecules parallel to the direction of the polarization. In the case of normal incidence the light field and the applied field are normal to each other. Thus if $\epsilon_a^{(S)} > 0$ the two fields weaken, while if $\epsilon^{(S)} < 0$ they reinforce each other.

To study the case $\epsilon_a^{(S)} > 0$ we chose octyl-cyano-biphenyl (OCB) which has a strong positive dielectric anisotropy (at $T = 35°C$ $\epsilon_\parallel = 12.5$; $\epsilon_\perp = 5.7$ [6]); the case $\epsilon_a^{(S)} < 0$ was studied on *p-n* methoxybenzilidene-*p*-butylaniline (MBBA) for which at T = 30°C $\epsilon_\parallel = 4.6$, $\epsilon_\perp = 5.0$.[7] Three different cells were investigated.

1. HOMEOTROPICALLY ALIGNED OCB

As discussed in detail in Ref. [5], in this case the alignment becomes instable at a threshold laser power. The applied field tries to stabize this alignment, consequently the threshold power increases as the applied field is increased. In Figure 1 an experimental curve is presented. In Figure 2 the beam divergence is shown as a function of the applied field at a fixed laser power. As it can be expected, the divergence monotonically decreases and becomes zero at a critical field strength.

2. HOMEOTROPICALLY ALIGNED MBBA

In this case as $\epsilon_a^{(S)} < 0$, the threshold power decreases as the applied field is increased, as shown in Figure 1. The threshold power becomes zero when the applied voltage is high enough to destabilize alone the alignment.

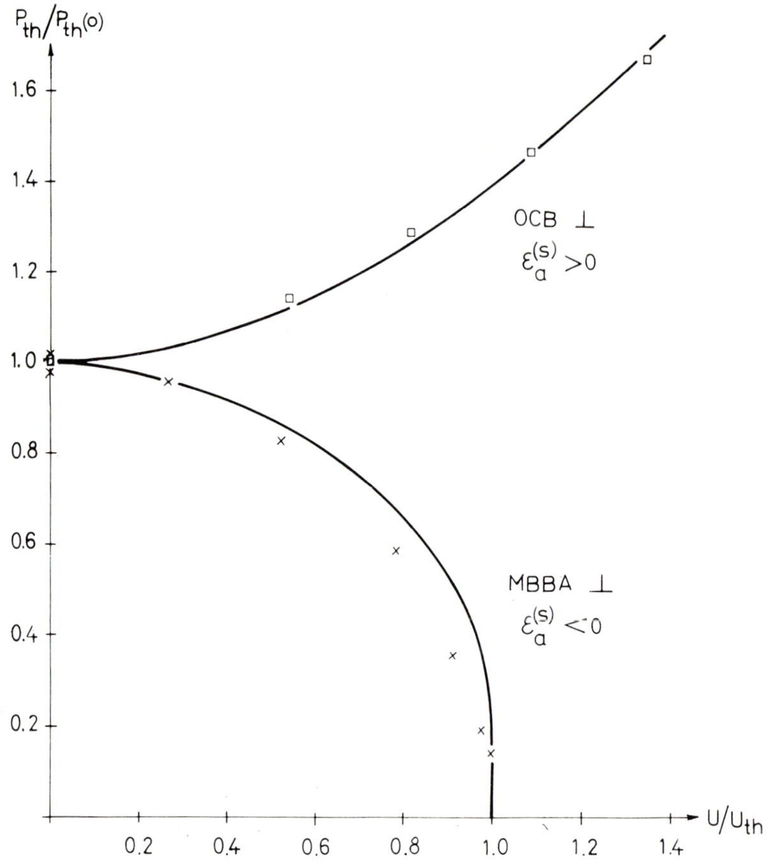

FIGURE 1 The threshold power as a function of the applied voltage. For OCB P_{th} $(O) = 58$ mW, $U_{th} = 0.93$ V; for MBBA P_{th} $(O) = 41.5$ mW, $U_{th} = 3.86$ V.

The divergence of the laser beam as a function of the applied field at a fixed power is shown in Figure 3. This curve, exhibiting a sharp maximum, can be explained qualitatively by remembering that the divergence of the laser beam is due to the inhomogeneity of the deformation within the laser spot. At low voltages the applied field increases this inhomogeneity. At high fields however the director becomes almost parallel to the surfaces in the *whole* cell. Thus the sample becomes again homogeneous and the ring system disappears.

The variation of the threshold power as a function of the applied field in the two cases discussed above can be given theoretically by a

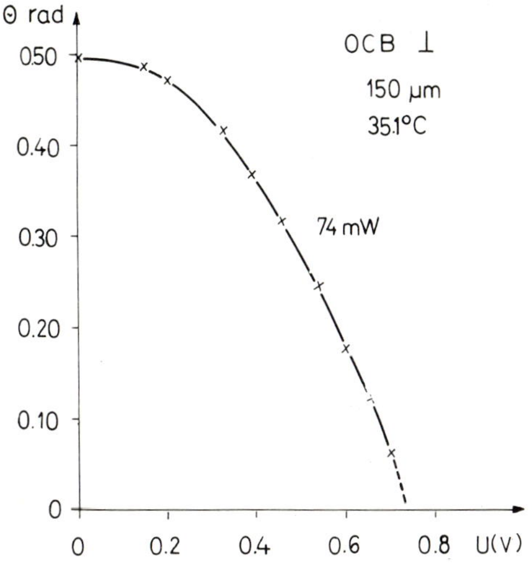

FIGURE 2 The beam divergence as a function of the applied voltage for homeotrop OCB.

slight modification of the theory given in Ref. [5]. As it is shown there the threshold power can be determined from the equation

$$\gamma \frac{J'_0(\gamma)}{J_0(\gamma)} = \kappa \frac{K'_0(\kappa)}{K_0(\kappa)} \tag{1}$$

where J_0 and K_0 denote the 0-order Bessel and modified Hankel functions. For zero applied field

$$\gamma^2 = P_{\text{th}}\, \epsilon_0 \epsilon_a \frac{n_o}{n_e^2} \frac{1}{\pi c K} - \frac{\pi^2}{l^2} r_0^2; \quad \kappa^2 = \frac{\pi^2}{l^2} r_0^2$$

where P_{th} is the threshold power, K an effective Frank elastic constant, n_o and n_e the ordinary and extraordinary refractive indices, $\epsilon_a = n_e^2 - n_o^2$; l and r_0 are the sample thickness and spot radius resp. Taking into account the torque due to the applied field we get Eq. (1) again, where now γ and κ are

$$\gamma^2(U) = \gamma^2(O) \mp \frac{r_0^2}{l^2} \frac{U^2}{U_{\text{th}}^2}; \quad \kappa^2(U) = \kappa^2(O) \pm \frac{r_0^2}{l^2} \frac{U^2}{U_{\text{th}}^2}$$

where the upper signs refer to positive $\epsilon_a^{(S)}$, the lower ones to negative $\epsilon_a^{(S)}$. U is the effective applied voltage and

$$U_{th} = \pi \sqrt{\frac{K}{\epsilon_0 |\epsilon_a^{(S)}|}} \tag{2}$$

In Figure 1 the solid lines correspond to the above theory. The fitted parameters are $P_{th}(O)$ and U_{th}. The agreement between the experimental curve and the theoretical one is satisfactory for both material. From $P_{th}(O)$ and U_{th} two separate values of the "effective" elastic constants can be calculated. The measurements provide for OCB $K = 0.63 \times 10^{-11}$ N and $K = 0.53 \times 10^{-11}$ N; for MBBA $K = 0.62 \times 10^{-11}$ N and $K = 0.54 \cdot 10^{-11}$ N resp. (for the refractive index data see Refs. [2,5]). The consistency of these data and their agreement with independently measured values of elastic constants[8,9] can be regarded satisfactory tak-

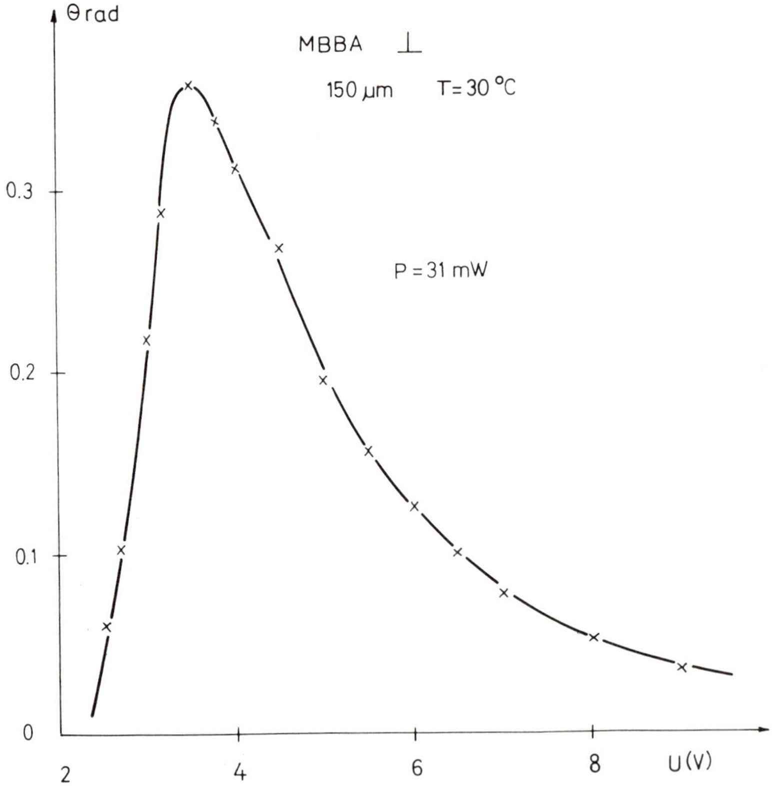

FIGURE 3 The beam divergence as a function of the applied voltage for homeotrop MBBA.

ing into account the simplifications in the theory and the uncertainities in the determination of the *absolute* value of the laser power.

3. PLANAR CELL OF OCB

As discussed in Ref. [2], in the case of planar orientation and normal incidence of the polarized light beam, deformation does not occur. The extraordinary component stabilizes the planar configuration, while the ordinary does not interact with it.

The applied field, due to the positive static dielectric anisotropy of OCB, destabilizes the planar orientation above a threshold voltage ($\approx$1 V [10]). Thus applying a voltage higher then 1 V, a ring system can be produced by a laser beam with extraordinary polarization, because the light field reorients the molecules within the spot. The beam divergence as a function of the applied field at different laser powers is given in Figure 4. In accordance with the above considerations the beam divergence goes to zero at approximately 1 V, independently from the laser

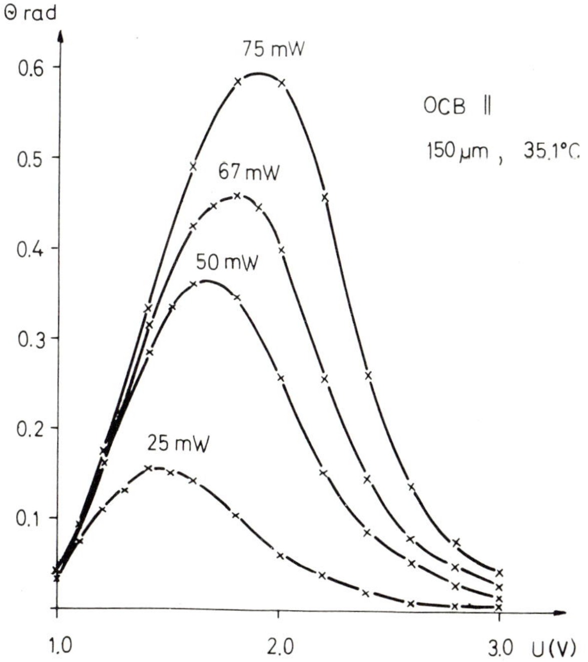

FIGURE 4 The beam divergence as a function of the applied voltage for planar OCB.

power. The decrease of the divergence at high voltages can be explained similarly as in the case of homeotropic MBBA. In the present case the high electric field removes the inhomogeneity of the deformation by orienting the molecules in the whole sample perpendicularly to the boundaries.

The result of the investigations described above fit well into the interpretation of the laser field induced reorientation of nematics, given in our earlier papers. The possibility of controlling the beam divergence by an external field may find some practical applications.

References

1. A. S. Zolot'ko, V. F. Kitaeva, N. Kroó, N. N. Sobolev and L. Csillag, *Pisma v ZhETF*, **34** (5), 263 (1981) (in Russian).
2. L. Csillag, I. Jánossy, V. F. Kitaeva, N. Kroó, N. N. Sobolev and A. S. Zolot'ko, *Mol. Cryst. Liq. Cryst.*, **78,** 173 (1981).
3. N. F. Pilipetski, A. V. Sukhov, N. V. Tabiryañ and B. Ya. Zel'dovich, *Optics Comm.*, **37** (4), 280 (1981).
4. S. D. Durbin, S. M. Arakelian and Y. R. Shen, *Optics Letters*, **6,** 411 (1981); *Phys. Rev. Lett.*, **47,** 411 (1981).
5. L. Csillag, I. Jánossy, V. F. Kitaeva, N. Kroó and N. N. Sobolev, *Mol. Cryst. Liq. Cryst.*, **84,** 425 (1982).
6. L. Bata and A. Buka, *Acta Phys. Pol.*, **A54,** 635 (1978).
7. I. P. Zhuk and L. E. Golovither, *Acta Phys. Pol.*, **A55,** 575 (1979).
8. P. P. Karat and N. V. Madhusudana, *Mol. Cryst. Liq. Cryst.*, **40,** 239 (1977).
9. P. de Gennes, The Physics of Liquid Crystals, Clarendon Press, Oxford, 1974, p. 66.
10. I. Jánossy and L. Bata, *Acta Phys. Pol.*, **A54,** 643 (1978).

External field enhanced optical bistability in nematic liquid crystals

Hiap Liew Ong[a)]
Department of Physics, Brandeis University, Waltham, Massachusetts 02254

(Received 9 January 1985; accepted for publication 18 February 1985)

The exact solution is obtained for describing the external field effects in the optically induced molecular reorientation and bistability in a homeotropic nematic liquid crystal. With a low external field applied transverse or parallel to the laser propagation direction, optical bistability can always be suppressed or enhanced and hence can be seen in all existing nematics at a low laser power (typically < 1 kW/cm^2). The fields corresponding to the tricritical point are obtained.

Since the first observation of optical bistability (OB) in 1976 by Gibbs, McCall, and Venkatesan,[1] a large number of investigations of this phenomenon have appeared in the literature.[2] OB offers the attractive advantages of parallelism and noninterfering propagation of optics over electronic switching devices. The study of optical nonlinearities via molecular reorientation of liquid crystals (LC's) is of particular interest because among fluids, LC's have the largest optical field induced refractive index changes. The reorienting torque produced by a cw optical field on LC's can result in an extremely strong collective molecular reorientation and hence large associated nonlinear effects. As a comparison: the light self-focusing nonlinear constants of nematic LC's (NLC's),[3–6] cholesteric LC's,[7] and smectic-C LC's[8] have, respectively, values larger by eight to ten, three, and nine orders of magnitude than that of carbon disulfide. The first quantitative observation of the optically induced molecular reorientation was reported in 1981 by Durbin, Arakelian, and Shen.[4] So far all the experimental results showed that it is a second-order structural transition in which the change in the LC spatial orientation is continuous with increasing optical field intensity.[3,4] Recently we obtained the exact solution for describing the NLC orientation and found that certain existing NLC's can have first-order transitions in which increasing optical field intensity results in discontinuous changes in the LC orientation.[5] This form of intrinsic OB does not use a resonant optical cavity and offers the attractive possibility of making large electroresponse devices with a low power sharp threshold.

Four current goals in the OB in NLC's are (1) to attain OB at room temperature, (2) to enlarge the width of the hysteresis cycle, (3) to reduce the power required for switching between two states, and (4) to increase the switching speed. In this letter, we discuss methods to achieve the first three goals. We study the external field effects on the OB in NLC's and obtain the exact solution and OB criterion. We predict that OB can always be enhanced or suppressed and hence is always possible for all existing NLC's. Typically, with an electric potential of a few volts or a magnetic field of 1 kG or less, OB occurs at a low power of < 1 kW/cm^2.

We consider a NLC cell of thickness d confined between the planes $z = 0$ and $z = d$ of a Cartesian coordinate system. The surfaces are treated to give homeotropic alignment, i.e., align NLC's parallel to the z axis. A laser beam is normally incident on the cell with the polarization parallel to the plane of incidence, which is the xz plane. The optical field will apply a torque on the NLC so as to reorient the NLC along the x axis. An additional magnetic (hereafter referred as MOB) or electric field (EOB) is applied to the sample with the orienting direction either conflicting (geometry I) or parallel (geometry II) with the optical reorienting direction. The applied field always tends to align the director so that the axis of maximum susceptibility is parallel to the field. Known NLC's are diamagnetic with positive diamagnetic anisotropy χ_a and also with positive dielectric anisotropy at optical frequencies. But for the electric field, the dielectric anisotropy can be either positive or negative depending on the material and frequencies. Two geometries are considered. In geometry I (II), the external field will orient the NLC parallel (normal) to the z axis and hence suppress (enhance) the optical orienting effects: the magnetic field is directed along the z axis (x axis), $\mathbf{H} = (0,0,H)$ [$\mathbf{H} = (H,0,0)$ for geometry II], whereas the electric field is always directed along the z axis, $\mathbf{E} = (0,0,E_z)$, and for geometry I (II), the dielectric anisotropy $\epsilon_a = \epsilon_\parallel - \epsilon_\perp$ is positive (negative), where $\epsilon_\perp$ and $\epsilon_\parallel$ are the dielectric constants perpendicular and parallel to the local director.

To determine the angle $\theta(z)$ between the director $\hat{n}(z) = (\sin\theta, 0, \cos\theta)$ and the z axis, we first express in terms of θ the total free-energy density which consists of the elastic deformation (F_d), optical field ($F_{\rm opt}$), and external field ($F_{\rm ext}$) energy densities. The magnetic field energy density $-\chi_a(\hat{n}\cdot\mathbf{H})^2/2$ can be written as $\pm(\chi_a H^2/2)\sin^2\theta$, where throughout this letter, the upper sign is always for geometry I and the lower sign for geometry II. The electric field satisfies $\operatorname{div}\mathbf{D} = 0$ from which D_z is a constant independent of z and $F_{\rm ext} = \mathbf{E}\cdot\mathbf{D}/8\pi = D_z^2/[8\pi\epsilon_\parallel(1 - w\sin^2\theta)]$, where $w = \epsilon_a/\epsilon_\parallel$. Recently, we showed that the optical energy density $-(\mathbf{E}\cdot\mathbf{D} + \mathbf{B}\cdot\mathbf{H})/8\pi$ can be expressed in terms of the intensity of the laser beam I as $F_{\rm opt} = -In_p/c$, where $n_p(\theta) = n_o/\sqrt{1 - u\sin^2\theta}$, $u = (n_e^2 - n_o^2)/n_e^2 > 0$, and n_o and n_e are the ordinary and extraordinary indexes of refraction.[5] The elastic deformation energy density due to splay and bend deformations is $F_d = k_{33}(1 - k\sin^2\theta)(d\theta/dz)^2/2$, where $k = (k_{33} - k_{11})/k_{33}$, and k_{11} and k_{33} are the splay and bend elastic constants.[9] The Euler equation resulting from the variation of the total free energy [$\mathscr{F} = \int_0^d (F_d + F_{\rm opt} + F_{\rm ext})dz$] takes the form

$$\frac{(1 - k\sin^2\theta)d^2\theta}{dz^2} - k\sin\theta\cos\theta\left(\frac{d\theta}{dz}\right)^2 + \frac{(2In_o/ck_{33})u\sin\theta\cos\theta}{(1 - u\sin^2\theta)^{3/2}} - \frac{CQ(z)}{k_{33}} = 0, \quad (1)$$

where $Q = \delta P/\delta\theta$, $C = \chi_a H^2$, and $P(\theta) = \pm\sin^2\theta$ for

[a)] Permanent address: IBM Thomas J. Watson Research Center, P.O. Box 218, Yorktown Heights, New York 10598.

MOB, and $C = D_z^2/(4\pi\epsilon_\parallel)$ and $P(\theta) = 1/(1 - w\sin^2\theta)$ for EOB.

By the symmetry of the problem, we look for symmetrical solutions satisfying $\theta(z) = \theta(d - z)$. Using the rigid boundary conditions $\theta(z=0) = \theta(z=d) = 0$, Eq. (1) can be integrated twice[5] to give for $d\theta/dz \neq 0$

$$z = \left(\frac{ck_{33}}{2I}\right)^{1/2} \times \int_0^\theta \left(\frac{1 - k\sin^2\theta}{n_p(\theta_m) - n_p(\theta) - R\,[P(\theta_m) - P(\theta)]}\right)^{1/2} d\theta, \quad (2)$$

for $0 \leqslant z \leqslant d/2$, where $\theta_m = \theta(z = d/2)$ and $R = C/(2I)$. For EOB, by div $\mathbf{D} = 0$, D_z is a constant independent of z but $E_z = D_z/[\epsilon_\parallel(1 - w\sin^2\theta)]$ depends on z through the dielectric constant. Therefore, the two constants D_z and $R = cD_z^2/(8\pi\epsilon_\parallel I)$ are functionals of the deformational profile. R is related to the applied voltage V through $V = \int_0^d E_z\, dz$ from which we obtain

$$V = 4\left(\frac{\pi k_{33} R}{\epsilon_\parallel}\right)^{1/2} \times \int_0^{\theta_m} P(\theta)\left(\frac{1 - k\sin^2\theta}{n_p(\theta_m) - n_p(\theta) - R\,[P(\theta_m) - P(\theta)]}\right)^{1/2} d\theta. \quad (3)$$

The maximum deformation angle θ_m can be determined by evaluating (2) at $z = d/2$. The spatial orientation of the NLC at a given field strength is then completely determined by Eq. (2) [and Eq. (3) for EOB]. Experimentally, one can observe the transition by several methods. A useful one, as discussed in Refs. 3–5, has been the measurement of the induced birefringence. It should be noticed that an OB element in the usual sense is one with a history-dependent output intensity, whereas for the OB considered here, the output intensity is equal to the input intensity and the OB exhibits a history-dependent output phase.

In the rising transition, there exists a rising threshold field below which no molecular reorientation can be induced. For geometry II, there exists a critical field above which the homeotropic state will be destabilized by the external field alone. The critical field is $H_0 = (\pi/d)\sqrt{k_{33}/\chi_a}$ for MOB and $V_0 = 2\pi\sqrt{\pi k_{33}/|\epsilon_a|}$ for EOB.[9] For simplicity, we define a reduced field strength $r = H/H_0$ for MOB and $r = V/V_0$ for EOB. Then $r \geqslant 0$. In the following discussion, we require $r < 1$ for geometry II. By expanding the solution for θ in Eqs. (2) and (3) up to and including terms $\sim O(\theta_m^2)$, we obtain that in the rising transition, $\theta = 0$ for $I < I_{\rm th}$, and for $I \geqslant I_{\rm th}$

$$\theta(z,I) \simeq \theta_m(I)\sin(\pi z/d) + O(\theta_m^3) \quad (4)$$

and

$$\theta_m(I) = [(I/I_{\rm th} - 1)/2B]^{1/2}, \quad (5)$$

where $B = B_0 \mp 9ur^2/16$ for MOB and $B = B_0 \mp (9u/16 - w/4)r^2$ for EOB, $I_{\rm th} = I_0(1 \pm r^2)$, $B_0 = (1 - k - 9u/4)/4$, and $I_0 = (ck_{33}/n_0u)(\pi/d)^2$.

From Eq. (5), we see that as $B > 0$, the transition is a second-order transition, in which $\theta = 0$ for $I < I_{\rm th}$ and θ changes continuously as $I \geqslant I_{\rm th}$, but $d\theta_m/dI$ is discontinuous at $I_{\rm th}$. However, if $B < 0$, which implies that $dI/d(\theta_m^2) < 0$, then small distortions are not stable and a first-order transition (OB) occurs: we include terms $\sim O(\theta_m^4)$ in the expansion and obtain

$$\theta_m^2(I) = (-B + \sqrt{B + 4GA})/2G, \quad (6)$$

where $A = \sqrt{I/I_{\rm th}} - 1$, $G = G_0 + [\pm(9u/4 + 63ku/4 + 639u^2/16)r^2 + 1539u^2r^4/32]/96$ for MOB, $G = G_0 + [\pm(9u/4 - 6w/5 + 63ku/4 - 42kw/5 - 513uw/10 + 639u^2/16 + 32w^2)r^2 + (1539u^2/32 + 519w^2/10 - 513wu/10)r^4]/96$ for EOB and $G_0 = (11/2 - k + 9u/4 + 63ku/4 - 9k^2/2 - 26lu^2/32)/96$. For $B > 0$, Eq. (6) can be approximated by Eq. (5). For $B < 0$, Eqs. (5) and (6) show that as I increases from zero, $\theta_m = 0$ for $I < I_{\rm th}$ and the state changes discontinuously at $I_{\rm th}$ from $\theta_m = 0$ to $\theta_m = \sqrt{-B/G}$ and then changes continuously as I increases. But if I decreases from above $I_{\rm th}$, the state assumes a finite amount of distortion even at $I < I_{\rm th}$. Upon reaching a lower falling threshold intensity $I'_{\rm th}$, the state changes discontinuously from $\theta_m = \sqrt{-B/2G}$ back to $\theta_m = 0$. The changes at both rising and falling threshold intensities are discontinuous and the rising and falling transitions assume the same deformation for $I < I_{\rm th}$ but different deformations for $I'_{\rm th} < I < I_{th}$; we thus obtain OB. The falling threshold intensity can be determined by the condition $dI/d(\theta_m^2) = 0$ and is given by $I'_{\rm th} = I_{\rm th}(1 - B^2/4G)^2$. There is a tricritical point separating the second and first-order transitions at $B = 0$ at the fields $r^* = H^*/H_0 = \sqrt{16|B_0|/9u}$ for MOB and $r^* = V^*/V_0 = \sqrt{16|B_0/(9u - 4w)|}$ for EOB. Here the transition is second order with $\theta_m = [(I/I_{\rm th} - 1)/4G]^{1/4}$ for $I > I_{\rm th}$. Thus at the rising threshold, the intensity dependence of θ_m is mean field like with the critical exponent 1/2 for $B > 0$ and 1/4 for $B = 0$.

From the above study, we conclude that the application of an external field parallel to the laser propagation direction can be a significant control parameter for enhancement of OB. For geometry I (II), the external field suppresses (enhances) the optical orienting effects, $I_{\rm th}/I_0 - 1$ increases (decreases) quadratically as r is increased. If $B < 0$, OB will occur. The enhancement of OB depends on B being decreased by the external field. For MOB, $\mathbf{H} = (0,0,H)$ [$\mathbf{H} = (H,0,0)$] enhances (suppresses) the OB. If $B_0 \geqslant 0$ ($B_0 < 0$), then the second-order (first-order) transition will become first order (second order) when $H > H^*$ ($H \geqslant H^*$). For EOB, if $0 < w < 3u/4$, the electric field enhances the OB, otherwise it is suppressed; if in addition, $B_0 \geqslant 0$, then the second-order transition will become first order if $V > V^*$.

At present, the purely optically induced molecular reorientation has been observed in a few well known NLC's, such as 5CB (pentyl-cyano-biphenyl), 8CB (octyl-cyano-biphenyl), and MBBA (*p-n*-methoxybenzilidene-*p*-bytylaniline).[3,4] The influence of a superposed electric field on $I_{\rm th}$ has been experimentally verified by Csillag *et al.*,[10] using 8CB ($w > 0$) and MBBA ($w < 0$). Since all the NLC's used in the experiments have positive values of B_0 and 8CB does not satisfy $0 < w < 3u/4$, OB was not observed, as expected. We predicted[5] in 1983 that OB could occur in a single component nematic PAA (*p*-azoxyanisole) at 110–130 °C. This prediction has not been verified experimentally. By examining data on more existing NLC's (up to a few hundred), we have

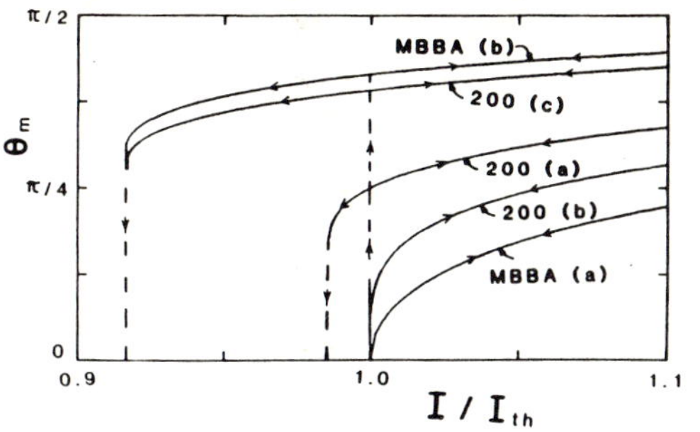

FIG. 1. Maximum deformation angle θ_m as a function of the reduced intensity I/I_{th} and magnetic field H/H_0 for RO-TN-200 and MBBA at temperature 22 °C and wavelength 6328 Å. For RO-TN-200 (abbreviation: 200) we put $k_{11} = 8.80\times 10^{-7}$ dyn, $k_{33} = 19.00\times 10^{-7}$ dyn, $n_o = 1.5345$, $n_e = 1.8100$, and $d = 250\ \mu$m. (a) $H = 0$. The transition is first order, i.e., optical bistable with $I_{th} = I_0 = 208.6$ W/cm^2, $I'_{th} = 0.986 I_{th} = 205.7$ W/cm^2. (b) $\mathbf{H} = (H^*, 0,0)$. $H^* = 0.518 H_0$ is the tricritical field. The OB is suppressed and the transition becomes second order with $I_{th} = 0.732 I_0 = 152.6$ W/cm^2. (c) $\mathbf{H} = (0,0,H_0)$. The OB cycle is enhanced with $I_{th} = 2I_0 - 417.1$ W/cm^2 and $I'_{th} = 0.917 I_{th} = 382.5$ W/cm^2. For MBBA we put $k_{11} = 6.95\times 10^{-7}$ dyn, $k_{33} = 8.99\times 10^{-7}$ dyn, $n_o = 1.5443$, $n_e = 1.7582$, $\chi_a = 10^{-7}$ cgs unit, and $d = 250\ \mu$m. (a) $H = 0$. The transition is second order with $I_{th} = I_0 = 120.7$ W/cm^2. (b) $\mathbf{H} = (0,0,2H_0)$. The OB is enhanced with $I_{th} = 5I_0 = 603.4$ W/cm^2 and $I'_{th} = 0.916 I_{th} = 553.0$ W/cm^2. $H_0 = 380$ G. The transition becomes first order in the field $\mathbf{H} = (0,0,H)$ with $H > H^* = 0.71 H_0$.

found that OB should occur in the following four NLC mixtures at room temperature without an external field (i.e., $B_0 < 0$): RO-TN-200, RO-TN-201, RO-TN-403 (from Hoffman–La Roche), and E7 (from British Drug House). Using the measured material parameters,[11] among these mixtures, RO-TN-200 has the lowest threshold intensity and largest hysteresis width $(1 - I'_{th}/I_{th})$. Because all eight NLC's have either $w < 0$ or $w > 3u/4$ at low frequency, the application of a low-frequency electric field will not enhance OB. But since the dielectric constants are strongly dependent on temperature and frequency, we can always find NLC's such that $0 < w < 3u/4$ at certain frequencies and thereby enhance OB.

However, OB can always be enhanced by a magnetic field. This is true for all known thermotropic NLC's and is illustrated in Fig. 1 for RO-TN-200 and MBBA. For RO-TN-200, using a magnetic field $\mathbf{H} = (H,0,0)$ with $H \gg H^* = \sqrt{16|B_0|/9u}\, H_0 = 0.518 H_0$, OB will be suppressed and the first-order transition becomes a second-order transition [Figs. 1(a) and 1(b)]. With $\mathbf{H} = (0,0,H)$ then the width of OB will always be enhanced as shown in Fig. 1(c) for $H = H_0$: the falling threshold intensity becomes $I'_{th} = 0.917 I_{th}$ (and $I_{th} - I'_{th} = 34.6$ W/cm^2 for $d = 250\ \mu$m) as compared to $I'_{th} = 0.986 I_{th}$ (and $I_{th} - I'_{th} = 2.9$ W/cm^2 for $d = 250\ \mu$m). Using the material parameters listed in Ref. 5, we find that the reduced tricritical magnetic fields H^*/H_0 are 1, 0.7, and 0.7 for 5CB, 8CB, and MBBA, respectively. In a typical experiment with $d = 250\ \mu$m, $H_0 \sim 280$ G for 5CB and 8CB, and $H_0 = 380$ G for MBBA, and I_0 varies from 100 to 200 W/cm^2. Thus using a low magnetic field (< 500 G) directed along the z axis, OB can also be seen in 5CB, 8CB, and MBBA at room temperature at a low power ($I_{th} < 500$ W/cm^2).

In conclusion, bistable optical devices using NLC's as the nonlinear medium should be viable candidates for the achievement of room temperature, low power, externally controllable optical switches which are easily fabricated and operated.

I gratefully acknowledge A. J. Hurd, R. B. Meyer, and T. Odagaki for useful discussions. This work is supported by the National Science Foundation under grant No. DMR-8210477-01.

[1] H. M. Gibbs, S. L. McCall, and T. N. C. Venkatesan, Phys. Rev. Lett. **36**, 1135 (1976).
[2] *Optical Bistability*, edited by C. M. Bowden, H. M. Gibbs, and S. L. McCall (Plenum, NY, 1984), Vol. 2.
[3] S. D. Durbin, S. M. Arakelian, M. M. Cheung, and Y. R. Shen, J. Phys. C (Paris) **44**, C2-161 (1983), and references therein.
[4] S. D. Durbin, S. M. Arakelian, and Y. R. Shen, Phys. Rev. Lett. **47**, 1411 (1981).
[5] H. L. Ong, Phys. Rev. A **28**, 2393 (1983); H. L. Ong and R. B. Meyer, in *Optical Bistability*, edited by C. M. Bowden, H. M. Gibbs, and S. L. McCall (Plenum, NY, 1984), Vol. 2, p. 333.
[6] I. C. Khoo, R. Normandin, and V. C. Y. So, J. Appl. Phys. **53**, 7599 (1982); M. M. Cheung, S. D. Durbin, and Y. R. Shen, Opt. Lett. **8**, 39 (1983).
[7] B. Ya. Zel'dovich and N. V. Tabiryan, Mol. Cryst. Liq. Cryst. **69**, 19 (1981); H. G. Winful, Phys. Rev. Lett. **49**, 1179 (1982).
[8] N. V. Tabiryan and B. Ya. Zel'dovich, Mol. Cryst. Liq. Cryst. **69**, 31 (1981); P. H. Lippel and C. Y. Young, Appl. Phys. Lett. **43**, 909 (1983); H. L. Ong and C. Y. Young, Phys. Rev. A **29**, 297 (1984).
[9] P. G. de Gennes, *The Physics of Liquid Crystals* (Oxford University, Oxford, 1974).
[10] L. Csillag, N. Eber, I. Janossy, N. Kroo, V. F. Kitaeva, and N. N. Sobolev, Mol. Cryst. Liq. Cryst. **89**, 287 (1982).
[11] M. Schadt and F. Muller, IEEE Trans. Electron Devices **ED-25**, 1125 (1978).

ERRATUM

On p. 822, second column, line 8 should read "Known thermotropic NLC's are diamagnetic and mostly with positive diamagnetic anisotropy..."

Equation (5) should read

$$\theta_m(I) \cong \sqrt{(I_{th}/I_0)(I/I_{th} - 1)/2B}\,.$$

After Eq. (6), the variable A should read $A = (I_{th}/I_0)(\sqrt{I/I_{th}} - 1) \cong (I_{th}/I_0)(I/I_{th} - 1)/2$.

On p. 823, second column, line 23, the falling threshold intensity should read $I'_{th} = I_{th}[1 - (I_0/I_{th})(B^2/4G)]^2$.

On p. 823, second column, lines 26 and 27 should read "Here the transition is second order with $\theta_m = [(I_{th}/I_0)(I/I_{th} - 1)/2G]^{1/4}$."

On p. 823, second column, lines 42 and 54, and on p. 824, first column, lines 8 and 12, $3u/4$ should read $9u/4$.

Observation of optical field induced first-order electric Freedericksz transition and electric bistability in a parallel aligned nematic liquid-crystal film

J. J. Wu
Institute of Electronics, National Chiao Tung University, Hsinchu, Taiwan 30050, Republic of China
Gan-Sing Ong
Institute of Electro-Optical Engineering, National Chiao Tung University, Hsinchu, Taiwan 30050, Republic of China
Shu-Hsia Chen
Department of Electrophysics and Institute of Electro-Optical Engineering, National Chiao Tung University, Hsinchu, Taiwan 30050, Republic of China

(Received 23 June 1988; accepted for publication 13 September 1988)

Optical field induced first-order electric Freedericksz transition and electric bistability in a homogeneously aligned nematic film is first observed. It is experimentally demonstrated that an applied optical field can transform the electric Freedericksz transition from second order to first order. The molecular reorientation as a function of electric field is then characterized by a hysteresis loop which exhibits the electric bistability. The results are in good agreement with theoretical predictions.

Recently, particular attention has been devoted to the first-order Freedericksz transition (FT), molecular reorientation, and bistability in nematic liquid crystals (NLC's). In the last few years, Ong has shown that all of the electric and magnetic FT's and most of the optical FT's in a nonconducting homogeneous or homeotropic NLC are second-order transitions.[1] However, applying a suitable static field, the second-order optical FT can be converted to first order in all existing NLC's.[2,3] Similarly, applying a suitable optical field, the second-order electric or magnetic FT can also be converted to first order.

The enhancement of a first-order optical FT by an additional magnetic field in a homeotropic NLC has been recently observed experimentally by Shen's group.[4] The electric field induced optical FT and optical field induced electric FT in a homeotropic NLC have been also observed and reported in our previous study.[5] In this letter, we report what we believe the first observation of optical field induced first-order electric FT and bistability in a homogeneously aligned NLC. The results are in good agreement with theoretical predictions.

To get a summary of Ong's theory concerning our experiment,[3] one can consider the geometry shown in the inset of Fig. 1. A homogeneously aligned NLC film with director parallel to the x axis occupies a layer $0\leqslant z\leqslant d$. An electric field of potential V is applied along the surface normal, and a linear polarized laser beam of intensity I with polarization parallel to the x axis is normally incident onto the NLC film. Assuming that $\epsilon_\parallel > \epsilon_\perp$ (under some proper frequency of ac electric field) and $n_\parallel > n_\perp$, where ϵ and n denote the dielectric constants and refractive indices, and refer to the director parallel and perpendicular to the director, respectively, the electric field will orient the molecular direction to the z axis and induce FT, whereas the optical field will orient the molecular direction toward the x axis and hence suppress the electric orienting effect. The orientation of the NLC's director can be described by the angle $\theta(z)$ between the director and the z axis.

To determine the angle $\theta(z)$, the total free energy F which consists of the elastic deformation, optical field, and electric field energy is expressed in terms of $\theta(z)$. The symmetry of the problem and the rigid boundary conditions require that the maximum deformation angle $\theta_m = \theta(z=d/2)$, and $\theta(z=0)=\theta(z=d)=0$. Variation of the total free energy leads to an Euler equation from which the solution to the deformation profile is obtained as

$$z=\int_0^\theta \left\{(1-k\sin^2\theta)\left[\frac{D_z^2}{4\pi\epsilon_\perp}\left(\frac{1}{1+w\sin^2\theta}-\frac{1}{1+w\sin^2\theta_m}\right) -\frac{2}{u}\left(\frac{\pi}{d}\right)^2\frac{I}{I_0}\left(\frac{1}{\sqrt{1+u\sin^2\theta}}-\frac{1}{\sqrt{1+u\sin^2\theta_m}}\right)\right]\right\}^{1/2} d\theta, \tag{1}$$

where $k=1-k_{33}/k_{11}$, k_{11} and k_{33} are the splay and bent elastic constants, $w=\epsilon_\parallel/\epsilon_\perp-1$, $u(n_\parallel/n_\perp)^2-1$, D_z is the z component of the displacement, and

$$I_0=(\pi/d)^2ck_{11}/(n_\parallel/u) \tag{2}$$

is a constant. In addition, because $V=\int_0^d E_z dz$, the relation of the maximum deformation angle and displacement to the applied voltage can be derived as

$$V=\frac{2D_z}{\epsilon_\perp}\int_0^{\theta_m}\left(\frac{1}{1+w\sin^2\theta}\right)\Bigg\{(1-k\sin^2\theta)\quad\Bigg[\frac{D_z^2}{4\pi\epsilon_\perp}\left(\frac{1}{1+w\sin^2\theta}-\frac{1}{1+w\sin^2\theta_m}\right)$$
$$-\frac{2}{u}\left(\frac{\pi}{d}\right)^2\frac{I}{I_0}\left(\frac{1}{\sqrt{1+u\sin^2\theta}}-\frac{1}{\sqrt{1+u\sin^2\theta_m}}\right)\Bigg]\Bigg\}^{1/2}d\theta. \quad (3)$$

The spatial orientation of NLC at a given electric potential and optical intensity is then completely described by Eqs. (1) and (3).

Analysis of θ_m in Eqs. (1) and (3) shows that there are two regimes: in the low optical intensity regime, the FT is second order and the forward threshold voltage $V_{\rm th}$ is equal to the backward threshold voltage $V'_{\rm th}$; but in the high optical intensity regime (above the tricritical intensity), the FT is first order, $V_{\rm th}\neq V'_{\rm th}$, and there exists a hysteresis loop of electric bistability. The tricritical intensity is given by $I^*=I_0[4(1-k+w)/(9u-4w)]$ and the threshold voltage is given by

$$V_{\rm th}=V_0\sqrt{1+I/I_0}, \quad (4)$$

where

$$V_0=2\pi\sqrt{\pi k_{11}/(\epsilon_\parallel-\epsilon_\perp)}. \quad (5)$$

Assuming that the NLC is at first perfectly homogeneously aligned, then applying the external optical and electric fields on it, we derive the following absolute induced phase shift:

$$\Delta\phi=\frac{2\pi n_\parallel d}{\lambda}-\frac{4\pi n_\perp}{\lambda}\int_0^{\theta_m}\left(\frac{1}{1+u\sin^2\theta}\right)\Bigg\{(1-k\sin^2\theta)$$
$$\Bigg[\frac{D_z^2}{4\pi\epsilon_\perp}\left(\frac{1}{1+w\sin^2\theta}-\frac{1}{1+w\sin^2\theta_m}\right)-\frac{2}{u}\left(\frac{\pi}{d}\right)^2$$
$$\times\frac{I}{I_0}\left(\frac{1}{\sqrt{1+u\sin^2\theta}}-\frac{1}{\sqrt{1+u\sin^2\theta_m}}\right)\Bigg]\Bigg\}^{1/2}d\theta, \quad (6)$$

where λ is wavelength of light.

In our experiment, the substance studied was a room-temperature NLC mixture ROTN-200 (from Haffmann–La Roche). It had been heated to an isotropic phase before it was used, and then sandwiched between two glass slides. The glass slides coated with an indium tin oxide transparent electrode was rubbed after they were treated with PVA (polyvinyl alcohol) to give a homogeneous molecular alignment. The sample film with a thickness of $260\pm5\,\mu$m was placed horizontally to minimize possible laser-induced convection. A cw Ar^+ laser, linearly polarized at 514.5 nm with a beam diameter 1.55 mm, was constricted to about 0.44 mm by an afocal optical system and was normally incident on the sample film. At the same time, an ac electric field at 560 kHz was applied normal to the film. The spatial Gaussian distributed optical intensity combined with the uniformly distributed electric field leads to the induced spatial self-phase modulation of the pump laser beam and results in a mulitple ring structure in the far-field pattern. The number of rings N is directly proportional to the relative induced phase shift $\Delta\Psi$ of the center and the margin of the beam in the medium, and then is given by $N=\Delta\Psi/2\pi$.

For the observation of electric FT, the number of rings was counted at a fixed optical field while changing the electric field by small steps.

Many samples have been studied, and some typical experimental results of second- and first-order electric FT's at fixed optical intensities are shown in Figs. 1–3. It can be seen in Fig. 1 that with low optical intensity ($I/I_0=1.19$) the data follow a smooth curve which exhibits second-order electric FT at $V_{\rm th}/V_0=1.4643$. However, in Figs. 2 and 3, with sufficiently high optical intensities ($I/I_0=3.53$ and 3.90), each of the data exhibits a hysteresis loop which characterizes the first-order FT and electric bistability. Though

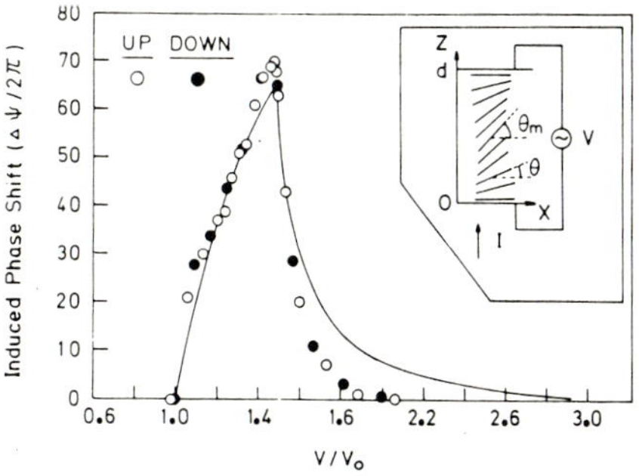

FIG. 1. Relative induced phase shift $\Delta\Psi/2\pi$ vs reduced voltage V/V_0 under the fixed optical intensity $I/I_0=1.19$. Open symbols were measured with increasing intensity; solid symbols, with decreasing intensity; solid line: numerical result. Inset shows geometry of molecular orientation in a homogeneous nematic liquid-crystal cell with electric and optical fields.

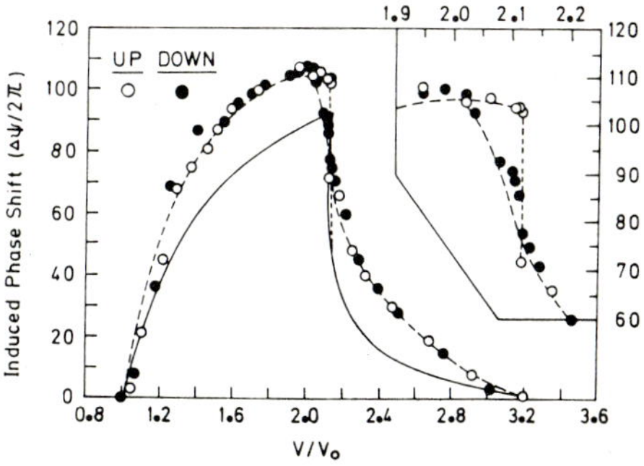

FIG. 2. Relative induced phase shift $\Delta\Psi/2\pi$ vs reduced voltage V/V_0 under the fixed optical intensity $I/I_0=3.53$. Open symbols were measured with increasing voltage; solid symbols, with decreasing voltage; solid curve: numerical result. Dashed curves were drawn to aid visualization of the data.

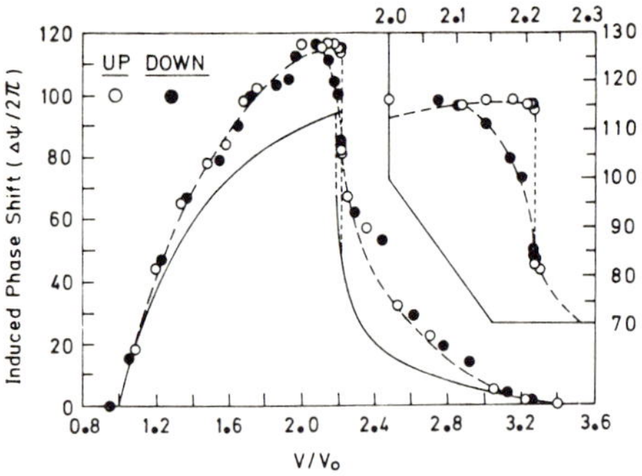

FIG. 3. Relative induced phase shift $\Delta\Psi/2\pi$ vs reduced voltage V/V_0 under the fixed optical intensity $I/I_0 = 3.90$. Open symbols were measured with increasing voltage; solid symbols with decreasing voltage; solid curve: numerical result. Dashed curves were drawn to aid visualization of the data.

the first-order transition phenomenon in our experiment seems unobvious in the backward transitions, it apparently exists in the forward transitions. The experimental data show an abrupt switching in the forward transitions at the threshold point $V_{th}/V_0 = 2.1166$ in Fig. 2 and $V_{th}/V_0 = 2.2132$ in Fig. 3.

In order to compare the experimental results in Figs. 1–3 with theory, the absolute phase shift $\Delta\phi$ versus the reduced electric field V/V_0 under some fixed optical intensity ($I/I_0 = 0$, 1.19, 3.53, and 3.90) has been calculated by using Eqs. (1), (3), and (6) with $n_\parallel = 1.795$, $n_\perp = 1.534$,[6] $\epsilon_\parallel = 8.8$, $\epsilon_\perp = 7.58$,[7] $k = -0.211$,[5] and the numerical results are shown in Fig. 4. The characteristics of these numerical results have been described in Ong's papers.[3] And the only difference is that he calculated θ_m as a function of the reduced electric voltage V/V_{th} and fixed optical field intensity I/I_0. However, in our experiment, the ring number N is not proportional to the absolute induced phase shift $\Delta\phi$

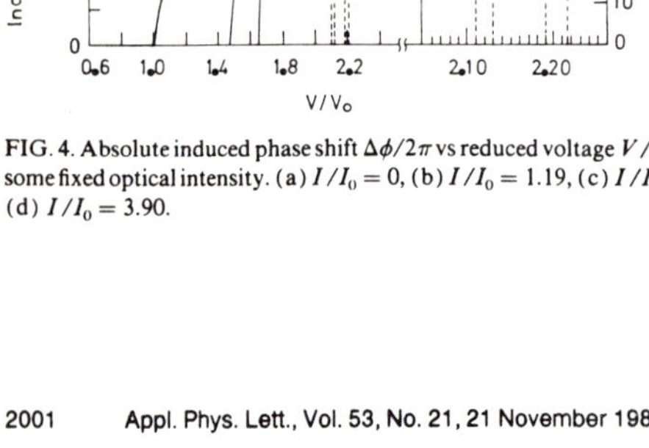

FIG. 4. Absolute induced phase shift $\Delta\phi/2\pi$ vs reduced voltage V/V_0 under some fixed optical intensity. (a) $I/I_0 = 0$, (b) $I/I_0 = 1.19$, (c) $I/I_0 = 3.53$, (d) $I/I_0 = 3.90$.

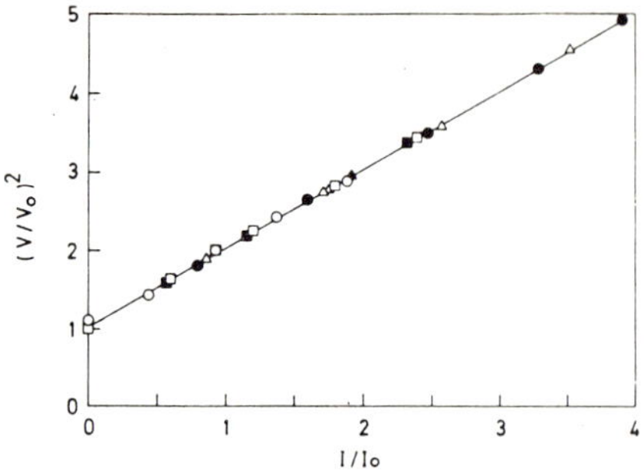

FIG. 5. Square of reduced electric potential vs reduced optical intensity with the theoretical line. Different samples are represented with different symbols.

but is proportional to the relative induced phase shift $\Delta\Psi$, so that the difference of $\Delta\phi$'s at the margin and center of the light beam was calculated. The numerical results are plotted in Figs. 1–3 with solid lines. As seen in these figures, the theory fits the data fairly well. The slight discrepancies seem acceptable since the material parameter of ROTN-200 may change after some stocking time.

We also compare the observed V_{th}^2/V_0^2 vs I/I_0 with the theoretical expression of V_{th}^2/V_0^2 in Eq. (4). Linear fitting for I/I_0 and V_{th}^2/V_0^2 was employed to find V_0 and I_0 in Eqs. (5) and (2). Averaging the values of all samples, we have $V_0 = 2.001$ V and $I_0 \cong 28.9$ W/cm^2. Here, we assume that $n_\parallel$, $n_\perp$, $\epsilon_\parallel$, and $\epsilon_\perp$ are not changed. It is obvious that the observed values of V_{th} in Fig. 5 agree well with the theoretical linear relation: $(V_{th}/V_0)^2 = (I/I_0) + 1$, in Eq. (4).

In conclusion, we have reported here what we believe to be the first observation of transformation of the electric FT in a homogeneous NLC film from second to first order by applying a fixed optical field. The accompanied hysteresis loop of electric bistability is very clear. Although a slight discrepancy exists between experimental results and numerical results, the phenomenon of first-order FT is unambiguously exhibited.

This work was supported by the Chinese National Science Council under contract No. NSC76-0208-M009-01.

[1]P. G. de Gennes, *The Physics of Liquid Crystals* (Clarendon, Oxford, 1947).
[2]H. L. Ong, Phys. Rev. A **28**, 2393 (1983); H. L. Ong and R. B. Meyer, in *Optical Bistability*, edited by C. M. Bowden, H. M. Gibbs, and S. L. McCall (Plenum, New York, 1984), Vol. 2, p. 333; H. L. Ong, Appl. Phys. Lett. **46**, 822 (1985); H. L. Ong, Phys. Rev. A **31**, 3450 (1985).
[3]H. L. Ong, Phys. Rev. A **33**, 3550 (1986); IEEE Trans. Electron Devices **ED-25**, 1195 (1986).
[4]A. J. Karn, S. M. Arakelian, Y. R. Shen, and H. L. Ong, Phys. Rev. Lett. **57**, 448 (1986).
[5]Shu-Hsia Chen and J. J. Wu, Appl. Phys. Lett. **52**, 1998 (1988).
[6]H. L. Ong, M. Schadt, and I. F. Chang, Mol. Cryst. Liq. Cryst. **132**, 45 (1986).
[7]M. Schadt and C. von Planta, J. Chem. Phys. **63**, 4379 (1975).

Undamped oscillations of NLC director in the field of an ordinary light wave

A. S. Zolot'ko, V. F. Kitaeva, N. Kroo, N. N. Sobolev, A. P. Sukhorukov, V. A. Troshkin, and L. Czillag

P. N. Lebedev Physics Institute, USSR Academy of Sciences
(Submitted 6 January 1984)
Zh. Eksp. Teor. Fiz. **87**, 859–864 (September 1984)

We have investigated experimentally some characteristic features of orientational aberration self-focusing of an ordinary light wave in a homotropically oriented NLC. A qualitative explanation of these features is offered.

It was noted in the first experiments aimed at observing the optically induced Fréedericksz effect[1] that the interactions of the fields of extraordinary and ordinary light waves with a homotropically oriented octyl-cyano-biphenyl (OCB) nematic liquid crystal (NLC) differ greatly. At oblique incidence of the light wave on an NLC, the characteristic pattern of the aberrational self-focusing that accompanies the reorientation of the NLC molecules in narrow light beams[2] is stable in the first case and unstable in the second, where the number of observed aberration rings changes periodically (oscillates) with time. The oscillations were observed at incidence angles $\alpha \lesssim 20°$. At larger angles α there was no aberration pattern at all, even at a laser radiation power $P \sim 300$ mW. The period of the oscillations and their amplitude depend on the incidence angle α and on the laser radiation power P. The nonstationary character of the self-focusing indicates that the NLC director also oscillates in the field of an ordinary light wave.

The purpose of the present study was a detailed investigation of the distinctive features of director reorientation in the field of an ordinary light wave, as well as an explanation of these features using simple models.

1. ORGANIZATION AND RESULTS OF EXPERIMENT

The object of the investigation was a homotropically oriented DCB crystal with $L = 150$ μm at a temperature 37 °C.

The cell with the crystal was placed in the focal spot of the beam of a cw argon laser (ILA-120 by Carl Zeiss, Jena or model 171 by Spectra Physics, USA) focused by a lens of focal length $f = 270$ mm; the spot radius was $w_0 \approx 44$ μm. The cell plane was vertical and perpendicular to the plane containing the unperturbed director $\mathbf{n}_0$ and the wave vector $\mathbf{k}$ of the light beam incident on the crystal. The incidence angle (the angle between $\mathbf{n}_0$ and $\mathbf{k}$, α_0, inside the crystal) was varied by rotating the cell around the vertical axis. The laser beam polarized in the vertical plane (ordinary wave) was displayed, after passing through the NLC, on a screen perpendicular to $\mathbf{k}$.

The picture on the screen has the following dependence on the incidence angle.

At large incidence angles $\alpha > 20°$ no changes take place in the laser beam passing through the crystal up to a laser-beam power $P \sim 300$ mW (no investigations were made at $P > 300$ mW to prevent damage to the crystal), i.e., no orientational self-focusing takes place.

At small incidence angles $1–2° < \alpha < 20°$ orientational self-focusing of the light beam sets in and the characteristic aberration pattern is observed on the screen. This pattern, however, is pulsating: the rings appear, "collapse" after some time, reappear, collapse again, etc. The effect has a threshold, viz., self-focusing occurs only if the light-beam power P exceeds a certain threshold P_{thr}. The polarization of the aberration rings is likewise not constant in time.

Detailed investigations reported in Ref. 3 have shown the following:

1) The threshold power P_{thr} depends on the angle α. Figure 1 shows the experimental plot of $P_{thr}(\alpha)/P_{thr}(0)$ (curve 1). It can be seen from Fig. 1 that the threshold power $P_{thr}(\alpha)$ increases monotonically with increasing α. As $\alpha \to 0$ the threshold power $P_{thr}(\alpha) \to P_{thr}(0)$.

2) The character of the oscillation of the number $N(t)$ of the aberration rings and of the beam divergence $\theta(t)$ depends on the angle α and on the laser radiation power P. Figure 2 shows by way of illustration the experimental dependences of N on the time t at different values of α and P.

Analysis of all the experimental $N(t)$ plots for $\alpha > 3°$ shows (as illustrated in Fig. 2) that the value N_1^{max} of the first

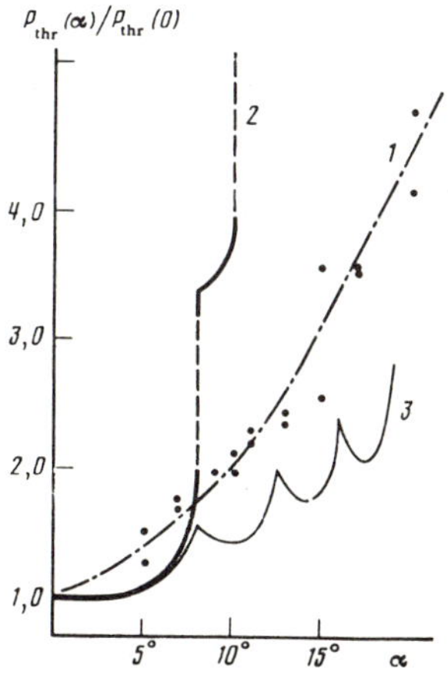

FIG. 1. Ratio of threshold power $P_{thr}(\alpha)$ of Fréedericksz transition in the field of an ordinary light wave obliquely incident on a crystal at an angle α to the threshold power $P_{thr}(0)$ at normal incidence: 1—experimental plot; 2—theoretical plot in the plane-wave approximation; 3—theoretical plot calculated by a variational method; $P_{thr}(0) = 50$ mW.

 0038-5646/84/090488-03$04.00

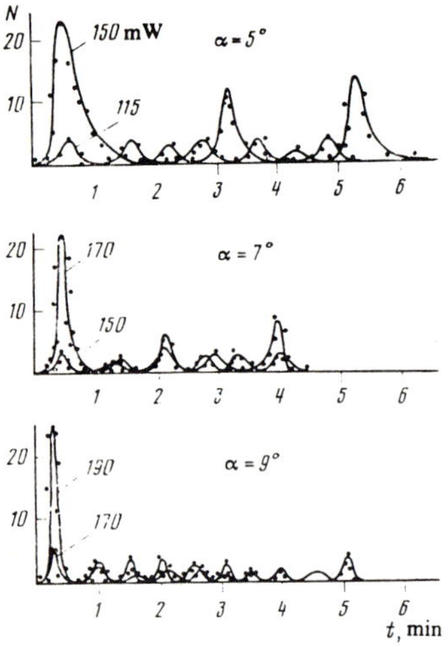

FIG. 2. Experimental time dependence of the number N of aberration rings at different values of α and P.

maximum of the function $N(t)$ at $P(\alpha)\sim(2\text{–}3)P(0)$ is several (2–10) times larger than $N_n^{\max}$ of all the succeeding maxima ($n > 1$). No regularity in the variation of the maxima $N_n^{\max}$, starting with the second ($n \geqslant 2$) could be observed. In some cases the values of $N_n^{\max}(n \geqslant 2)$ are practically equal for different n, and in others they differ from one another, sometimes considerably (by 2–3 times), but usually they do not exceed $N_1^{\max}$. With decreasing power $P(\alpha) N_1^{\max}$ ceases to be distinguishable from $N_n^{\max}$ of the remaining maxima for each of the α.

An aperiodic nonmonotonic time variation of N is observed at $\alpha \sim 1°$. N reaches a maximum 2–3 min after the start of the illumination, decreases to zero slowly, within 6–8 min, and thereafter no more aberration rings appear. A stationary aberrational pattern of self-focusing orientation is always observed at $\alpha = 0$ regardless of the position of the incident-light polarization plane. The number N of the aberration rings is then determined primarily by the light-beam power.

The measured temporal characteristics of the oscillatory regime of orientational aberrational self-focusing were the time $t_1^{\min}$ from the start of the crystal illumination to the instant of first collapse of the aberration pattern, and the oscillation period of the aberration-ring oscillations.

It was established that starting with the second ($n = 2$) maximum all the succeeding maxima ($n > 2$) follow one another at almost equal time intervals $T_n(\alpha, P) = t_{n+1}^{\max} - t_n^{\max}$($t_n^{\max}$ corresponds to $N_n^{\max}$). This allows us to introduce an oscillation period defined here as

$$T_{\mathrm{av}}(\alpha, P)=\frac{1}{n-1}\sum_{i=2}^{n} T_i,$$

where $n-1$ is the number of averaged periods.

Figure 3 shows the experimentally obtained dependence of $T_{\mathrm{av}}(\alpha,P)$ on α (The period T_{av} for each angle α at

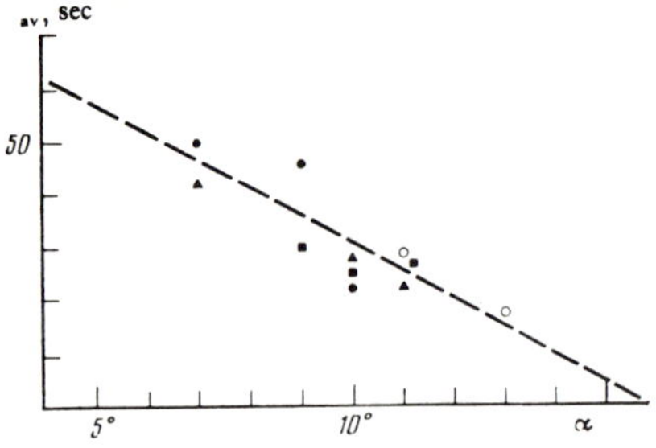

FIG. 3. Experimental plots of the average oscillation period T_{av} vs the crystal rotation angle: ●—$P = 135$ mW, ▲—170 mW, ■—190 mW, ○—210 mW.

constant P was obtained by averaging the time intervals between 5 to 10 maxima). The period T_{av} is practically independent of the beam power P.

As for the time $t_1^{\min}$, it was found to decrease with increasing α [at constant $P(\alpha)$]. The explicit dependence of $t_1^{\min}$ on P in the investigated P interval could not be found.

3) The polarization of the beams that form the aberration pattern in orientational self-focusing in an ordinary light wave differs from that of the incident radiation[1)] and varies strongly with time. This can be deduced from observations of the aberration rings. Recall that the aberration rings are not quite regular in shape. They are elongated in a direction perpendicular to the polarization of the incident radiation, and the "elongation" of the rings allows us to track the polarization of the incident light beam and all its variations as the beam passes through the NLC.

Observations of the aberration pattern in our case show that after a time $t_1^{\min}$ the polarization in the transmitted beam changes from its vertical position in the incident beam and at the instant $t = 0$, to horizontal at the instant $t = t_1^{\min}$. At the instant $t_1^{\max}$ the polarization plane of the beam makes an angle $\sim 45°$ with the vertical. It remains practically horizontal for all the succeeding maxima ($n \geqslant 2$).

2. DISCUSSION OF EXPERIMENTAL RESULTS

The observed features of the orientational self-focusing of an ordinary light wave can be first interpreted qualitatively as follows.

If an ordinary light wave with $\mathbf{E}\perp\mathbf{n}_0$ is obliquely incident on an NLC, the vectors $\mathbf{k}$ and $\mathbf{n}_0$ are not collinear. When the director is tilted away from the horizontal plane, two waves will therefore propagate in the interior of the crystal, ordinary $\mathbf{E}_o$ and extraordinary $\mathbf{E}_e$. This alters greatly the polarization of the incident wave and leads to a substantial change of the interaction between the field and the director. A consequence of these complicated processes is an oscillatory regime of the self-focusing and an increase of the self-focusing threshold with increasing incidence angle. Let us explain this in somewhat greater detail.

The external bulk force $\mathbf{G}$ exerted on the director by the electric field $\mathbf{E} = \mathbf{E}_o + \mathbf{E}_e$ is[4]

$$\mathbf{G}=\frac{\Delta\varepsilon}{4\pi}(\mathbf{n}\mathbf{E}_e)\mathbf{E}_e+\frac{\Delta\varepsilon}{4\pi}(\mathbf{n}\mathbf{E}_e)\mathbf{E}_o. \qquad (1)$$

With increasing angle between the director **n** and the vector **k** in the course of the change of director orientation, the anisotropy increases and the energy of the ordinary wave incident on the crystal is effectively transferred to the extraordinary one. When the anisotropy becomes very large, the phase difference between the ordinary and extraordinary waves changes greatly and the second term of Eq. (1) oscillates rapidly over the length of the crystal. Clearly, the force corresponding to this term cannot cause a significant reorientation of the director. That is to say, when considering the action of a strong optical electric field on the director, only the field of the extraordinary wave need be considered. Since the director is forced to become parallel to the field of the extraordinary electric wave, the moment of the forces exerted on the director by the electric field is perpendicular to the plane containing $\mathbf{E}_e$ and **n**, i.e., to the plane containing **k** and **n**. But the director is acted upon also by elastic forces that tend to return it to the initial position $\mathbf{n}_0$. The moment of these forces is thus perpendicular to the plane containing **n** and $\mathbf{n}_0$. Since the vectors **k** and $\mathbf{n}_0$ are not collinear in our case, the moments of the electric and elastic forces are also noncollinear, and cannot therefore balance each other.

The resultant torque causes the director to precess towards the horizontal plane. This decreases the fraction of the light-field energy going into the extraordinary wave, making the angle between **n** and **k** smaller. This manifests itself in experiment as an approach of the aberration-ring polarization to horizontal and a decrease in the number of the aberration rings. The director approaches gradually the unperturbed state. The entire process is then repeated. Since it starts out then from a somewhat different position, the amplitude and period of this oscillation differ from those of the preceding one.

With increasing angle α_0, the increase of the anisotropy leads to a more substantial change of the polarization of the incident radiation, and this decreases the amplitude and period of the oscillations.

A quantitative description of oscillating self-focusing of an oblique beam in an NLC encounters the following difficulties. In our experiments the laser beam had a focal spot (transverse radius) comparable with the thickness L of the NLC sample (the characteristic parameter $^2g = \sqrt{2}L/\pi w_0$ is equal to 1.5). The limited transverse dimension of the beam plays therefore a rather important role in the director reorientation. Thus, even at normal incidence ($\alpha = 0$) the field-intensity threshold is 6.3 times the corresponding value for a plane wave. This is why the theory developed by a number of workers for plane unbounded waves[5,6] cannot be applied directly to our experimental results. An attempt can be made to take the influence of the limited transverse dimension into account either by introducing a phenomenological factor g in the exact plane-wave solution for the threshold[2,6] (curve 2 of Fig. 1) or by finding a correction by the Ritz method[7] (curve 3 of Fig. 1). It can be seen that both approaches lead to a qualitatively correct description of the threshold curve. For a better quantitative agreement, numerical methods must apparently be invoked.

A theoretically even more complicated problem is that

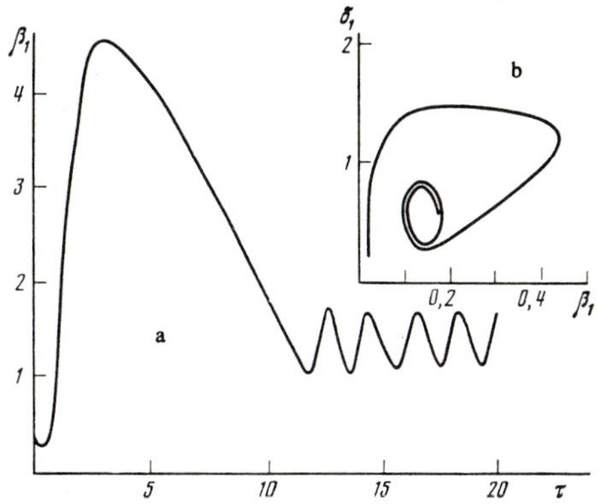

FIG. 4. Oscillations of director **n**, plotted with β_1 and τ (a) and with β_1 and δ_1 (b) as coordinates, in the center of the crystal at $\alpha_0 \approx 1°$ and at a power 5 times the threshold; β_1 is the change of the angle between the director **n** and the wave vector **k**; δ_{12} is the angle between the horizontal plane and the director projection on the plane perpendicular to $\mathbf{n}_0$; τ is the dimensionless time.

of the oscillating self-focusing observed in our experiment. We have taken here the first step towards the development of a mathematical model of self-focusing. A numerical analysis of the system of Maxwell's equations that describe the reorientation of the director in the electric field of a light wave, carried out in the plane-wave approximation with allowance for only the lowest spatial harmonics of the director-deflection angles ($\beta(t)$ and $\delta(t)$ in Fig. 4), shows a limit cycle, i.e., the presence of undamped oscillations. Their calculated period, $T \sim 60$ sec for $\alpha_0 \approx 1°$ and for a beam power 5 times the threshold, is in satisfactory agreement with experiment. The model of director reorientation is made additionally complicated by the need for taking into account higher longitudinal modes and by the complex transverse structure that differs, generally speaking, from Gaussian. In our opinion, however, the limit cycle is preserved also in the complicated model. Since different limit cycles (different oscillation periods) are to be expected for different spatial modes, the director oscillations and the self-focusing periods can become quasiperiodic.

[1)]In contrast to the polarization effects described in Ref. 2, we are dealing here with the change of the polarization of the aberration pattern as a whole.

[1]A. S. Zolot'ko, V. F. Kitaeva, N. Kroo, N. N. Sobolev, and L. Csillag, Pis'ma Zh. Eksp. Teor. Fiz. **32**, 170 (1980) [JETP Lett. **32**, 158 (1980)].
[2]A. S. Zolot'ko, V. F. Kitaeva, N. N. Sobolev, *et al.* Zh. Eksp. Teor. Fiz. **81**, 993 (1981); **83**, 1368 (1982) [Sov. Phys. JETP **54**, 496 (1981); **56**, 786 (1982)]. A. S. Zolot'ko, F. V. Kitaeva, V. A. Kuyumchyan, N. N. Sobolev, and A. P. Sukhorukov, Pis'ma Zh. Eksp. Teor. Fiz. **36**, 66 (1982) [JETP Lett. **36**, 80 (1982)].
[3]A. S. Zolot'ko, V. F. Kitaeva, N. Kroo, *et al.*, FIAN Preprint No. 225, 1983.
[4]S. Chandrasekhar, Liquid Crystals, Cambridge Univ. Press., Chap. 3.
[5]B. Ya. Zel'dovich, S. K. Merzlikin, N. F. Pilipetskiĭ, A. V. Sukhov, and N. V. Tabiryan, Pis'ma Zh. Eksp. Teor. Fiz. **37**, 568 (1983) [JETP Lett. **37**, 676 (1983)].
[6]M. S. Arakelyan and Yu. S. Chilingaryan, Calculus of Variation and Integral Equations, Nauka, 1966.

Translated by J. G. Adashko

Collective Rotation of Molecules Driven by the Angular Momentum of Light in a Nematic Film

E. Santamato
Dipartimento di Fisica Nucleare, Struttura della Materia e Fisica Applicata di Napoli, 80125 Napoli, Italy

B. Daino, M. Romagnoli, and M. Settembre
Fondazione Ugo Bordoni and Istituto Superiore di Poste e Telecomunicazioni, 00142 Roma, Italy

and

Y. R. Shen
Department of Physics, University of California, Berkeley, California 94720
(Received 10 September 1986)

It is experimentally demonstrated that a circularly polarized laser beam normally incident on a homeotropically aligned nematic film can induce a collective precession of the molecules in the film if the laser intensity is above the threshold for the Fréedericksz transition. The effect is shown to result from a transfer of angular momentum from the laser beam to the medium.

PACS numbers: 61.30.Gd, 42.65.−k, 64.70.Md

We report here an interesting new nonlinear optical phenomenon—self-induced time-dependent polarization rotation in a liquid-crystal medium. It occurs as a result of angular momentum transfer from the radiation field to the medium, causing the molecules to precess around the beam propagation direction. An elliptically polarized light beam of frequency ω, intensity I, and ellipticity $S_3=(I_+-I_-)/I$ carries an average angular momentum of $-\Phi\hbar S_3$ per unit time and unit area, where $\Phi=I/\hbar\omega$ is the photon flux and I_+ and I_- are the intensities of the right- and left-circularly polarized components of the beam, respectively.[1] The beam angular momentum can be transferred to a transparent medium if the latter is anisotropic, as some photons must emerge with their spin components reversed. The continuous transfer of angular momentum results in a torque on the medium. The torque is usually very small and is insufficient to put a macroscopic body into continuous rotation.[2] Its effect on the molecules in a liquid crystalline medium is, however, easily observable. In this Letter, we present the first observation of such an effect in a thin nematic film of 4-cyano-4′-pentyl-biphenyl (5CB). The rotation of molecules is manifested by the rotation of the polarization ellipse of the output beam. The process can also be recognized as a type of stimulated scattering. Our experimental results are in good agreement with the theoretical predictions based on angular momentum conservation.

The experimental arrangement is shown in Fig. 1. A circularly polarized argon-laser beam ($\lambda=0.515\ \mu$m) propagating along the z axis was focused at normal incidence on a homeotropically aligned nematic 5CB sample 65 μm thick by use of a lens of 10-cm focal length. The polarization state of the light beam emerging from the sample was monitored with use of a heterodyne polarimeter scheme.[3] As shown in Fig. 1, a fraction of the incoming laser beam was split off and frequency shifted by an acoustic-optic modulator driven at 10 MHz and used as the reference beam. This reference beam, linearly polarized, was then divided by a beam splitter followed by two crossed polarizers into a horizontal (H) and a vertical (V) component of equal intensities. The signal beam, after transmission through the sample, was elliptically polarized in general, and was also divided into H and V components by the same beam-splitter polarizers. The H and V components of both beams fell separately on two photodiodes and generated a heterodyne beat signal at 10 MHz from each of them. The two beat signals from the photodiodes were then sent to the x and y terminals of a sampling oscilloscope for visual display of the output polarization ellipse, or to an IBM-PC computer for data

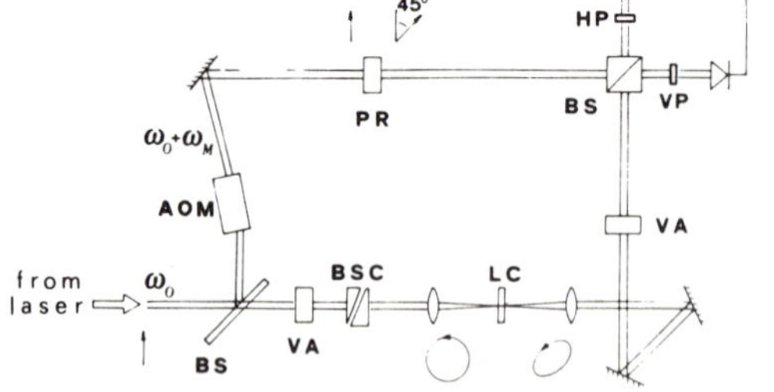

FIG. 1. The experimental arrangement: BS, beam splitter; VA, variable attenuator; BSC, Babinet-Soleil compensator; LC, liquid crystal; AOM, acousto-optic modulator; PR, polarization rotator; P, Glan prism polarizer.

processing. A Babinet-Soleil compensator and a polarization rotator were used for calibration of the apparatus.

Below a threshold pump power intensity I_{th}, no effect was observed and the transmitted light remained circularly polarized. Above the threshold, however, it became elliptically polarized and the polarization ellipse rotated continuously, as could be seen on the oscilloscope screen. The period of the ellipse rotation ranged from 40 to 50 s. This optical-field-induced transition appeared to be first order. As the laser intensity decreased from $I > I_{th}$ to $I < I_{th}$, the ellipticity of the output beam varied, but switched back to circular polarization only when I was significantly below I_{th}. Figure 2 describes the ellipticity change Δs_3 of the beam in traversing the sample as a function of the normalized input laser intensity. The hysteresis loop *ABCD* characterizes the first-order transition. This is a clear demonstration of an intrinsic (mirrorless) optical bistability effect in the absence of any static field.

With the input intensity changing slowly, the hysteresis loop is well reproducible. The upward transition *BC* in Fig. 2 is much faster than the downward transition *DA*: The switching times are $t_{up} \simeq 1$ s and $t_{down} \simeq 15$ s, respectively. The observed threshold power for up transition at point *B* was $P_{th} = 110$ mW. With a beam cross section of 5×10^{-5} cm^2 at the sample, the corresponding threshold intensity was $I_{th} = 2.2$ kW/cm^2. This transition should result from the field-induced molecular reorientation usually known as the Fréedericksz transition. We found that the observed threshold for a circularly polarized input beam was twice that for a linearly polarized input beam, as expected from theory.[4] However, no polarization rotation of the output was observed in the latter case.

All data points in Fig. 2 refer to an output whose polarization ellipse rotated continuously with a nearly constant angular velocity and no appreciable change in its form.[5] The branch *CD* is stable and reversible. No diffraction ring was observed in the far field except for a weak halo. This indicates that the birefringence induced in the sample is so small that the phase difference between the ordinary and extraordinary waves accumulated in transversing the medium is always less than π. When the input beam was changed from left to right circular polarization, the direction of the ellipse rotation was reversed accordingly, but the speed of the rotation remained unchanged.

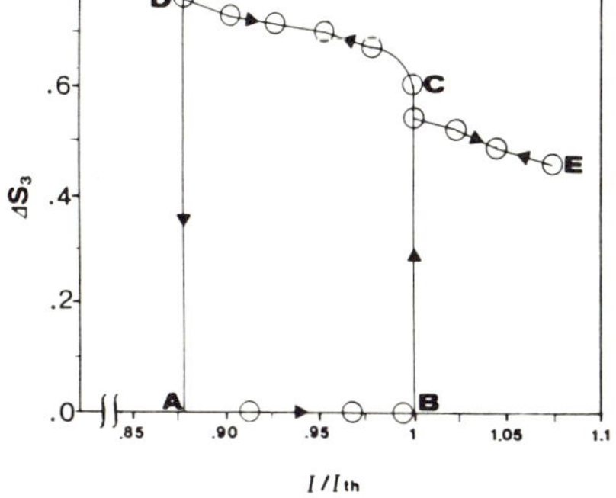

FIG. 2. Ellipticity change Δs_3 of the laser beam traversing the sample as a function of the normalized light intensity. The branch *CD* is characterized by a stable uniform rotation of the output polarization ellipse.

The branch *CE* is also reversible, but is metastable, in the sense that after a certain time (ranging from 10 to 60 min) the system switches spontaneously in 1–10 s to a different regime, characterized by a continuous variation of the form of the output polarization ellipse as well as by the appearance of a large number (up to 10) of diffraction rings in the far field. The number of rings is an indication of the magnitude of the induced birefringence in the sample.[6] During switching, the output polarization ellipse was continuously changing in form, until a steady-state regime was reached where the ellipse rotated again uniformly but with a much longer period of about 12 min. This slow-rotation regime could be reached only if the laser power is varied sufficiently slowly, $(\partial/\partial t)I/I_{th} < 2.5 \times 10^{-3}$/min. Otherwise, a dynamic regime would set in with the diffraction rings undergoing oscillating motion of expansion and contraction. If the laser intensity was beyond *E* in Fig. 2, then the system switched directly to the dynamic regime without going through any metastable or steady-state regime.

As the results for $I > I_B$ in Fig. 2 are not yet well understood, we shall, in this paper, focus our attention on the results for $I < I_B$. The fact that the ellipse rotation occurred with a circular rather than linear input polarization suggests that the effect arises from angular momentum transfer between the radiation field and the medium. While details of the theory are yet to be worked out, we present here a simple calculation that can explain the main observations.

Let us first give a qualitative description. The molecular reorientation of the homeotropic alignment is initiated only if the incoming laser intensity is above the threshold for the Fréedericksz transition. With circularly polarized input, the director $\hat{\mathbf{n}}$ would tend to be reoriented in such a way that it is tilted away from the $\hat{\mathbf{z}}$ axis but random in the azimuthal plane. However, in order to keep the elastic energy low, $\hat{\mathbf{n}}$ is likely to be tilted along a single direction. Some residual anisotropy would define that direction. The medium then appears birefringent to the incoming beam and, consequently, the beam polarization becomes elliptical. Transfer of angular momentum from the beam to the medium causes $\hat{\mathbf{n}}$ to precess around $\hat{\mathbf{z}}$, and hence the rotation of the polarization ellipse following $\hat{\mathbf{n}}$. With the beam becoming elliptically polarized, the threshold for the Fréedericksz transition is reduced. Therefore, even if the incoming I is now decreased to a value below I_{th} for circularly polarized light, the director $\hat{\mathbf{n}}$ may still point away from $\hat{\mathbf{z}}$ and precess around $\hat{\mathbf{z}}$ until the elliptically polarized light in the medium can no longer sustain the molecular reorientation, leading to the downward transition DA in Fig. 2. If $I > I_{th}$, then tranformation of the beam polarization from circular to elliptical tends to tilt $\hat{n}(z)$ farther away from $\hat{z}$ and enhances the birefringence.[7] As the birefringence increases, the beam polarization changes accordingly, and in turn, modifies $\hat{n}$ again. Because the reorientation of $\hat{n}$ responds rather slowly to the change of beam polarization, this could lead to a libration of $\hat{n}$ in its polar angle, and hence an oscillation in the observed birefringence.

We now consider some quantitative features of the problem. The threshold I_{th} for the Fréedericksz transition with circularly polarized light is given by

$$I_{th} = 2(\pi/L)^2(ck_{33}/n_o)[n_e^2/(n_e^2-n_o^2)], \tag{1}$$

where L is the thickness of the nematic film, k_{33} is the bend elastic constant, and n_0 and n_e are the ordinary and extraordinary refractive indices of the medium, respectively. With $L = 65$ μm and taking $k_{33} = 4.4\times10^{-7}$ dyne, $n_o = 1.53$, and $n_e = 1.70$ for 5CB,[8] we find $I_{th} = 2.1$ kW/cm^2, which is in good agreement with the observed value.

The angular momentum per unit area lost by the beam to the medium in traversing the nematic film is $(I/\omega)\Delta s_3$, where $\Delta s_3 = s_3(L) - s_3(0)$ is the ellipticity change of the beam polarization. This angular momentum transfer appears as a dynamic torque on the medium, causing the director $\hat{n}$ to precess, and is balanced by the torque arising from the viscous force:

$$I\Delta s_3/\omega = \gamma_1 \int_0^L \sin^2\theta \frac{\partial\phi}{\partial t} dz, \tag{2}$$

where γ_1 is a viscosity coefficient[9] and θ and ϕ are the polar and azimuthal angles of $\hat{n}(z,t)$, respectively. We have neglected in Eq. (2) the inertia term and the backflow effect.

We now assume, for the CD section in Fig. 2, that $\partial\phi/\partial z = 0$, $\partial\theta/\partial t = 0$, and $\partial\phi/\partial t = \Omega$ which describes the constant rotation of the director and the output polarization ellipse. Equation (2) then yields

$$\Omega = I\Delta s_3/(\omega\gamma_1 \int_0^L \sin^2\theta\, dz). \tag{3}$$

We recognize, however, that the phase difference between the ordinary and extraordinary waves at $z = L$ is

$$\chi = \pm\pi/2 + (\omega/c)\int_0^L [n_o - n_o n_e/(n_e^2\cos^2\theta + n_o^2\sin^2\theta)^{1/2} dz \simeq \pm\pi/2 - (\omega n_o/2c)(1 - n_o^2/n_e^2)\int_0^L \sin^2\theta(z)dz, \tag{4}$$

where $\pm\pi/2$ refers to the phase difference at $z = 0$ for the right- and left-circularly polarized waves, respectively, and the approximation is valid for small θ. Equation (4) indicates that the output at $z = L$ is elliptically polarized. Since $s_3(L) = \sin\chi$ and $s_3(0) = \pm 1$, we find, by eliminating $\int_0^L \sin^2\theta dz$ in Eq. (3) using Eq. (4),

$$\Omega = \pm \frac{I}{I_{th}} \left(\frac{\pi}{L}\right)^2 \frac{k_{33}}{\gamma_1} \frac{|\Delta s_3|}{\sin^{-1}[(|\Delta s_3|/2)^{1/2}]}. \tag{5}$$

It is seen that Ω changes sign when the circular polarization of the input beam is reversed, as observed in the experiment.

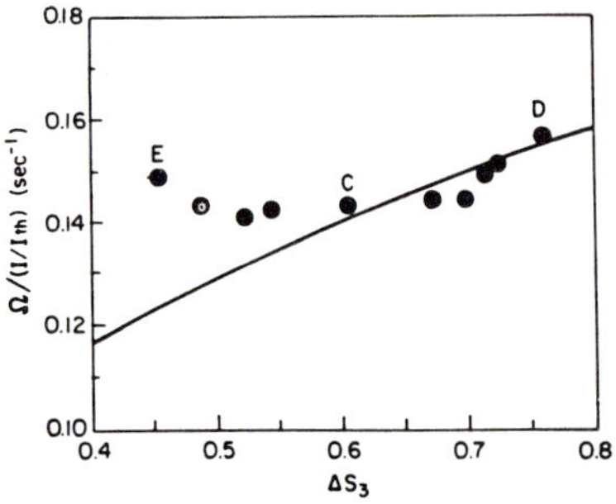

FIG. 3. Angular velocity Ω (normalized against I/I_{th}) of the output polarization ellipse as a function of the ellipticity change Δs_3 across the sample. The continuous line is obtained from Eq. (5).

From the measured I/I_{th} and Δs_3 given in Fig. 2 and $\gamma_1 = 0.38$ P taken from the literature,[8] we can find Ω from Eq. (5). The results are shown in Fig. 3 in comparison with the measured Ω. Note that the agreement is fairly good for the CD branch, where the approximations we used in the derivation are reasonable. For the metastable CE branch, the approximations are no longer valid, and hence the agreement is poor. The transient dynamic behavior of the problem is generally more complex. We are still in the process of finding a satisfying solution to explain all our observations quantitatively.

We also realize that the induced molecular precession requires a deposition of the beam energy in the medium. Since the medium is transparent, this can only happen if part of the beam is downward shifted in frequency. Indeed, the rotation of the polarization ellipse means that the two circularly polarized components of the elliptical polarization have different frequencies ω and ω', respectively, with $\omega - \omega' = 2\Omega$. Using a heterodyne technique, we were able to measure directly the ω' component in the output. In this respect, we can also regard the present nonlinear optical effect as a stimulated light-scattering process in which a new frequency component at ω' is generated. Details of our experiment and theoretical description will be reported elsewhere.

In summary, we have demonstrated that via the optical Fréedericksz transition, a circularly polarized input beam can activate a precession of the director in a homeotropic nematic film by transferring part of its angular momentum to the medium. This yields an output beam with a continuously rotating polarization ellipse. The phenomenon is intrinsically bistable. At input intensities above the transition threshold, the induced birefringence in the medium may break into oscillation. The observations can be understood physically, but the details are yet to be worked out.

This work was supported by Ministero della Publica Istruzione and Istituto Superiore di Poste e Telecomunicazioni, Roma Italy (E.S., B.D., M.R., and M.S.), and by National Science Foundation Solid State Chemistry Grant No. DMR8414053 (Y.R.S).

[1]See, for example, J. M. Jauch and F. Rohrlich, *The Theory of Photons and Electrons* (Addison-Wesley, Cambridge, 1955).

[2]Nevertheless, this torque has been measured in static conditions with a sophisticated torsional balance by R. A. Beth, Phys. Rev. **50**, 115 (1936).

[3]C. F. Buhrer, L. R. Bloom, and D. H. Baird, Appl. Opt. **2**, 839 (1963); R. Calvani, R. Caponi, and F. Olsternino, Opt. Commun. **54**, 63 (1985).

[4]B. Ya. Zel'dovich and N. V. Tabiryan, Zh. Eksp. Teor. Fiz. **82**, 1126 (1982) [Sov. Phys. JETP **55**, 99 (1982)].

[5]A small (~3%) oscillation of the output polarization ellipticity was always present. This effect is due to an imperfect, rigid rotation of the liquid-crystal molecules that has been neglected in our simplified mode. The values of Δs_3 reported in Figs. 2 and 3 are obtained by averaging over this oscillation.

[6]S. D. Durbin, S. M. Arakelian, and Y. R. Shen, Opt. Lett. **6**, 411 (1981).

[7]The same mechanism is responsible for the sharp switchon of *BC* in Fig. 2.

[8]The mechanical constants and refractive indices of 5CB are taken, respectively, from K. Skarp, S. T. Lagerwall, and B. Stebler, Mol. Cryst. Liq. Cryst. **60**, 215 (1980), and from R. G. Horn, J. Phys. (Paris) **39**, 105 (1978).

[9]See, for example, P. G. de Gennes, *The Physics of Liquid Crystals* (Oxford Univ. Press, Oxford, 1974), pp. 156–158.

Enhanced nonlinear birefringence in hybrid aligned nematics

G. Barbero and F. Simoni
UNICAL Liquid Crystals and Optics Group, Dipartimento di Fisica, Università della Calabria, Cosenza, Italy

(Received 7 April 1982; accepted for publication 7 July 1982)

We show that birefringence can be optically induced in hybrid aligned nematic (HAN) liquid crystals at normal incidence for light fields much lower than the Fredericksz threshold. In this case no threshold exists and the nonlinear optical effect at normal incidence is of the same order as that obtained with homeotropic (HOM) cells for the maximum useful angle. An analysis of the effect of the anchoring energy W_s is worked out showing that weak anchoring on the walls enhances the nonlinear effect by one order of magnitude or more.

PACS numbers: 42.10.Qj, 42.65.Jx, 61.30.Gd

The nonlinear optical properties of nematic liquid crystals have recently been studied by various authors.[1-4] It has been shown that an intensity-dependent refractive index is due to the optical reorientation of the molecules.

In both the homeotropic and planar configurations an incidence angle $\theta \neq 0°$ is necessary to induce the nonlinear effect. Moreover, in the homeotropic (HOM) case an incidence angle $\theta = 0°$ can be used if the field amplitude is higher than a threshold value $E_{th} = (2\pi/d)(\pi K/\Delta\epsilon)^{1/2}$ corresponding to a Fredericksz-like transition.

In this letter we show that if a hybrid aligned (HAN) cell is used an intensity-dependent refractive index can be obtained even with $\theta = 0°$, for optical field amplitude $E_{op} < E_{th}$. The effect is of the same order of magnitude as that found for homeotropic samples for the maximum useful angle θ ($25° + 30°$). This idea seems to be of practical application for those optical devices that might use self-focusing to get bistability.[5] So an analysis of the effect of the anchoring energy for both HAN and HOM cell is relevant, to show that the effect can be enhanced if weak anchoring is induced on the walls.

Let us consider a HAN cell as depicted in Fig. 1, with the z axis as direction of propagation of the light, taking into account the intrinsic birefringence of the medium to get the effective dielectric constant seen by the field component normal to z[6]:

$$\epsilon_{eff} = \frac{\epsilon^2 - (\Delta\epsilon/2)^2}{\epsilon + (\Delta\epsilon/2)\cos 2\varphi(z)}, \tag{1}$$

where $\epsilon = (\epsilon_\perp + \epsilon_\parallel)/2$, $\Delta\epsilon = \epsilon_\parallel - \epsilon_\perp$, $\epsilon_\parallel$ and $\epsilon_\perp$ are the dielectric constants parallel and perpendicular to the molecular director, and $\varphi(z)$ is the angle between the director and the z axis. For the HAN cell the total free-energy density is

$$F_0(\varphi,\varphi_z) = (K/2)\varphi_z^2 - (\Delta\epsilon/8\pi)E_{op}^2\sin^2\varphi. \tag{2}$$

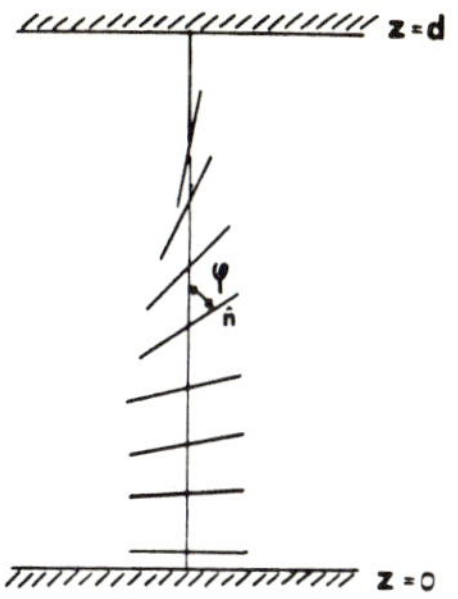

FIG. 1. Schematic representation of the hybrid aligned nematic (HAN) cell with no applied field. The rod lines show the alignment of the director **n** at different points in the cell.

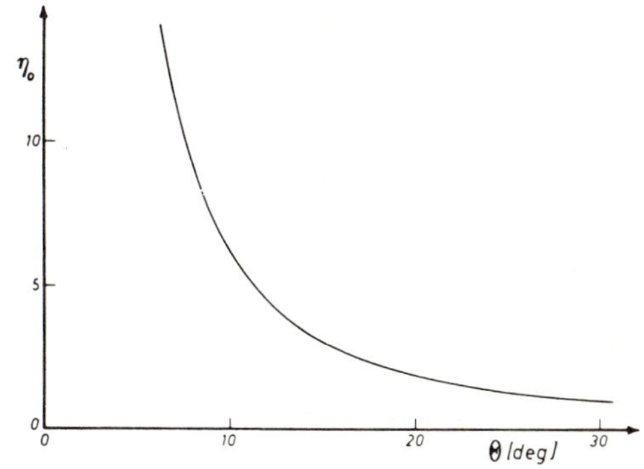

FIG. 2. Plot of η_0 vs ϑ, where ϑ is the internal incidence angle for the homeotropic (HOM) cell.

Here $K = K_{11} = K_{33}$ is the elastic constant and the boundary conditions are $\varphi(0) = 0, \varphi(d) = \pi/2$, where d is the cell thickness. Subscript z means a differentiation with respect to z. We follow the standard method[7] in the hypothesis of strong anchoring, with $E_{\rm op} \ll E_{\rm th}$ and obtain

$$\varphi = \varphi_0 + \varphi_2(E_{\rm op}/E_{\rm th})^2, \qquad (3)$$

where $\varphi_0 = (\pi/2)(z/d), \varphi_2 = (1/2)\sin 2\varphi_0$.

Equation (3) is obtained using a perturbative expansion[8]

$$\varphi = \sum_0^\infty \kappa\varphi_\kappa (E_{\rm op}/E_{\rm th})^2.$$

Then we insert (3) in (1) and keep only the change in the dielectric constant linear in the intensity[3]; we get

$$\epsilon_2^{\rm HAN} = \Delta\epsilon \frac{\epsilon^2 - (\Delta\epsilon/2)^2}{[\epsilon + (\Delta\epsilon/2)\cos 2\varphi_0]^2}\varphi_2 \sin 2\varphi_0. \qquad (4)$$

Equation (4) shows that no threshold exists even for optical field $E_{\rm op} \ll E_{\rm th}$, when the angle of incidence on the boundaries is $\theta = 0°$. It is interesting to compare this result with that of a HOM cell. The calculation is worked out in the same way, but with $\theta \neq 0°$ and the boundary condition $\varphi(0) = \varphi(d) = 0$.

We have the following expression[10]:

$$\epsilon_2^{\rm HOM} = \Delta\epsilon \frac{\epsilon^2 - (\Delta\epsilon/2)^2}{[\epsilon + (\Delta\epsilon/2)\cos 2\vartheta\,]^2}\sin 2\vartheta\, \varphi_2^{\rm HOM}, \qquad (5)$$

where

$$\varphi_2^{\rm HOM} = (\pi/2d)^2(-z^2 + zd)\sin 2\vartheta.$$

On comparing Eq. (4) with Eq. (5) we find that they are identical from a formal point of view. In Eq. (4) φ_0 plays the role of the incidence angle ϑ of Eq. (5), so we realize why no threshold exists for normal incidence in hybrid aligned cells.

Now we define

$$\eta_0(\vartheta) = \langle \epsilon_2^{\rm HAN}\rangle / \langle \epsilon_2^{\rm HOM}(\vartheta)\rangle, \qquad (6)$$

where different angles of incidence are considered only for the HOM case, $\langle\ \rangle$ means averaging on the cell thickness. The function $\eta_0(\vartheta)$ is plotted in Fig. 2, using the parameters $\epsilon = 2.8$ and $\Delta\epsilon = 0.8$ (typical for MBBA).[11] From this it appears that $\eta_0(\vartheta)$ decreases from ∞ (at $\theta = 0°$) to a value close to 1 for the maximum useful angle.

The above calculations were performed in the case of strong anchoring on the walls, i.e. when the anchoring energy $W_s \to \infty$. If we drop this approximation Eq. (2) becomes $F(\varphi,\varphi_z) = F_0(\varphi,\varphi_z) + F_s(\varphi)$, where $F_0(\varphi,\varphi_z)$ is still given by Eq. (2) and $F_s(\varphi)$ represents the free-energy density due to the nematic-substrate interaction,[12] and hence

$$F_s(\varphi) = (W_s/2)[\sin^2\varphi(0)\ \delta(z) + \cos^2\varphi(d)\delta(z-d)], \qquad (7)$$

$\varphi(0)$ and $\varphi(d)$ being the tilt angles at the boundaries and δ standing for Dirac's function. We work out the calculation as in the previous case and we get the same initial differential equation, but now the boundary conditions are[13]

$$2L\varphi_z - \sin 2\varphi = 0 \qquad (8)$$

at $z = 0$ and $z = d$. Here $l = K/W_s$ is the de Gennes–Kleman extrapolation length.

Using the perturbative expansion we get Eq. (3) once again, but now we have

$$\varphi_0(L) = c_1^{(0)}(L)z + c_2^{(0)}(L),$$
$$\varphi_2(L) = (1/2)(\pi/2c_1^{(0)}(L)d)^2 \sin 2\varphi_0(L) + c_1^2(L)z + c_2^{(2)}(L), \qquad (9)$$

where $c_i^{(j)}(L)$ is obtained by Eq. (8).

Combining Eqs. (9) and (4) we easily get an expression for $\epsilon_2^{\rm HAN}(z,L)$ that allows us to study the effect of the anchoring energy on the intensity-dependent dielectric constant.

The same calculation can be carried out for the HOM case to get $\epsilon_2^{\rm HOM}(z,L)$, taking into account that the surface interaction term of Eq. (7) is now

$$F_s(\varphi) = (W_s/2)[\sin^2\varphi(0)\ \delta(z) + \sin^2\varphi(d)\ \delta(z-d)]$$

and the boundary conditions are

$$(2L\varphi_z - \sin 2\varphi)_{z=0} = 0, \quad (2L\varphi_z + \sin 2\varphi)_{z=d} = 0.$$

We again average ϵ_2 on the cell thickness and define

$$\eta_{(L)}^{\rm HAN} = \frac{\langle \epsilon_2^{\rm HAN}(L)\rangle}{\langle \epsilon_2^{\rm HAN}(0)\rangle}, \quad \eta_{(L)}^{\rm HOM} = \frac{\langle \epsilon_2^{\rm HOM}(L)\rangle}{\langle \epsilon_2^{\rm HOM}(0)\rangle},$$

where $\epsilon_2^{\rm HOM}$ is calculated for $\theta = 24°$.

Figure 3 shows the result obtained with the same parameters as before. In Fig. 3 we see that by increasing L (i.e., lowering W_s) the nonlinear effect is enhanced in both cases, but the enhancement is more striking for the HAN cell, with a gain close to 30 for $L = d$. With a common surfactant[14] such as lecthin and Ryschenkow's technique[15] $L = d$ or

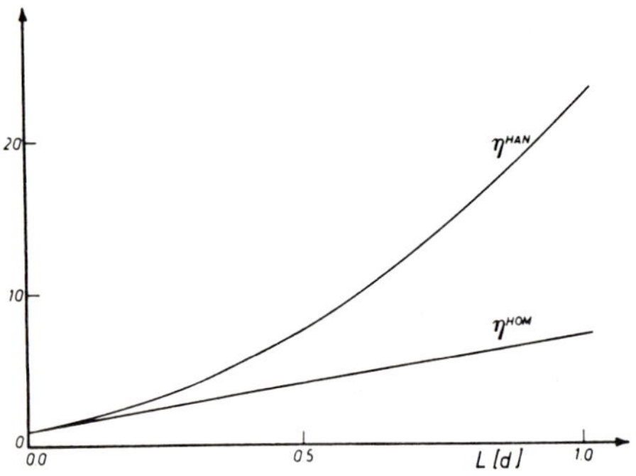

FIG. 3. Plot of η^{HAN} and η^{HOM} ($\vartheta = 24°$) vs the extrapolation length L.

higher can be attained with cell thickness of the order of 10 μ. Furthermore, the gain in the nonlinear dielectric constant can be calculated using a HAN cell, rather than HOM cell: $\bar{\eta} = \langle \epsilon_2^{\mathrm{HAN}}(L) \rangle / \langle \epsilon_2^{\mathrm{HOM}}(0) \rangle$; in this case a value of up to ~ 30 is reached for $L = d$.

In conclusion, we have shown that the use of a HAN cell is convenient to get nonlinear change in the dielectric constant for the following reasons: (a) an incidence angle $\vartheta = 0$ can be used even with very low optical fields; (b) the nonlinear effect can be greatly enhanced inducing a weak anchoring on the boundaries using a suitable surfactant on the walls.

[1] S. D. Durbin, S. M. Arakelian, and Y. R. Shen, Phys. Rev. Lett. **47**, 1411 (1981).

[2] N. F. Pilipetski, A. V. Sukhov, N. V. Tabiryan, and B. Y. Zel'Dovich, Opt. Commun. **37**, 280 (1981).

[3] I. C. Khoo, Phys. Rev. A **23**, 2077 (1981).

[4] I. C. Khoo, S. L. Zuang, and S. Sherpard, Appl. Phys. Lett. **39**, 937 (1981); and references therein.

[5] A. E. Kaplan, Opt. Lett. **6**, 360 (1981).

[6] R. M. Herman and R. J. Serinko, Phys. Rev. A **19**, 1757 (1981).

[7] In order to determine the tilt angle distribution $\varphi(z)$ inside the cell we minimize the total free energy $\mathcal{F}$. In the case of strong anchoring hypothesis, $\mathcal{F}$ is obtained as an ordinary fucntional: $\mathcal{F}_0(\varphi) = \int_0^d F_0(\varphi,\varphi_z)dz$, $F_0(\varphi,\varphi_z)$ being given by Eq. (2), and furthermore, $\delta\varphi(0) = \delta\varphi(d) = 0$, with fixed boundaries. By minimizing $\mathcal{F}(\varphi)$, the Euler–Lagrange standard equation is assumed: $\partial F_0/\partial\varphi - (\partial F_0/\partial\varphi_z)'_z = 0$. From Eq. (2) in the HAN case $\varphi_{zz} + (1/2)(\pi E_{\mathrm{op}}/dE_{\mathrm{th}})^2 \sin 2\varphi = 0$ is obtained, while a similar procedure gives $\varphi_{zz} + (1/2)(\pi E_{\mathrm{op}}/dE_{\mathrm{th}})^2 \sin 2(\vartheta + \varphi) = \rangle$, in the HOM case.

[8] G. Barbero and A. Strigazzi, Mol. Cryst. Liq. Cryst. (in press).

[9] Actually ϵ_2 is defined through the relation $\epsilon = \epsilon(E_{\mathrm{op}} = 0) + \epsilon_2(E_{\mathrm{op}}/E_{\mathrm{th}})^2$.

[10] The paper by Khoo *et al.* published in Appl. Phys. Lett. **39**, 937 (1981) reports an expression of $\delta\epsilon(I)$ quite approximat. Fortunately the same crude approximations cancel each other out so that an excellent agreement between theory and experiment can be claimed. We would point out that incorrect parameters are used for MBBA ($\Delta\epsilon = 0.4$ instead of $\Delta\epsilon = 0.8$ and $\epsilon_{\parallel}/\epsilon_{\perp} \sim 1$, where $\epsilon_{\parallel}/\epsilon_{\perp} \sim 0.75$) and a factor $\cos\varphi$ appears instead of $\cos^2\varphi$ [where $\cos(30°) = 0.86$].

[11] I. Haller, H. A. Huggins, and M. J. Freiser, Mol. Cryst. **16**, 53 (1972).

[12] G. Barbero and A. Strigazzi, Nuovo Cimento **64B**, 101 (1981).

[13] In the case of low anchoring energy the total free energy is given by a generalized functional: $\mathcal{F}(\varphi) = \int_0^d F_0(\varphi,\varphi_z)dz + \mathcal{F}_s(\varphi)$ where $\mathcal{F}_s$ represents the nematic-substrate interaction. By minimizing $\mathcal{F}(\varphi)$ the Euler–Lagrance standard equation is still obtained, and the transversality conditions are found to be $\{\mathbf{v}\cdot\nabla F_0 + \partial\mathcal{F}'_s/\partial\varphi\}_{z=0,d} = 0$, $\mathbf{v}$ being the surface normal unit vector. By means of the last relation, $\mathcal{F}_s$ form gives both boundary conditions (8) for the HAN cell and the analogous one for the HOM cell.

[14] S. Naemura, J. Phys. (Paris) **40**, Colloq. 3, 514 (1979); D. Riviere, Y. Levy, and E. Guyon, J. Phys. Lett. **40**, 215 (1979); A. Hochbaum and M. M. Labes, J. Appl. Phys. **53**, 2998 (1982).

[15] G. Ryschenkow and M. Kleman, J. Chem. Phys. **64**, 404 (1976).

Interaction of light waves with nonuniformly oriented liquid crystals

B. Ya. Zel'dovich and N. V. Tabiryan

Institute of Problems in Mechanics, Academy of Sciences of the USSR, and Special Design Bureau, Academy of Sciences of the Armenian SSR

(Submitted 22 November 1984; resubmitted 26 August 1985)
Zh. Eksp. Teor. Fiz. **90**, 141–152 (January 1986)

The orientation interaction is considered for light waves and liquid crystals with a uniform initial director orientation (such crystals include twisted nematics and cholesterics, and crystals with "hybrid" or "hyperhybrid" structures). Conditions are found for critical behavior to occur for a nonthreshold interaction. A light-induced Frederiks transition is predicted for cells with a hybrid nematic orientation, in which radiation absorption may reorient the director.

1. INTRODUCTION

Much recent work has been done on the interaction of light waves with the mesophases of liquid crystals (LC). Most of these results (particularly the experimental data) are limited to liquid crystals with a uniform initial orientation. Only a few papers have dealt with light wave interaction with initially nonuniformly oriented mesophases. Specifically Refs. 1–4 investigated the deformation or change in pitch of spiral cholestric LC structures in response to a light field, and the associated nonlinear optical phenomena were considered; self-focusing of light in a cell containing a nematic LC with a hybrid orientation was considered theoretically in Refs. 5 and 6; finally, light-induced bleaching of an unconfined nematic liquid crystal (NLC) was observed experimentally in Ref. 7 (the bleaching was accompanied by a straightening out of the initially nonuniform directors).

In this paper we theoretically predict and analyze several novel phenomena specific to the interaction of light with nonuniform mesophases. We also discuss how the twisting of the director field influences the light-induced Frederiks transition (LFT).

2. FUNDAMENTAL EQUATIONS

The equations governing the interaction of external fields with LC's can be derived by minimizing the free energy: $\delta\int F dV=0$. The free energy density F [erg/cm^3] of a nematic or cholesteric LC in a light field is given by

$$F=\frac{1}{2}K_1(\operatorname{div}\mathbf{n})^2+\frac{1}{2}K_2(\mathbf{n}\operatorname{rot}\mathbf{n}+q)^2+\frac{1}{2}K_3[\mathbf{n}\operatorname{rot}\mathbf{n}]^2-\frac{\varepsilon_{ik}}{16\pi}E_iE_k^*, \qquad (1)$$

where the K_i (in dynes) are the Franck constants, $\mathbf{n}=\mathbf{n}(\mathbf{r})$ is the director, E is the complex amplitude of the (monochromatic) light wave, and

$$\mathbf{E}_{\text{real}}(\mathbf{r},t)=0.5[\mathbf{E}\exp(-i\omega t+i\mathbf{kr})+\mathbf{E}^*\exp(i\omega t-i\mathbf{kr})];$$

$\varepsilon_{ik}=\varepsilon_\perp\delta_{ik}+\varepsilon_a n_i n_k$ is the dielectric permittivity tensor of the NLC at the light frequency. We have $q=0$ and $q=2\pi/h$ for nematics and cholesterics, respectively, where h is the equilibrium pitch of the cholesteric helix.

For situations when the director (unperturbed by the external fields) is uniformly oriented ($n=n^0=\text{const}$), it is helpful to write the variational equations in the form

$$\Pi_{ik}\left[\frac{\delta F}{\delta n_i}-\frac{\partial}{\partial x_j}\frac{\delta F}{\delta(\partial n_i/\partial x_j)}\right]=-\frac{\delta R}{\delta(\partial n_i/\partial t)}, \qquad (2)$$

where the factor $\Pi_{ik}=\delta_{ik}-n_in_k$ ensures the normalization $n^2=1$, $\eta[\Pi]$ is the orientation viscosity coefficient, and R [erg/cm$^3\cdot$s] is the dissipation function density. The right-hand side of (2) describes the relaxation processes.[2] To estimate the relaxation time it suffices to take R of the form

$$R=0,5\eta(\partial\mathbf{n}/\partial t)^2,$$

i.e., we may neglect the effects of hydrodynamic motion in the nematic LC on the director orientation.

However, in the general case when $\mathbf{n}^0=\mathbf{n}^0(\mathbf{r})$ even the linearized equations (including only terms of first order in the director perturbation $\delta\mathbf{n}=\mathbf{n}-\mathbf{n}^0$) are quite complicated. It is simpler to write out the equilibrium equations in terms of two variable angles specifying the director orientation—i.e., we write

$$\mathbf{n}(\mathbf{r})=[\mathbf{e}_x\cos\varphi(z)+\mathbf{e}_y\sin\varphi(z)]\sin\theta(z)+\mathbf{e}_z\cos\theta(z), \qquad (3)$$

where $\mathbf{e}_x,\mathbf{e}_y,\mathbf{e}_z$ are the unit vectors of a cartesian coordinate system with z axis normal to the walls of the NLC cell. If we assume that the system is uniform in the x,y plane then the variational equations take the form

$$(K_1\sin^2\theta+K_3\cos^2\theta)\frac{\partial^2\theta}{\partial z^2}-\frac{1}{2}(K_3-K_1)\sin 2\theta\left(\frac{\partial\theta}{\partial z}\right)^2-\sin\theta\cos\theta[K_3-2(K_3-K_2)\sin^2\theta]\left(\frac{\partial\varphi}{\partial z}\right)^2+K_2q\sin 2\theta\frac{\partial\varphi}{\partial z}$$
$$+\frac{\varepsilon_a}{16\pi}\{\sin 2\theta[\cos^2\varphi|E_x|^2+\sin^2\varphi|E_y|^2-|E_z|^2+\sin\varphi\cos\varphi(E_xE_y^*+E_x^*E_y)]$$
$$+\cos 2\theta[\cos\varphi(E_xE_z^*+E_x^*E_z)+\sin\varphi(E_yE_z^*+E_y^*E_z)]\}=\eta\frac{\partial\theta}{\partial t}, \qquad (4a)$$

 0038-5646/86/010080-07$04.00

$$\sin^2\theta(K_2\sin^2\theta+K_3\cos^2\theta)\frac{\partial^2\varphi}{\partial z^2}$$

$$+\sin 2\theta[K_3-2(K_3-K_2)\sin^2\theta]\frac{\partial\theta}{\partial z}\frac{\partial\varphi}{\partial z}$$

$$-K_2q\sin 2\theta\frac{\partial\theta}{\partial z}+\frac{\varepsilon_a}{16\pi}\{\sin^2\theta[\sin 2\varphi(|E_y|^2-|E_x|^2)$$

$$+\cos 2\varphi(E_xE_y^*+E_x^*E_y)]$$

$$+\sin\theta\cos\theta[\cos\varphi(E_yE_z^*+E_y^*E_z)-\sin\varphi(E_xE_z^*+E_x^*E_z)]\}$$

$$=\eta\sin^2\theta\frac{\partial\varphi}{\partial t}. \quad (4b)$$

Equations (4) must be solved simultaneously with the wave equation, which we write in the form

$$\Delta\mathbf{E}+(\omega/c)^2\hat{\varepsilon}(z)\mathbf{E}-\text{grad div }\mathbf{E}=0. \quad (5)$$

They simplify considerably in the "one-constant" approximation, i.e., if $K_1=K_2=K_3=K$. However, in this approximation Eq. (2) becomes

$$K\frac{\partial^2 n_i}{\partial z^2}-Kn_i\mathbf{n}\frac{\partial^2\mathbf{n}}{\partial z^2}-2Kq(\text{rot }\mathbf{n})_i+2Kqn_i\mathbf{n}\text{ rot }\mathbf{n}$$

$$+\frac{\varepsilon a}{16\pi}\{\mathbf{n}\mathbf{E}[E_i^*-n_i\mathbf{n}\mathbf{E}^*]+\mathbf{n}\mathbf{E}^*[E_i-n_i\mathbf{n}\mathbf{E}]\}=\eta\frac{\partial n_i}{\partial t}, \quad (6)$$

which is also relatively tractable. In what follows we will use whichever form of the equilibrium equation happens to be more convenient.

When no external fields are present, Eqs. (4) and (6) describe how the equilibrium director configuration $\mathbf{n}^0(\mathbf{r})$ depends on the boundary conditions. In particular, if the director on the surface $z=0$ is rigidly oriented normal to the plate (i.e., along the z axis) but points along the x axis on the plane $z=L$, $\mathbf{n}^0(z)$ will lie in the x,z plane everywhere inside the NLC if no external fields are present. Setting $\varphi=0$ and $E=0$ in Eqs. (4), we get the simple expression

$$\theta^{(0)}(z)=(\pi/2)(z/L) \quad (7)$$

for $\theta^{(0)}(z)$ in the one-constant approximation. Cells of this type are said to be "hybrid." Cells for which $\theta^{(0)}(z)=pz$, $p>\pi/2L$, will be called hyperhybrid cells.

3. NONTHRESHOLD FREDERIKS EFFECT IN NONUNIFORM NEMATICS

The nonthreshold Frederiks effect occurs when the extraordinary wave (e-wave) makes an oblique angle with the director. It was shown theoretically and experimentally in Ref. 8 that this effect can be detected even in extremely weak light fields.

In this section we analyze how light interacts with an LC with a twisted initial director orientation and find configurations for which the effects of director nonuniformity are particularly pronounced.

1. Twisted nematics. We have

$$\mathbf{n}^{(0)}(z)=\mathbf{e}_x\cos\varphi(z)+\mathbf{e}_y\sin\varphi(z),\quad \varphi=pz \quad (8)$$

for an unperturbed nematic. If we assume that an e-wave propagates with wave vector

$$\mathbf{k}=(\omega/c)(\mathbf{e}_x\sin\alpha\cos\beta+\mathbf{e}_y\sin\alpha\sin\beta+\mathbf{e}_z\cos\alpha)\varepsilon_e^{1/2} \quad (9)$$

in the cell and make the approximation

$$(\omega/c)(\varepsilon_e^{1/2}-\varepsilon_\perp^{1/2})L\gg 1,\quad \varepsilon_a/\varepsilon_\perp\ll 1,$$

we can write the wave polarization as

$$\mathbf{e}=\mathbf{E}/E=[\mathbf{k}[\mathbf{kn}]]/|[\mathbf{k}[\mathbf{kn}]]|, \quad (10)$$

where $[\mathbf{kn}]$ denotes the vector product. Substitution of Eqs. (8)–(10) into (4) yields an expression for the magnitude of the light-induced component $nz=\cos\theta$ of the director, where $\theta=\pi/2-\gamma$, $\gamma\ll 1$. To first order in the wave intensity, the steady-state equation for γ takes the form

$$\frac{d^2\gamma}{dz^2}+\xi\gamma-\frac{\varepsilon_a|E|^2\sin\alpha\cos\alpha}{8\pi K_1}\cos(\beta-pz)=0, \quad (11)$$

where $\xi=(2K_2-K_3)p^2/K_1$. The behavior is particularly interesting for $\xi>0$, because in this case the system becomes unstable to perturbations which bend the director out of the x,y plane by twisting the NLC director by the angle

$$\varphi_k=\pm\pi[K_1/(2K_2-K_3)]^{1/2}$$

(see, e.g., Ref. 9). Thus, if we take

$$2K_2>K_3\quad\text{and}\quad\varphi(z=L)=\varphi_L=\varphi_k(1-\Delta/\pi),$$

where $\Delta\ll 1$, the solution of (11) satisfying the boundary conditions $\gamma(z-0)-\gamma(z-L)-0$ is found to be

$$\gamma=\left[\varepsilon_a|E|^2\sin 2\alpha\cos\frac{\varphi_L}{2}\cos\left(\beta-\frac{\varphi_L}{2}\right)L^2\sin\frac{\pi z}{L}\right]$$

$$\times[8\pi K_1\Delta(\varphi_L^2-\pi^2)]^{-1}. \quad (12)$$

The magnitude of the response γ depends in an extremely complicated way on the interaction geometry. In general, γ becomes anomalously large as $\Delta\to 0$. It is easy to check that in the one-constant case ($K_1=K_2=K_3$), for which instability develops when the director is twisted by an angle $\varphi_k=\pi$, the behavior of γ exhibits no anomalies at $\beta=0$. We must have $\beta\neq 0$ for instability to be observable in the orientation interaction of a light wave with a nematic liquid crystal.

2. Hyperhybrid cells. We take the unperturbed director distribution to be

$$\mathbf{n}^{(0)}=\mathbf{e}_x\sin\theta(z)+\mathbf{e}_z\cos\theta(z),\quad \theta(z)=pz \quad (13)$$

and consider the linearized equation for the y-component n_y. Using (9) and (13), we find that

$$\frac{d^2n_y}{dz^2}+p^2n_y-\frac{\varepsilon_a|E|^2\sin\alpha\sin\beta}{8\pi K}$$

$$\times(\cos\alpha\cos pz+\sin\alpha\cos\beta\sin pz)=0 \quad (14)$$

from (6) in the steady-state case. The solution of (14) satisfying $n_y(z=0)=n_y(z=L)=0$ is

$$n_y=\frac{\varepsilon_a|E|^2\sin\alpha\sin\beta}{16\pi Kp}\{(z-L)\cos\alpha\sin pz$$

$$+\sin\alpha\cos\beta(L\,\text{ctg}\,pL\sin pz-z\cos pz)\}. \quad (15a)$$

As $\Delta=\pi-pL\to 0$, the component n_y again exhibits critical behavior:

$$n_y \approx -\frac{\varepsilon_a |E|^2 L^2 \sin^2\alpha \sin\beta \cos\beta}{16\pi^2 K\Delta} \sin\frac{\pi z}{L}. \qquad (15b)$$

We must have $\cos\beta \neq 0$ in order for criticality to occur (the latter arises because as $pL \to \pi$, the deformed NLC structure becomes unstable to fluctuations that take the director out of the x,z plane).

The deformed states of an NLC with $pL \lesssim \pi$ correspond to local minima of the free energy. These states can relax to an absolute energy minimum through the generation of disclinations. These questions are both interesting and extremely difficult to analyze; in particular, it would be of interest to see if light-induced generation of disclinations can occur in the cases considered above. Moreover, the types of instability mentioned above for nonuniform nematic liquid crystals are not exhaustive—other instabilities involving reversible adhesion of molecules on the substrate surfaces can also occur.[10]

We also mention the following interesting effect, which occurs, e.g., in planar-homeotropic (hybrid) oriented NLC's and is not directly related to nonlinear optical effects. According to (10), in the adiabatic approximation with $\beta \neq 0$, the polarization vector of the light wave transmitted by the cell will be rotated by 90° relative to the incident wave polarization. A wave with polarization $\mathbf{e} = \mathbf{e}_x$ incident in the z,y plane will be an ordinary wave (o-wave) if it is incident from the homeotropic wall ($z = 0$) and an e-wave if it is incident from the planar wall ($z = L$).

We also observe that in contrast to uniform LC's, in general the perturbation of the director in nonuniformly oriented cells is not greatest at the center of the cell.

3. Deformed cholesteric. Assume that Eq. (8) describes the initial structure of the director, but that the cell is filled with a cholesteric liquid crystal with equilibrium pitch $h = 2\pi/q$. For $p = q$ we have an ordinary Grandjean structure. For $p \neq q$ elastic stresses are present which in principle should be revealed when the cholesteric interacts with light. In the Mauguin limit, the equations describing this interaction are given by (11) except that

$$\xi = \frac{2K_2 - K_3}{K_1} p^2 - \frac{2K_2}{K_1} qp. \qquad (16)$$

We can use (16) to find conditions for $\xi > 0$, i.e., for instability to be possible. Thus, for $p \ll q$ the condition $\xi > 0$ requires that sign $(qp) < 0$, i.e., the direction of the mechanical twisting must be opposite to the cholesteric twisting. The critical twisting angle $\varphi_{cr} = p_{cr} L$ is

$$|\varphi_{cr}| = \pi^2 K_1 / 2K_2 |q| L.$$

The magnitude of the light-induced component n_z and its dependence on the experimental configuration are given by an expression analogous to (12) with ξ given by (16).

When the Mauguin condition is violated, it is much more difficult to describe the lightwave-cholesteric interaction because the natural modes of the LC are elliptically polarized waves whose properties depend in a complicated way on the angle of incidence of the wave and on the ratio of the wavelength divided by the pitch of the helices. However, this problem can be solved without difficulty by using the results in Ref. 11.

Under our assumptions, the phase shift of the e-wave can be found from the expression

$$\delta\Phi = \frac{\omega}{2c\varepsilon_e^{1/2}\cos\alpha} \int_0^L \delta\varepsilon_{ik} e_i e_k \, dz$$

$$= -\frac{\omega\varepsilon_a}{c\varepsilon_e^{1/2} k^2 \cos\alpha} \int_0^L (\mathbf{k}\mathbf{n}^{(0)})(\mathbf{k}\delta\mathbf{n})\, dz. \qquad (17)$$

The minus sign in (17) does not mean that the nonlinear phase shift is negative—indeed, the orientational nonlinearity always results in self-focusing of the light, i.e., $\delta\Phi > 0$. This can be seen by calculating $\delta\Phi$ for some specific cases by substituting the corresponding expressions for $\mathbf{n}^{(0)}(z)$ and $\delta\mathbf{n}(z)$ into (17).

4. FREDERIKS TRANSITION IN HYBRID-ORIENTED NEMATICS

If the light wave is incident normally on a homeotropically oriented cell, the director will become reoriented if the light intensity exceeds a certain threshold (this is called a light-induced Frederiks transition, or LFT). There may also be a threshold intensity for director reorientation when an o-wave is incident on a cell with a uniform planar orientation, but only if the adiabatic condition is violated.[12] This leads naturally to the question of what happens when an o-wave is normally incident on a hybrid cell (Fig. 1).

In order to find the threshold intensity for director reorientation, we must linearize and solve Eq. (6) to first order in the perturbation of the director n_y. In the one-constant approximation, we obtain

$$\eta \frac{\partial n_y}{\partial t} + K \frac{\partial^2 n_y}{\partial z^2} + K p^2 n_y$$

$$+ \frac{\varepsilon_a}{16\pi} [2n_y |E_y|^2 + (n_x E_x + n_z E_z) E_y^* + (n_x E_x^* + n_z E_z^*) E_y] = 0, \qquad (18)$$

where $p = \pi/2L$ and E_y is the strength of the light field incident on the nematic liquid crystal. Because of the perturbation of the director, the field components E_x and E_z are nonzero: $E_x, E_z \propto \varepsilon_a n_y$.

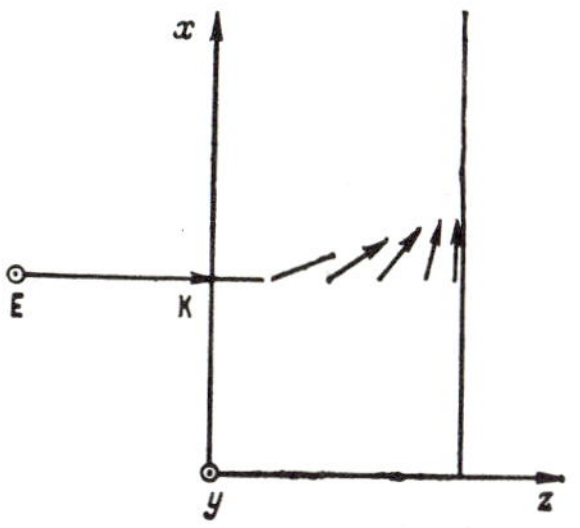

FIG. 1. Light-induced Frederiks transition in a hybrid cell with director initially aligned in the x, z plane. The light wave is incident normally on the cell (along the z axis) and is polarized along the y axis, i.e., perpendicular to the plane of the figure.

The field amplitudes E in (17) can be found from the Maxwell equations

$$\frac{d^2E_x}{dz^2}+\left(\frac{\omega}{c}\right)^2[(\varepsilon_\perp+\varepsilon_a n_x^2)E_x+\varepsilon_a n_x n_y E_y+\varepsilon_a n_x n_z E_z]=0, \quad (19a)$$

$$\frac{d^2E_y}{dz^2}+\left(\frac{\omega}{c}\right)^2[\varepsilon_a n_x n_y E_x+(\varepsilon_\perp+\varepsilon_a n_y^2)E_y+\varepsilon_a n_y n_z E_z]=0, \quad (19b)$$

$$\varepsilon_a n_x n_z E_x+\varepsilon_a n_z n_y E_y+(\varepsilon_\perp+\varepsilon_a n_z^2)E_z=0. \quad (19c)$$

Using (19c) to eliminate E_z, we can recast (19a) and (19b) to first order in the director perturbation as

$$\frac{d^2E_y}{dz^2}+k_\perp^2E_y=0, \quad (20a)$$

$$\frac{d^2E_x}{dz^2}+\left(\frac{\omega}{c}\right)^2\frac{\varepsilon_\parallel\varepsilon_\perp}{\varepsilon_\perp+\varepsilon_a n_z^2}E_x=-\left(\frac{\omega}{c}\right)^2\frac{\varepsilon_a\varepsilon_\perp n_x n_y}{\varepsilon_\perp+\varepsilon_a n_z^2}E_y, \quad (20b)$$

where $k_\perp=(\omega/c)\varepsilon_\perp^{1/2}$. The solution of (20a) is $E_y=E_0\times\exp(ik_\perp z)$.

We assume a solution of the form

$$E_x=E_0A(z)e^{ik_\perp z} \quad (21)$$

for (20b), where the function $A(z)$ varies slowly over a characteristic length $2k_\perp^{-1}$. Equation (20b) then gives

$$\frac{dA}{dz}-\frac{i}{2}k_\perp\frac{\varepsilon_a n_x^2}{\varepsilon_\perp+\varepsilon_a n_z^2}A=\frac{i}{2}k_\perp\frac{\varepsilon_a n_x n_y}{\varepsilon_\perp+\varepsilon_a n_z^2}E_0. \quad (22)$$

If A is smooth even over distances $(\varepsilon_a k_\perp)^{-1}$, then (22) and (19c) imply that $A\sim-n_y/n_x$ and $E_z\approx0$.

The o-wave thus adiabatically follows the rotation of the director and the right-hand side of (18) is equal to zero. On the other hand, if $A(z)$ is not assumed to satisfy the above condition, (22) has the solution

$$A(z)=\frac{i}{2}\varepsilon_a k_\perp\int_0^z\frac{n_x(z')n_y(z')}{\varepsilon_\perp+\varepsilon_a n_z^2(z')}\exp[i\psi(z)-i\psi(z')]dz',$$

$$\psi(z)=\frac{1}{2}k_\perp\varepsilon_a\int_0^z\frac{n_x^2}{\varepsilon_\perp+\varepsilon_a n_z^2}dz. \quad (23)$$

From (19c) and (23),

$$E_z\approx-\frac{\varepsilon_a n_z E_0}{\varepsilon_\perp+\varepsilon_a n_z^2}[n_y+n_xA(z)]e^{ik_\perp z}, \quad (24)$$

and Eq. (18) takes the form

$$\eta\frac{\partial n_y}{\partial t}=K\frac{\partial^2 n_y}{\partial z^2}+\left[Kp^2+\frac{\varepsilon_a|E_0|^2}{8\pi}\left(1-\frac{\varepsilon_a n_z^2}{\varepsilon_\perp+\varepsilon_a n_z^2}\right)\right]n_y$$

$$-\frac{\varepsilon_a^2|E_0|^2k_\perp}{16\pi}\left(1-\frac{\varepsilon_a n_z^2}{\varepsilon_\perp+\varepsilon_a n_z^2}\right)n_x\int_0^z\frac{n_x(z')n_y(z')}{\varepsilon_\perp+\varepsilon_a n_z^2(z')}$$

$$\times\sin[\psi(z)-\psi(z')]dz'. \quad (25)$$

If no field is present ($|E_0|^2=0$), Eq. (25) describes the damping of the small perturbations:

$$n_y(z,t)=\sum_{m=1}^\infty\sin\frac{m\pi z}{L}\exp(-\Gamma_m t), \quad (26a)$$

$$\Gamma_m=\frac{K}{\eta}\left(\frac{\pi}{L}\right)^2\left(m^2-\frac{1}{4}\right). \quad (26b)$$

The term proportional to the coefficient $-1/4$ in (26b), and the corresponding term in (25), describes the destabilizing influence of the initial nonuniform structure. It is easy to show that when

$$Ld\theta/dz=\theta(L)-\theta(0)\geqslant\pi$$

the planar structure is unstable (see also Ref. 9). In our case $\theta(L)-\theta(0)=\pi/L$ and this destabilization is offset by stabilization at the walls $z=0$ and $z=L$, where the director is rigidly oriented.

We will discuss the light-induced Frederiks transition for the case when the light is incident from the homeotropically orienting boundary; then $\theta(z)=\pi z/2L$. The lightwave-LC interaction is strongest where the phase difference $\psi(\Delta z)$ between the o- and the e-waves is $\lesssim1$. The width Δz of the strong-interaction region is typically a small fraction of the total cell width L, so that for $z\lesssim\Delta z$ we need only retain the first nonvanishing term in the expression for $\psi(z)$:

$$\psi(z)=\left(\frac{d\theta}{dz}\right)^2\frac{k_\perp\varepsilon_a}{6\varepsilon_\parallel}z^3=\frac{\pi^2}{24}\frac{k_\perp\varepsilon_a}{\varepsilon_\parallel L^2}z^3\equiv Bz^3. \quad (27)$$

For $z\lesssim\Delta z$ we can also set $n_z\approx1$ and $n_x\approx\pi z/2L$.

To simplify the notation the subscript y in $n_y(z,t)$ will be omitted in what follows. We define the dimensionless coordinate v by

$$v=B^{1/3}z,\quad \psi(z)=v^3, \quad (28)$$

so that $\psi(z)=1$ corresponds to $v=1$ and $\psi(z)=2\pi$ to $v=(2\pi)^{1/3}\approx1.85$. The cell boundary $z=L$ corresponds to the following (large) dimensionless value:

$$v(z=L)=M=(\pi^2\varepsilon_a k_\perp L/24\varepsilon_\parallel)^{1/3}.$$

For example, we have $M\approx6$ for radiation of wavelength $\lambda=0.6\ \mu$m in vacuum for a cell with $\varepsilon_\parallel\approx3$, $\varepsilon_a\approx1$, and $L=100\ \mu$m.

We also introduce the parameter ρ which gives the incident power density relative to the LFT threshold for an ordinary homeotropic cell of the same width L:

$$|E_0|^2=\rho(8\pi\varepsilon_\parallel K_3/\varepsilon_a\varepsilon_\perp). \quad (29)$$

In these variables the linearized small-perturbation equation takes the form

$$\frac{\partial n(v,t)}{\partial t}=\frac{K}{\eta}\left(\frac{\pi}{L}\right)^2\left(\frac{M}{\pi}\right)^2\left\{\frac{\partial^2 n}{\partial v^2}+\left(\frac{\pi}{2M}\right)^2(1+4\rho)n-3\rho\left(\frac{\pi}{M}\right)^2v\int_0^v v'n(v')\sin(v^3-v'^3)dv'\right\}\equiv\hat{L}n, \quad (30a)$$

$$n(v=0,t)=n(v=M,t)=0. \quad (30b)$$

We solve (30) by separation of variables by substituting $n(v,t)=n(v)\exp(-\Gamma t)$ into (30a). The problem then reduces to an eigenvalue problem for Γ. Because the operator on the right-hand side of (30a) is nonhermitian, the eigenvalues Γ are in general complex, $\Gamma = \Gamma' + i\Gamma''$. At threshold, the incident wave intensity p is such that the real part of Γ becomes negative: $\Gamma' \leqslant 0$; since Γ'' may be nonzero, the director perturbation may oscillate. The situation here is similar to the Frederiks transition in a uniform NLC cell induced by o-wave light.[12,13]

We have not yet succeeded in solving Eqs. (30) in closed form. Some qualitative estimates for the threshold LFT intensity in a hybrid cell can be derived as follows. We first estimate the width Δz of the strong-interaction region; taking $\psi(2\pi) \approx 2\pi$, we find from (27) that $\Delta z = (2\pi/B)^{1/3}$, or $\Delta z/L \approx (2\pi)^{1/3}/M$. The LFT threshold in a hybrid cell of width L is of course lower than for a homeotropic cell of width Δz:

$$P < \frac{c\varepsilon_{\parallel}K_3}{\varepsilon_a\varepsilon_{\perp}^{1/2}}\frac{\pi^2}{(\Delta z)^2} = \left(\frac{\pi}{L}\right)^2 \frac{c\varepsilon_{\parallel}K_3}{\varepsilon_a\varepsilon_{\perp}^{1/2}}\frac{M^2}{(2\pi)^{2/3}}. \qquad (31)$$

For the above parameter values, $M \approx 6$ and $M^2/(2\pi)^{2/3} \approx 10$.

In fact, in some cases the quantity $(\Delta z L)^{1/2}$ can be regarded as the "effective width" of the hybrid cell—this is true, e.g., for the light-induced Frederiks transition in the field of a wave localized within a distance Δz from the NLC surface.

When the light wave is incident from the plane-orienting boundary of the cell, we must take $\theta(z) = (\pi/2L)(L - z)$ in (25). The form of the resulting equation then differs considerably from (30). This implies that the LFT thresholds may differ for light incident from opposite directions on a cell with a nonuniform initial NLC orientation.

In some cases the integral in (30), or even the entire field term, can be regarded as a small perturbation, as is the case when $\theta(z) = pz$, $pL \ll 1$. In the opposite case when $\theta(z) \lesssim pz$, $p \lesssim \pi/L$, the director distribution becomes very sensitive to perturbations out of the x,z plane. Clearly, the LFT threshold is low in this case, $p \ll 1$. However, there is little point in analyzing the various cases enumerated above in the absence of specific experimental data. Moreover, in addition to the hybrid cell it would also be of great interest to investigate light-induced effects in cells in which a static magnetic field, say, is used to produce a nonuniform director orientation. This would make it possible to study these effects under conditions when the initial director field can be changed continuously.

5. THERMAL ORIENTATION EFFECTS IN A HYBRID CELL

The liquid crystal mesophases provide various opportunities for thermal reorientation; for instance, heating alters the pitch of cholesteric helices. Reference 14 considered nonlinear optical properties of C smectics associated with changes in the molecular orientation angle during heating. In this section we show that thermal orientation effects are also present in nonuniformly oriented nematics.

We assume that an o-wave of intensity below the LFT threshold is incident on a hybrid cell. It turns out that thermal effects can reorient the director even in this case. We will analyze this effect by using the equation

$$(K_1\sin^2\theta + K_3\cos^2\theta)\frac{d^2\theta}{dz^2} - \frac{1}{2}(K_3 - K_1)\sin 2\theta\left(\frac{d\theta}{dz}\right)^2 = 0 \qquad (32)$$

for the equilibrium director distribution in a hybrid cell. This equation follows from (4a) by setting $\alpha = 0$ and using $\mathbf{E}\cdot\mathbf{n} \equiv \mathbf{En} = 0$. We assume that the light field affects the nematic LC through the temperature-dependence of the Franck constants. Equation (32) has the implicit solution

$$\mathrm{E}(\theta, u^{1/2}) = \mathrm{E}\left(\frac{\pi}{2}, u^{1/2}\right)\left(1 - \frac{z}{L}\right), \qquad (33)$$

which satisfies the boundary conditions $\theta(z = 0) = \pi/2$ and $\theta(z = L) = 0$; here E is the elliptic integral of the second kind and $u = 1 - K_1/K_3$. We can obtain an explicit expression for $\theta(z)$ if $u \ll 1$; to first order in u,

$$\theta(z) = \frac{\pi}{2}\left(1 - \frac{z}{L}\right) - \frac{u}{8}\sin\frac{\pi z}{L}. \qquad (34)$$

The change in θ due to heating is given by

$$\delta\theta = -\frac{\delta u}{8}\sin\frac{\pi z}{L}. \qquad (35)$$

Particularly large changes $\delta u = -\delta(K_1/K_3)$ should be expected near the nematic-smectic-A transition, for which the constant K_3 increases rapidly.[15]

If an e-wave is incident on a hybrid NLC cell, all three interaction mechanisms (orientation, thermal, and thermo-orientation) will occur simultaneously. We will use Eq. (17) and the corresponding expressions for the perturbation of the permittivity tensor to estimate the relative contributions of these mechanisms to the nonlinear phase shift. For the orientation and thermo-orientation mechanisms, $\delta\varepsilon_{ik}e_ie_k = \varepsilon_a \sin 2\theta^{(0)}\delta\theta$. We readily find an expression for $\delta\theta$ for the orientation mechanism from Eq. (4a) by setting $\varphi = 0$, $E_x \approx E = \text{const}$, and $E_z \approx 0$ (for definiteness we consider a normally indicent e-wave). For the thermal mechanism of nonlinearity,

$$\delta\varepsilon_{ik}e_ie_k = [(\partial\varepsilon_{\perp}/\partial T) + (\partial\varepsilon_a/\partial T)(\mathbf{n}^{(0)}\mathbf{e})^2]\delta T.$$

The heating δT in the light field can be estimated as $\delta T \approx \sigma P\tau_t/C_p$, where σ is the absorption coefficient in cm^{-1}, τ_t [s] is the relaxation time, and C_p [erg/cm^3·deg] is the specific heat per unit volume. If the width of the beam is greater than the cell width L then $\tau_t \approx (L/\pi)^2/\chi$, where χ [cm^2/s] is the thermal diffusivity.

The above discussion leads to the following expressions for the nonlinear phase shifts $\delta\Phi_0$, $\delta\Phi_{t0}$, and $\delta\Phi_t$ due to the orientation, thermo-orientation, and thermal mechanisms:

$$\delta\Phi_{T0} = \frac{\partial(K_1/K_3)}{\partial T}\frac{\omega\varepsilon_a\sigma L^3|E|^2}{256\pi^3 C_P\chi}, \qquad (36)$$

$$\delta\Phi_0 = \frac{\omega\varepsilon_a^2 L^3|E|^2}{64\pi^3 c n_e K}, \qquad (37)$$

$$\delta\Phi_T = \left(\frac{\partial\varepsilon_{\parallel}}{\partial T} + \frac{\partial\varepsilon_{\perp}}{\partial T}\right)\frac{\omega\sigma L^3|E|^2}{32\pi^3 C_P\chi}. \qquad (38)$$

We see at once that the phase shifts depend differently on the anisotropy of the permittivity: $\delta\Phi_0 \propto \varepsilon_a^2$, $\delta\Phi_{t0} \propto \varepsilon_a^1$, $\delta\Phi_t \propto \varepsilon_a^0$. We obtain $|\delta\Phi_{t0}/\delta\Phi_0| \approx |\delta\Phi_{t0}/\delta\Phi_t| \approx 6$ if we take $|\partial(K_1/K_3)/\partial T| \sim 10^{-1}\ \text{deg}^{-1}$, $\sigma \sim 5\ \text{cm}^{-1}$, $C_p \sim 10^7$

erg/cm³·deg, $\chi \sim 10^{-4}$ cm²/s, $n_e \sim 1.6$, $\varepsilon_a \sim 0.5$, $K \sim 5 \cdot 10^{-7}$ dyn, and $|\partial \varepsilon_{\parallel}/\partial T + \partial \varepsilon_{\perp}/\partial T| \sim 10^{-3}$ deg^{-1}.

The thermal relaxation time ($\tau_t \approx 2.5 \cdot 10^{-2}$ s for $L = 50$ μm) is significantly less than the relaxation time for the director orientation τ_{0r}. The latter can be estimated from Eqs. (4) or (6) and is $\tau_{0r} \sim \eta L^2/\pi^2 K \sim 5$ s for $\eta \sim 1$ Poise (see, e.g., Ref. 8).

"Giant" optical nonlinearity of an NLC mesophase caused by light-induced changes in the molecular conformation was discovered in Ref. 16. These changes alter the molecular polarizabilities and the macroscopic properties of the nematic (the order parameter Q and the phase transition temperature[17]). Presumably, the Franck constants are also affected. Well away from the nematic-smectic-A phase transition, the Franck constants depend on temperature as $K_i \propto Q^2$, so that $K_i/K_j = \text{const}$. On the other hand, it is not clear *a priori* how the ratios K_i/K_j will behave during light-induced conformational changes. A study of the absorption and orientation nonlinearity might shed some light on this question.

6. THRESHOLD THERMO-ORIENTATION NONLINEARITY

Cholesteric LC's in cells with a rigid homeotropic orientation at the walls were considered in Refs. 18–21, where it was shown that for cholesteric pitches h exceeding a critical value $h_{cr} = 2LK_2/K_3$, the LC assumes a uniform homeotropic orientation with director $n_x = n_y = 0$, $n_z = 1$. When h is less than h_{cr} (equivalently, if the wave vector q of the cholesteric LC exceeds the critical value $q_{cr} = \pi K_3/LK_2$), the director takes on a nonuniform orientation and the resulting structure is similar to that for a spring which is stretched at both ends.[21] Let us now examine what happens when $q < q_{cr}$ and q increases with T. Clearly, the lightwave intensity must exceed a threshold value if the LC is to be heated enough so that $q > q_{cr}$ and the director becomes realigned. If we take the change in the wave vector of the cholesteric helix to be

$$\delta q = (\partial q/\partial T)\delta T = (\partial q/\partial T)\sigma L^2 P/\pi^2 C_P \chi$$

(see Sec. 5), we get the result

$$P_{\text{thr}} = \frac{q_{cr} - q}{q_{cr}} \frac{\pi^2 \chi C_P}{\sigma L^2} \left(\frac{1}{q} \frac{\partial q}{\partial T} \right)^{-1} \qquad (39)$$

for the threshold intensity. If $L = 10$ μm, $(q_{cr} - q)/q_{cr} \sim 10^{-1}$, $q^{-1}(\partial q/\partial T) \sim 10$ K^{-1}, and the remaining parameter values are as above, we obtain $P_{\text{thr}} \approx 2$ W/cm².

The resulting structure can be shown to be optically active, so that we are in fact dealing with nonlinear optical activity. By suitably selecting the absorption coefficient of the cholesteric, the difference $q - q_{cr}$, and the polarization of the light wave, it is quite easy to ensure that the direct orientational effects within the light beam have no influence on the director orientation.[22] In general the magnitude of the thermal reorientation can be calculated numerically by using the formulas derived in Ref. 21. It was shown there that if the Franck constants satisfy $K_1 - 3(K_3 - K_2) < 0$, then hysteresis is present in the dependence $\theta_m(q/q_{cr} - 1)$, where θ_m is the angle of greatest director deflection (this occurs at the center of the cell) and $q/q_{cr} - 1$ is the relative increase above threshold. In addition to heating, light-induced conformational changes in the LC may also influence q.

7. CONCLUSIONS

We have considered several examples which illustrate the diversity of the interesting effects associated with lightwave-liquid-crystal interaction in cells with a nonuniform initial director distribution. Many such cells can be constructed. We have already seen that their qualitative properties depend not only on the boundary conditions but also on the specific physical properties of the LC material. This fact should be stressed, since in cells with a uniform director orientation the differences in the Franck constants, say, from one LC to another lead merely to quantitative differences. In many cases, even the description of the equilibrium structure for "nonuniform" cells encounters serious mathematical difficulties. Conservation laws[23] may provide a powerful technique for solving problems of this type. For instance, a theorem of E. Noether was used in Ref. 23 to derive analytic expressions for the equilibrium structure of complex configurations such as homeotropic-planar oriented cholesterics and cholesterics in magnetic fields with a homeotropic orientation at the walls.

Bistability is quite common in nonuniform cells (see, e.g., Ref. 9), and even the linear optical properties of such cells are far from trivial.[24] It is here that the qualitative differences between the effects of lightwaves and static fields on LC's are most pronounced.

We hope that the above discussion will stimulate more experimental work on the interaction of light with mesophases in cells with nonuniform LC orientations.

1. N. V. Tabiryan and B. Ya. Zel'dovich, Molec. Cryst. Liq. Cryst. **69**, 19 (1981).
2. B. Ya. Zel'dovich and N. V. Tabiryan, Zh. Eksp. Teor. Fiz. **82**, 167 (1982) [Sov. Phys. JETP **55**, 99 (1982)].
3. N. F. Pilipetskiĭ, A. V. Sukhov, and B. Ya. Zel'dovich, Molec. Cryst. Liq. Cryst. **92**, 157 (1983).
4. H. G. Winfull, Phys. Rev. Lett. **49**, 1179 (1982).
5. B. Ya. Zel'dovich and N. V. Tabiryan, Zh. Eksp. Teor. Fiz. **79**, 2388 (1980) [Sov. Phys. JETP **52**, 1210 ((1980).
6. G. Barbaro and F. Simoni, Appl. Phys. Lett. **41**, 504 (1982).
7. S. R. Nersiyan, V. B. Pakhalov, N. V. Tabiryan, and Yu. S. Chilingaryan, in: Interaction of Laser Radiation with Liquid Crystals, Vol. 1, EGU, Erevan (1982), p. 197.
8. B. Ya. Zel'dovich, N. F. Pilipetskiĭ, A. V. Sukhov, and N. V. Tabiryan, Pis'ma Zh. Eksp. Teor. Fiz. **31**, 287 (1980) [JETP Lett. **31**, 263 (1980)].
9. R. N. Thurston, J. de Phys. **43**, 117 (1982).
10. G. Barbero and R. Barberi, J. de Phys. **44**, 609 (1983).
11. R. S. Akopyan, B. Ya. Zel'dovich, and N. V. Tabiryan, Zh. Eksp. Teor. Fiz. **83**, 1170 (1982) [Sov. Phys. JETP **56**, 1024 (1982)].
12. B. Ya. Zel'dovich and N. V. Tabiryan, Zh. Eksp. Teor. Fiz. **82**, 1126 (1982) [Sov. Phys. JETP **55**, 656 (1982)].
13. B. Ya. Zel'dovich, S. K. Marzlikin, N. F. Pilipetskiĭ, *et al.*, Pis'ma Zh. Eksp. Teor. Fiz. **37**, 568 (1983) [JETP Lett. **37**, 676 (1983)].
14. B. Ya. Zel'dovich and N. V. Tabiryan, Kvantovaya Elektron. **7**, 770 (1980) [Sov. J. Quantum Electron. **10**, 440 (1980)].
15. S. Chandrasekhar, *Liquid Crystals*, Cambridge Univ. Press (1977).
16. S. G. Odulov, Yu. A. Reznikov, O. K. Sarbeĭ, *et al.*, Ukr. Fiz. Zh. **25**, 1922 (1980).
17. S. G. Odulov, Yu. A. Reznikov, M. S. Soskin, and A. I. Khizhnyak, Zh. Eksp. Teor. Fiz. **85**, 1988 (1983) [Sov. Phys. JETP **58**, 1154 (1983)].
18. M. Brehm, H. Finkelmann, and H. Stegemeier, Phys. Chem. **78**, 883 (1974).
19. F. Fischer, Z. Natur. **31a**, 41 (1976).

[20]B. Ya. Zel'dovich and N. V. Tabiryan, Pis'ma Zh. Eksp. Teor. Fiz. **34**, 428 (1981) [JETP Lett. **34**, 406 (1981)].
[21]B. Ya. Zel'dovich and N. V. Tabiryan, Zh. Eksp. Teor. Fiz. **83**, 998 (1982) [Sov. Phys. JETP **56**, 557 (1982)].
[22]R. S. Akopyan, B. Ya. Zel'dovich, and N. V. Tabiryan, Kristallogr. **28**, 973 (1983) [Sov. Phys. Crystallogr. **28**, 576 (1983)].
[23]R. S. Akopyan and B. Ya. Zel'dovich, Zh. Eksp. Teor. Fiz. **83**, 2137 (1982) [Sov. Phys. JETP **56**, 1239 (1982)].
[24]V. V. Zheleznikov, V. V. Kocharovskiĭ, and Vl. V. Kocharovskiĭ, Usp. Fiz. Nauk **141**, 257 (1983) [Sov. Phys. Usp. **26**, 877 (1983)].

Translated by A. Mason

Orientational effect of an ordinary wave on a nematic with a hybrid orientation

B. Ya. Zel'dovich, N. F. Pilipetskiĭ, and A. V. Sukhov

Institute of Problems in Mechanics, Academy of Sciences of the USSR, Moscow

(Submitted April 9, 1986)
Kvantovaya Elektron. (Moscow) **14**, 202–203 (January 1987)

An experimental investigation was made of aberration self-focusing due to orientational deformation of a nematic with a hybrid orientation subjected to the field of an incident ordinary optical wave. When the radiation was incident on a sample from the homeotropic substrate side, the reorientation occurred at intensities higher than the threshold for a photoinduced Fréedericksz transition in a homeotropic cell of the same thickness, and it depended strongly on the angle of incidence of the ordinary wave on the sample. The effect was not observed when light was incident from the planar substrate side.

Reorientation of a nematic liquid crystal above a certain threshold value of the electric field of an optical wave, known as the photoinduced Fréedericksz transition (PFT)—recently discovered and investigated both experimentally[1] and theoretically[2,3]—exhibits a certain number of special features which are absent in the case of the rf analog of the Freedericksz effect.

Among the most interesting of these features are, in particular, a strong dependence of the threshold of the PFT induced in a homeotropic cell by an obliquely incident ordinary (o) wave on the angle of incidence of this wave and the oscillatory time dependence of the above-threshold distribution of the orientation in this geometry.[1,4,5] These effects are due to a considerable change in the state of polarization of the exciting radiation in a medium with a perturbed initial orientation, particularly by the lattice components of the orientational deformation in the field of interfering o and e (extraordinary) waves.

In view of this information, it seemed of interest to investigate the PFT in cells with an initially inhomogeneous orientation, particularly in a hybrid cell where the state of polarization of the exciting radiation is perturbed much more strongly than in a homeotropic cell of the same thickness. The PFT in a hybrid cell in the field of an ordinary wave has already been considered theoretically.[6] Unfortunately, even linearized equations for the near-threshold perturbation of the orientation in a hybrid cell have no analytic solutions, so that this treatment is limited to purely qualitative estimates.

Our aim was to investigate experimentally the orientational effect of an ordinary optical wave on a nematic with a hybrid orientation. We used the experimental geometry shown schematically in Fig. 1. An exciting o wave $\mathbf{E}_0 = \mathbf{e}_y E_0$ was directed from the side of a substrate with the homeotropic easy orientation axis $\mathbf{n}(0)$ oriented at an angle α to the substrate [wave vector $\mathbf{k} = k(\mathbf{e}_z \cos\alpha + \mathbf{e}_x \sin\alpha)$]. It should be pointed out that the sign of the birefringence α (and, consequently, of the angle of incidence α_0) was very important, as demonstrated later. We assumed that this sign was positive in the geometry of Fig. 1. In our experiments we used a hybrid cell with a nematic 4-*n*-pentyl-4-cyanobiphenyl (5CB) which was 140 μ thick; the homeotropic orientation on the input substrate was ensured by treatment with chromium stearyl chloride, whereas the planar orientation on the input substrate resulted from treatment in polyvinyl alcohol. The cell was slightly wedge-shaped (the wedge angle was in the YZ plane), which made it possible to determine easily the sign of α or, which was equivalent, the directional bending of the unperturbed distribution of the orientation. It was easy to show that constant-thickness fringes which appeared in transmitted light in crossed polarizers had a smaller spatial period in the $\alpha > 0$ case (in the geometry of Fig. 1) than for $\alpha' = -\alpha$.

The exciting radiation was in the form of a focused argon laser ($\lambda = 0.4880\,\mu$) beam consisting of a single transverse mode with a transverse size of the constriction (at midamplitude) of $a = 180\,\mu$ and with a power which could be varied continuously from zero to ~ 1 W.

The main results of our experiments were as follows. In a transmitted beam we observed the usual, for the PFT, aberration self-focusing rings. The polarization of these rings was linear and orthogonal to the incident polarization, i.e., it corresponded to the e wave in the liquid crystal.

Figure 2 shows the experimental dependences of the number N of self-focusing rings on the laser radiation power W obtained for different angles of incidence α_0 (the arrow on the abscissa corresponds to the specially measured threshold power for the PFT in a homeotropic cell with the same thickness and created in the same transverse cross section of the beam under normal incidence conditions).

For normal incidence ($\alpha_0 = 0$) the reorientation threshold was approximately three times higher than the PFT threshold in a homeotropic cell of the same thickness. It is clear from Fig. 2 that such reorientation (and the value of N) depended strongly on the angle of incidence, namely it increased on reduction in α_0. Moreover, in addition to the threshold reorientation, a weak subthreshold reorientation also took place. This could be explained qualitatively as follows: the power density of the exciting wave in our experiments was ~ 3 kW/cm^2, so that even weak ($\sim 1\%$) spontaneous scattering of the incident o wave into an e wave created an e wave of ~ 30 W/cm^2 power density, which was quite

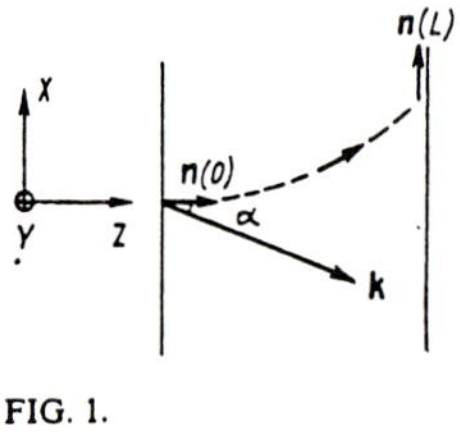

FIG. 1.

 0049-1748/87/010120-02$04.10

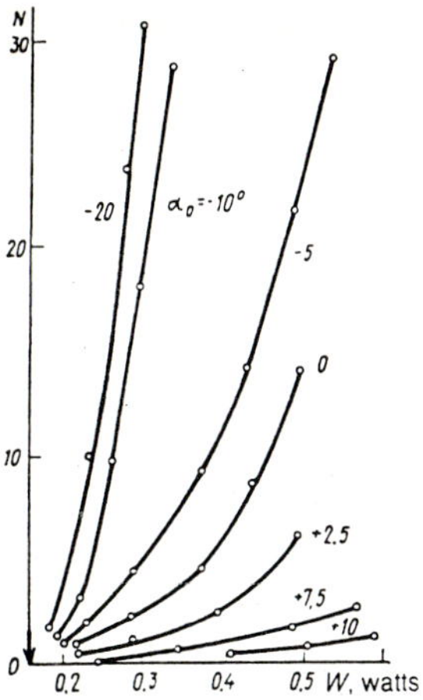

FIG. 2.

sufficient for weak threshold-free reorientation in the XZ plane (Fig. 1; see also Ref. 7).

It should also be mentioned that in the course of propagation of radiation from the side of the substrate to the planar easy orientation axis it was found, for exactly the same conditions, that reorientation was not observed in the investigated range of powers (up to 1.2 W) for any value of α_0. The possibility of such nonreciprocity was pointed out in Ref. 6.

Our experimental investigation of the orientational effect or wave on a nematic with a hybrid orientation thus revealed a number of specific features of the reorientation process in the adopted geometry and these features would require a theoretical analysis.

The authors are grateful to R. S. Akopyan and N. V. Takiryan for valuable discussions, and to V. A. Nikishkin for his help in the experiments.

[1]A. S. Zolot'ko, V. F. Kitaeva, N. Kroo, N. N. Sobolev, and L. Čillag (Csillag), Pis'ma Zh. Eksp. Teor. Fiz. **32**, 170 (1980) [JETP Lett. **32**, 158 (1980)].

[2]B. Ya. Zel'dovich, N. V. Tabiryan, and Yu. S. Chilingaryan, Zh. Eksp. Teor. Fiz. **81**, 72 (1981) [Sov. Phys. JETP **54**, 32 (1981)].

[3]B. Ya. Zel'dovich and N. V. Tabiryan, Zh. Eksp. Teor. Fiz. **82**, 1126 (1982) [Sov. Phys. JETP **55**, 656 (1982)].

[4]A. S. Zolot'ko *et al.*, Preprint No. 225 [in Russian], Lebedev Physics Institute, Academy of Sciences of the USSR, Moscow (1983).

[5]B. Ya. Zel'dovich, S. K. Merzlikin, N. F. Pilipetskiĭ, A. V. Sukhov, and N. V. Tabiryan, Pis'ma Zh. Eksp. Teor. Fiz. **37**, 568 (1983) [JETP Lett. **37**, 676 (1983)].

[6]B. Ya. Zel'dovich and N. V. Tabiryan, Zh. Eksp. Teor. Fiz. **90**, 141 (1986) [Sov. Phys. JETP **63**, 80 (1986)].

[7]G. Barbero and F. Simoni, Appl. Phys. Lett. **41**, 504 (1982).

Translated by A. Tybulewicz

 0049-1748/87/010121-02$04.10

Optically induced twist Fréedericksz transitions in planar-aligned nematic liquid crystals

E. Santamato,* G. Abbate,† and P. Maddalena*
Dipartimento di Fisica Nucleare, Struttura della Materia e Fisica Applicata, Padiglione 20, Mostra d'Oltremare, I-80125 Napoli, Italy

Y. R. Shen
Department of Physics, University of California, Berkeley, California 94720
(Received 10 April 1987)

The optical-field-induced Fréedericksz transition for a twist deformation by a normally incident laser beam in a planar-aligned nematic liquid crystal is studied. The Euler equation for the molecular director and the equations describing the evolution of the beam polarization in the birefringent medium are solved simultaneously in the small-perturbation limit. The stability of the undistorted state is investigated. An alternate series of stable and unstable bifurcations is found. This phenomenon has no analog in the Fréedericksz transition induced by dc electric and magnetic external fields.

In recent years, nonlinear interaction between a normally incident linearly polarized light beam and a homeotropically oriented nematic liquid-crystal film has received a great deal of attention.[1] The existence of a characteristic threshold intensity for induced molecular reorientation was experimentally well demonstrated.[2] The underlying physical mechanism for such an effect, known as the optical Fréedericksz transition, is essentially the same as in the corresponding dc Fréedericksz transition.[3] The geometry dictates that the polarization of the light beam remains unchanged in traversing the cell, even with molecular reorientation. There are a number of other dc Fréedericksz transitions with different geometries to which one may also find optical analogue. In most cases, however, the underlying physical processes of the dc- and optical-field-induced transitions are very different because the beam polarization varies in propagating through the medium. In this paper, we consider optical Fréedericksz transition in a planar-aligned cell induced by a light beam linearly polarized in a direction perpendicular to the molecular alignment. The induced molecular reorientation yields a change of birefringence seen by the beam, and consequently, the beam polarization changes continuously as it propagates through the cell. We must consider, simultaneously, the local action of the *elliptically polarized* light beam on the liquid-crystal molecules and the change of the beam polarization due to birefringence arising from molecular reorientation. It turns out that even the threshold behavior for the induced transition is characteristically unique. This is what we would like to focus on in the present paper. The transition is second order. The present case shows little resemblance to the corresponding dc case. The characteristics of our case are intimately related to the exchange of angular momentum between the light beam and the liquid-crystal medium.

Let us consider a planar nematic cell of thickness L, with the molecular alignment originally along $\hat{\mathbf{x}}$. A light beam, linearly polarized along $\hat{\mathbf{y}}$, is normally incident on the cell along $\hat{\mathbf{z}}$. If the optical field is sufficiently strong to reorient the liquid-crystal molecules, then the local molecular orientation is described by the unit director $\hat{\mathbf{n}}=(\cos\phi,\sin\phi,0)$, and the local beam polarization is described by the Stokes unit vector $\hat{\mathbf{s}}=(s_1,s_2,s_3)$ on the Poincaré sphere. Here, both $\hat{\mathbf{n}}$ and $\hat{\mathbf{s}}$ are functions of z and time t. The quantity s_3 denotes the polarization ellipticity: $s_3=(I_R-I_L)/(I_R+I_L)$, I_R and I_L being the intensities of the right- and left-handed circular components of the beam, respectively.

It has been shown that the evolution of the beam polarization of an optical wave propagating in an inhomogeneous uniaxial medium is approximately governed by the precession equation[4,5]

$$\partial\hat{\mathbf{s}}/\partial z=\mathbf{\Omega}\times\hat{\mathbf{s}}\ , \tag{1}$$

where, in the present case,

$$\mathbf{\Omega}=(2\pi/\lambda)\Delta n\,(\cos(2\phi),\sin(2\phi),0)\ ,$$

λ is the optical wavelength and $\Delta n=n_e-n_0$ is the difference between the extraordinary and the ordinary refractive indices of the medium. Equation (1) can be derived from Maxwell's equations in the slowly-varying envelope approximation in the limit of low birefringence $\Delta n/n$. For higher birefringence, one needs a more rigorous approach, in which the beam polarization should be described by a set of pseudo-Stokes parameters.[5] For conventional liquid crystals, $\Delta n/n\simeq 0.15$, and Eq. (1) is sufficiently accurate.

According to the angular momentum conservation law, the angular momentum lost by the light beam per unit time in traversing the medium must be equal to the optical torque exerted on the liquid-crystal molecules. The angular momentum carried by a beam with intensity I and ellipticity S_3 is $(IS_3/\hbar\omega)\hbar=(I\lambda/2\pi c)S_3$ per unit area and unit time. Then, the angular momentum per unit volume and unit time lost by the light beam between planes z and $z+dz$ is $(I\lambda/2\pi c)\partial S_3/\partial z$, which is also the optical torque per unit volume acting on the molecules. With the inertia term neglected, the equation of motion for the molecular reorientation is obtained by equating the optical torque to the viscous and elastic torques:

$$-\gamma_1(\partial\phi/\partial t)+k_{22}(\partial^2\phi/\partial z^2)+(I/c)(\lambda/2\pi)(\partial S_3/\partial z)=0\ , \tag{2}$$

where γ_1 is the viscosity coefficient and k_{22} the elastic constant for twist.

Equations (1) and (2) form a set of coupled nonlinear partial differential equations for the quantities $\hat{s}$ and ϕ, which must be solved with the initial and boundary conditions

$$\phi(z,0)=0\ ,\quad \phi(0,t)=\phi(L,t)=0\ ,\quad \hat{s}(0,t)=\hat{s}_0\ , \tag{3}$$

where $\hat{s}_0$ is the Stokes vector for the input light. For linear polarization along the y axis, we have $\hat{s}_0=(-1,0,0)$. It is easily seen that $\hat{s}=\hat{s}_0=(-1,0,0)$ and $\phi=0$ is a solution of Eqs. (1) and (2).[6] This corresponds to the propagation of an ordinary wave in the undistorted sample. This solution, however, becomes unstable when a critical intensity threshold I_{th} is reached. The bifurcation point can be found by using Lyapunov's first method for test of stability extended to continuous media as follows.

Equation (1) is linearized by taking ϕ, S_2, and S_3 as small quantities. With the introduction of the following normalized quantities:

$$u=z/L\ ,\quad \tau=t(k_{22}/\gamma_1 L^2)\ ,\quad \tilde{I}=(I/ck_{22}\Delta n)(\lambda/2\pi)^2\ ,\quad \tilde{L}=2\pi\,\Delta n\,L/\lambda\ , \tag{4}$$

the *linearized* set of equations becomes, for $0\le u\le 1$,

$$-\partial\phi/\partial\tau+(\partial^2\phi/\partial u^2)+\tilde{I}\tilde{L}(\partial S_3/\partial u)=0\ ,\quad \partial S_1/\partial u=0\ ,\quad \partial S_2/\partial u=-\tilde{L}S_3\ ,\quad \partial S_3/\partial u=\tilde{L}(S_2+2\phi)\ , \tag{5}$$

with $S_1\simeq S_1(0,t)=-1$. Although the set of Eq. (5) is not self-adjoint, we can still take a trial solution of the form

$$\phi(u,\tau)=\psi(u)\exp(\alpha\tau)\ ,\quad S_{2,3}=\sigma_{2,3}(u)\exp(\alpha\tau)\ . \tag{6}$$

Insertion of Eq. (6) into Eq. (5) yields a set of equations that can be solved by the standard method for eigenvalue problems. We report here only the results of the calculation. The dimensionless eigenvalue α is obtained from the transcendental equation

$$[(x_{22}-\tilde{L}^2)/(x_1^2-\tilde{L}^2)]\sin(x_2)/x_2=\sin(x_1)/x_1\ , \tag{7}$$

where $x_1(\alpha)$ and $x_2(\alpha)$ are the two roots of the biquadratic algebraic equation

$$(x^2+\alpha)(x^2-\tilde{L}^2)-2\tilde{I}\tilde{L}^2x^2=0\ . \tag{8}$$

The results show that α always appears to be real instead of complex. A plot of α as a function of $\tilde{I}$ for $\tilde{L}=5$ is presented in Fig. 1. At a given intensity I, there exists a discrete set of eigenvalues α_n $(n=0,1,2,\dots)$ with $\alpha_0\ge\alpha_1\ge\alpha_2\ge\cdots$.

According to Lyapunov's criterion for stability, the solution $\hat{s}=(-1,0,0)$, $\phi=0$ is stable if, and only if, all α_n have a negative real part. This is certainly not the case here. Let $\tilde{I}_m$ $(m=0,1,2,\dots)$ be the zeroes of $\alpha(I)$, as shown in Fig. 1. We have $\tilde{I}=\tilde{I}_m$ (m even) correspond to the points where the solution $\hat{s}=(-1,0,0)$, $\phi=0$ becomes unstable with further increase of $\tilde{I}$. The lowest threshold for instability is given by $\tilde{I}_0$. As seen in Fig. 1, there is a series of alternate stable and unstable intervals as the light intensity increases. In the regions $\tilde{I}<\tilde{I}_0$ and $\tilde{I}_m^*\le\tilde{I}\le\tilde{I}_{m+1}$ (m odd) the undistorted state $\phi=0$ is stable, such that any spontaneous fluctuations will be damped out in time. For $\tilde{I}_m\le\tilde{I}\le\tilde{I}_{m+1}$ (m even), however, the eigenvalue α_0 is positive, and the fluctuations will grow exponentially. The initial growth of the instability is governed by a time constant proportional to $1/\alpha_0$. For $\tilde{I}_m\le\tilde{I}\le\tilde{I}_m^*$ (m odd), the state $\phi=0$ is also unstable, but the initial growth of the fluctuations has a double-exponential character with two time constants proportional to $1/\alpha_0$ and $1/\alpha_1$, respectively. At the points $\tilde{I}=\tilde{I}_m^*$ (m odd) the eigenvalues α_0 and α_1 coalesce. In any case, in the unstable regions, deviation from the undistorted state will grow in time, until the nonlinear terms, neglected in Eq. (5), become significant and drive the system to a steady-state saturation regime. This behavior has no analog in the corresponding dc Fréedericksz transition.

The threshold intensities $\tilde{I}_m$, at which the solution $\hat{s}=(-1,0,0)$, $\phi=0$ switches from stable to unstable can be obtained as a function of $\tilde{L}$ by setting $\alpha=0$ in Eq. (8). We find $x_1=0$ and $x_2=\tilde{L}\sqrt{1+2\tilde{I}}$, so that Eq. (7) reduces to

$$2\tilde{I}_{th}\left[\frac{\sin(\tilde{L}\sqrt{1+2\tilde{I}_{th}})}{\tilde{L}\sqrt{1+2\tilde{I}_{th}}}\right]=-1\ , \tag{9}$$

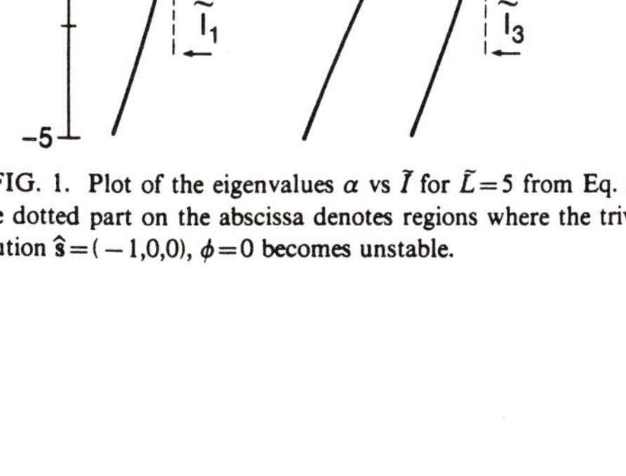

FIG. 1. Plot of the eigenvalues α vs $\tilde{I}$ for $\tilde{L}=5$ from Eq. (7). The dotted part on the abscissa denotes regions where the trivial solution $\hat{s}=(-1,0,0)$, $\phi=0$ becomes unstable.

from which $\tilde{I}_{th}(\tilde{L})$ can be found. This is plotted in Fig. 2. For any fixed $\tilde{L}$, we have a discrete set of values $\tilde{I}_{th}=\tilde{I}_m$ ($m=0,1,2,\ldots$), corresponding to the zeros of α [for example, the zeroes of α in Fig. 1 correspond to the intersecting points of the $\tilde{L}=5$ line with the curves $\tilde{I}_{th}(\tilde{L})$ in Fig. 2]. The solid curves in Fig. 2 describe $\tilde{I}_m(\tilde{L})$ with even m and the dashed curves $\tilde{I}_m(\tilde{L})$ with odd m. Only the former denotes the threshold intensities at which the system switches from stable to unstable. The lowest threshold for a given $\tilde{L}$ corresponds to $\tilde{I}_{th}=\tilde{I}_0$. The result can be better understood if we know the molecular orientation (eigenfunction) associated with the present eigenvalue problem.

For a given L, we find the twisted molecular reorientation corresponding to the eigenvalue α_n to be

$$\phi_n(u)=A\left[\frac{\sin[x_1(\alpha_n)u]}{\sin[x_1(\alpha_n)]}-\frac{\sin[x_2(\alpha_n)u]}{\sin[x_2(\alpha_n)]}\right],\quad 0\le u\ (\equiv z/L)\le 1 \qquad (10)$$

where A is a constant. This eigenfunction $\phi_n(u)$ vanishes at $u=0$ and L, and exhibits a number of sinusoidal-like oscillations in between. A plot of the eigenfunction $\phi_0(u)$ is shown in Fig. 3 for $\tilde{I}$ just above the minimum threshold $\tilde{I}_0$ and $\tilde{L}=2$, 5, and 10. We see that the molecular distortion from the unperturbed state is an oscillating function having a number of zeros, which is of the order of $\tilde{L}$. This is also generally true for ϕ_n. Since $\tilde{L}$ is close to (L/λ), the number of oscillations is usually very high, unless very thin samples are used. For this reason, the free energy associated with $\phi_0(u)$ is relatively large and the corresponding threshold intensity I_0 is expected to be high, as compared with the optical Freédericksz transition in a homeotropic sample having the same thickness. Also, $\tilde{I}_0$ is larger for larger $\tilde{L}$, as seen in Fig. 2. The smallest $\tilde{I}_0$ occurs at $\tilde{L}\cong 2$, where the molecular distortion ϕ_0 has only half an oscillation between $z=0$ and L (Fig. 3). For $\tilde{L}<2$, $\tilde{I}_0$ is proportional to $1/L^2$ for sufficiently small $\tilde{L}$. This latter behavior is similar to that

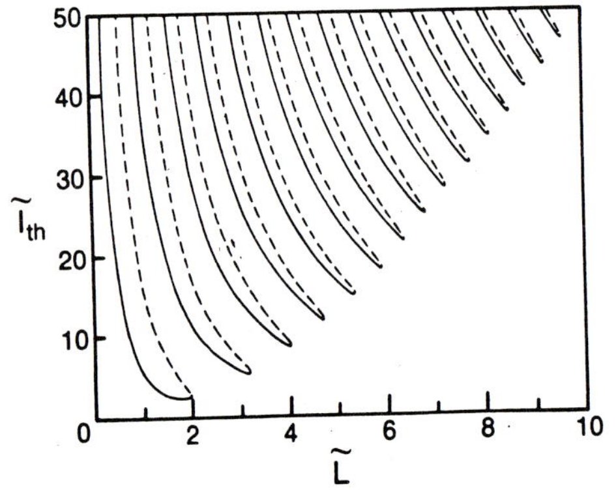

FIG. 2. Threshold intensity $\tilde{I}_{th}$ vs sample thickness $\tilde{L}$. The intersecting points of an $\tilde{L}$=constant line with the solid curves yield the threshold intensities where the solution switches from stable to unstable, for that given $\tilde{L}$. Dashed lines correspond to unstable solutions.

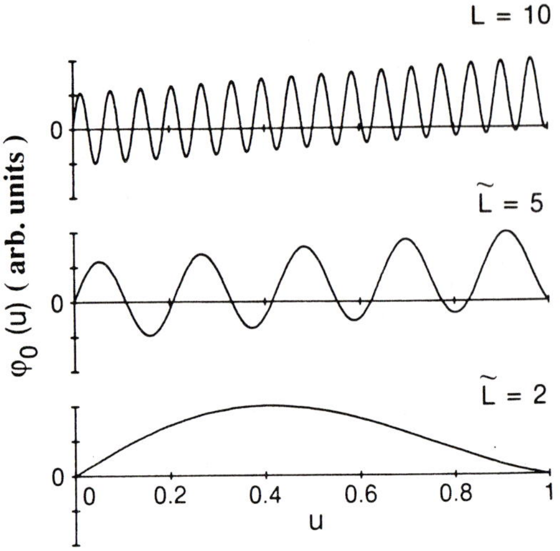

FIG. 3. Director reorientation $\phi_0(u)$ ($0\le u\le 1$) as a function of position u in the sample at the minimum threshold intensity $\tilde{I}=\tilde{I}_0$ for three different sample thicknesses $\tilde{L}=2$, 5, and 10.

of the corresponding dc case and also to that of the optical Fréedericksz transition in a homeotropic sample, as one would expect from the distortion profile.

For a typical nematic sample ($n_0=1.5$; $n_e=1.7$; $k_{22}/k_{33}\simeq 0.6$) with a thickness of 100 μm and $\lambda=0.5$ μm, we find $\tilde{L}=250$ and I_{th} (homeotropic)/I_0 (planar)$\cong 1.5\times 10^{-4}$. Only for very thin samples with L being of the order of unity, the two thresholds are of comparable magnitude, but then they are both very high ($\simeq 1$ MW/cm^2). We note, however, that in the present case, if Δn is smaller, then for a given sample thickness, both $\tilde{L}$ and the threshold intensity I_0 are smaller. Considering the case with $\Delta n=10^{-3}$ and $k_{22}=2.5\times 10^{-7}$ dyne, we find, for $L=160$ μm (corresponding to $\tilde{L}=2$) and $\lambda=0.5$ μm, a threshold intensity $I_0=30$ kW/cm^2, which could be obtained in reality by a focused CW laser beam.

The above analysis gives no information about the character of the transition, i.e., whether it is first order or second order. To answer this question, we have solved numerically the set of nonlinear equations (1) and (2) in the stationary case ($\partial\phi/\partial t=0$ and $\partial\hat{s}/\partial t=0$). In Fig. 4 the ellipticity S_3 of the output beam is plotted as a function of the input intensity $\tilde{I}$. The stable branches of stationary solutions are drawn in solid lines, while the unstable branches are dashed. An alternate series of stable and unstable bifurcations is found, according to Fig. 1. The curve in Fig. 4 clearly shows that the transition is second order.

When the intensity $\tilde{I}$ gradually increases from zero, the system remains undistorted with $\hat{s}=(-1,0,0)$ and $\phi=0$ until the first branching point $\tilde{I}=\tilde{I}_0$ is reached. At this point, a second-order optical twist Fréedericksz transition occurs and the output ellipticity begins to deviate from zero, following the solid line in Fig. 4. To reach the second steady-state branch in Fig. 4, the intensity $\tilde{I}$ should be suddenly switched to the stable region and then

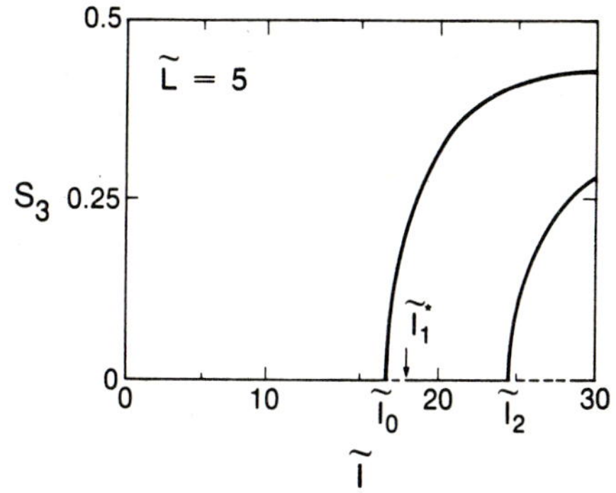

FIG. 4. Ellipticity s_3 of the transmitted light beam through the sample as a function of $\tilde{I}$ for $\tilde{L}=5$. The dashed part on the horizontal abscissa corresponds to the unstable solutions shown in Fig. 1.

gradually increased. At the second branching point $\tilde{I}=\tilde{I}_2$, the system again undergoes a second-order Fréedericksz transition and moves onto the second steady-state branch. If $\tilde{I}$ is slowly decreased instead, then as it reaches the end of the stable interval $\tilde{I}=\tilde{I}_1^*$, the system should switch to the first steady-state branch via a first-order transition. Note that in the stable regions (e.g., $\tilde{I}_1^* < \tilde{I} < \tilde{I}_2$), the system is intrinsically bistable, with the two stable solutions given by the unperturbed state $\hat{s}=(-1,0,0)$, $\phi=0$ and the nonlinearly perturbed state on the steady-state branch. If $\tilde{I}$ is suddenly switched to an unstable region, the system would vary in time until the corresponding nonlinearly perturbed state on the steady-state branch is reached.

The present theory applies also to smectic-*c* liquid crystals in Rapini's *N*-configuration.[7] In this case, the following formal substitutions should be used:

$$\Delta n = n_0 \left[\frac{n_e}{(n_e^2\cos^2\theta + n_0^2\sin^2\theta)^{1/2}} - 1 \right],$$

$$\tilde{I} = \frac{(I/ck_{22}\Delta n)(\lambda/2\pi)^2}{(1-k\cos^2\theta)\sin^2\theta},$$

$$k = (k_{33}-k_{22})/k_{22},$$

where k_{33} is the bending elastic constant and θ is the tilt angle characteristic of the smectic material. Samples with small θ are expected to have lower threshold $\tilde{I}_{th}$, which is roughly scaled by the factor $\sin^2\theta$.

This work was partially supported by Consiglio Nazionale delle Ricerche and Ministero della Pubblica Instruzione, Italy, and partially by National Science Foundation, Solid State Chemistry, Grant No. DMR8414053.

*Also at Centro Interuniversitario di Elettronica Quantistica e Plasmi, I-80125 Napoli, Italy.

†Also at Centro Interuniversitario di Struttura della Materia, I-80125 Napoli, Italy.

[1]See, for example, H. L. Ong, Phys. Rev. A **28**, 2393 (1983), and references therein.

[2]S. D. Durbin, S. M. Arakelian, and Y. R. Shen, Phys. Rev. Lett. **47**, 1411 (1981).

[3]V. Fréedericksz and V. Zolina, Trans. Faraday Soc. **29**, 919 (1933). For a review on the dc-field-induced Fréedericksz transition in nematics, see H. J. Deuling, in *Liquid Crystals*, edited by L. Liebert (Academic, New York, 1978), p. 77.

[4]H. Kubo and R. Nagata, J. Opt. Soc. Am. **73**, 1719 (1983).

[5]E. Santamato and Y. R. Shen, J. Opt. Soc. Am. A **4**, 356 (1987).

[6]Another solution is $\hat{s}=(+1,0,0)$; $\phi\equiv0$. This solution corresponds to a light linearly polarized along the molecular director. This solution is always stable.

[7]A. Rapini, J. Phys. (Paris) **33**, 237 (1972).

Optically induced molecular reorientation in a smectic-*C* liquid crystal

Hiap Liew Ong and Charles Y. Young*
Department of Physics, Brandeis University, Waltham, Massachusetts 02254
(Received 19 July 1983)

The Euler equations for describing the director in the optically induced molecular reorientation of a smectic-*C* liquid crystal in an external magnetic field are presented. The optical alignment effect is shown to be localized and not to produce point defects. Analytic expressions are given explicitly in the small-distortion regime, which, generally, have the form of zeroth-order Bessel functions. Threshold behavior exists if the polarization of the optical beam is normal to the magnetic field. For a magnetic field of the order of 1 kG, the threshold power varies from 3 to 120 mW for a typical smectic-*C* sample with a laser spot size of about 1 to 100 μm, but is independent of the thickness of the sample. For polarizations oblique to the magnetic field, there is no threshold and the amplitude of rotation of the director depends on spot size, laser power, and the angle between laser polarization and external magnetic field. The transient response of molecular reorientation to the laser switch on is shown to have exponential time dependence for a normally polarized incident beam with an incident intensity greater than the threshold intensity. The response time of such a reorientation is of the order of milliseconds, depending on incident laser intensity. We propose that, experimentally, the molecular reorientation can be quantitatively measured by the reflectivity or transmissivity of a normally incident probe beam. Optical reflectivity and transmissivity from a typical smectic-*C* film are also calculated.

I. INTRODUCTION

Recently the nonlinear optics of liquid crystals has received a great deal of attention.[1] Shelton and Shen initiated the study of the normal and umklapp optical third-harmonic generation in cholesteric liquid crystals (CLC).[1–3] The orientational optical nonlinearity of CLC was recently considered by Tabiryan and Zel'dovich[4] in 1981 and by Winful[5] in 1982. Tabiryan and Zel'dovich showed that when a light wave propagates along the helical axis, self-focusing should not occur for circularly polarized light, but for linearly polarized light the nonlinear dielectric constant ϵ_2 is about 8×10^{-8} cm^3/erg. The elliptic-function solutions in the Bragg regime have been obtained by Winful and the results show that optically induced changes in the pitch of the cholesteric helix lead to a bistable reflection even in the absence of external reflectors. In the nematic-liquid-crystal (NLC) phase, the optically induced Freedericksz transition has received even more attention. The effect of the optically induced molecular reorientation in the NLC was explained qualitatively by Zolot'ko *et al.* in 1980.[6] A quantitative theory was later constructed by Zel'dovich, Tabiryan, and Chilingaryan using the geometrical-optics approximation.[7] In 1981, Durbin, Arakelian, and Shen reported the first observation of the optically induced Freedericksz transition in nematic 5CB (4-cyano-4'-pentylbiphenyl) and showed that the results were in quantitative agreement with their theoretical predictions using the infinite-plane-wave approximation.[8] In 1981 Khoo also presented an approximate solution and made a quantitative experimental verification of the associated nonlinear optical processes.[9] Recently the exact solution describing the orientation of the NLC molecule was obtained by Ong.[10] These studies show that for certain NLC's with large dielectric and elastic anisotropies, the transition can be first order accompanied by hysteresis, and that the optical nonlinearity of NLC's is larger by eight to ten orders of magnitudes than that of carbon disulfide (CS_2). The term "gigantic optical nonlinearity" (GON) has been coined by Zel'dovich for such large nonlinearities.

Contrary to cholesterics and nematics, not much work has been done regarding the effects of external fields on smectics. The possibility of a Freedericksz transition in a smectic liquid crystal by an external dc magnetic field has been considered by Helfrich,[11] Rapini,[12] Hurault,[13] and Meirovitch *et al.*[14,15] The results show that all transitions requiring a distortion of the layers are probably not observable because the distortion is a very weak function of the field. Such a transition has therefore been called a "ghost" by Rapini and has not been observed experimentally. However, those magnetic-field-induced transitions which involve the rotation of the director about the normal to the smectic-*C*-liquid-crystal (Sm*C*) layers should be observable, as discussed by Rapini[12] and by Meirovitch *et al.*[14] Indeed, magnetic fields have been used to align the azimuthal angle of the Sm*C* experimentally.[14,15] The effects of weak anchoring between the smectics and the surfaces on the dc-field-induced Freedericksz transition have also been studied by Meirovitch *et al.*[14]

In contrast, using a linearly polarized light source with the electric field directed in the plane of the layer at an angle to the initial orientation, it is possible to reorient the azimuthal component of the director of the Sm*C* molecules. This transition involves only a rotation of the director about the normal to the layers and does not involve any distortion in the layer. Indeed, the optically induced molecular reorientation in the SmC was recently directly observed by Lippel and Young using a linearly polarized laser beam incident on a freely suspended film of a SmC.[17] They reported that large reorientations have been observed with an incident optical power of less than 50 mW and have presented a simple theory explaining the reorientation effect. The theoretical study of the orientational optical nonlinearity of SmC induced by an optical field in the presence of an external orienting dc magnetic field has been considered by Tabiryan and Zel'dovich who showed that the effective nonlinear dielectric constant ϵ_2 is ~ 0.2 cm^3/erg for the self-focusing light and $\chi_{\text{eff}} \sim 0.6\times10^{-6}$ cm^3/erg for the four-wave interaction.[18] These nonlinear constants are, respectively, nine and five orders larger than those of CS_2. However, the solution to the spatial orientation of the director of the SmC has not been found.

It is the purpose of this paper to present the Euler equations for describing the director in the optically induced molecular reorientation in a SmC sample. The sample is assumed initially oriented by a homogeneous dc magnetic field in the SmC layer so that without the optical field, the azimuthal angle of the SmC is well aligned in the magnetic field direction. A polarized light beam is then normally incident on the sample. If the polarization is at an angle to the magnetic field, the azimuthal angle of the director will vary under the action of the optical field. We discuss the general properties of the solution for the director and give analytic expressions explicitly in the small-distortion linearized regime. The results show that the optical alignment effect is localized and does not produce point defects. Generally, the solutions in the small-distortion regime have the form of zeroth-order Bessel functions. Using the continuity condition at the boundary imposed by the spot size of the optical field, the amplitudes of the deformations can be determined. In regions far away from the optical field, the azimuthal angle approaches its asymptotic orientation exponentially $\sim e^{-qr}/\sqrt{qr}$ where q is the inverse magnetic field coherence length. If the polarization is normal to the orienting magnetic field, there exists a characteristic threshold intensity below which no molecular reorientation can be induced. The threshold power depends on the applied magnetic field and the laser spot size, but is independent of the sample thickness. For a magnetic field of the order of 1 kG, the threshold power varies from 3 to 120 mW for a typical SmC sample with a laser spot size of about 1 to 100 μm in radius. We also discuss the dynamics of the transition. The transient response of molecular reorientation to the laser switch on is shown to have exponential time dependence for a normally polarized incident beam with an incident intensity greater than the threshold intensity. We propose that, experimentally, the molecular reorientation can be quantitatively measured by the reflectivity or transmissivity of a normally incident probe beam. The reflected and transmitted power of the probe beam covering a known area are derived.

In the following sections, we first discuss the free energy density and the Euler equations (Sec. II). A section on the solution describing the orientation of the director in the small-distortion regime follows (Sec. III), including a general discussion of the solutions at the origin and at infinity. Finally, proposed methods to observe the optical nonlinearity effects experimentally are discussed in Sec. IV.

II. FREE ENERGY DENSITY AND EULER EQUATIONS

The elastic theory of SmC has been considered by the Orsay group[19] and by Rapini.[12] Both approaches use the Oseen description of smectic A, neglecting all changes in internal parameters such as density, interlayer distance, and tilt angle. The Orsay group used the Lagrangian description for the elastic strains with a vector $\vec{\Omega}(\vec{r})$ describing the local rotation of the director. In order to be consistent with the elastic theory for NLC's, we shall use the Eulerian description developed by Rapini based on the director.

We introduce two unit vectors $\hat{k}$ and $\hat{n}$ to describe the SmC structure. $\hat{k}$ is normal to the layers and $\hat{n}$ is along the "long axis" of the molecules as shown in Fig. 1. In

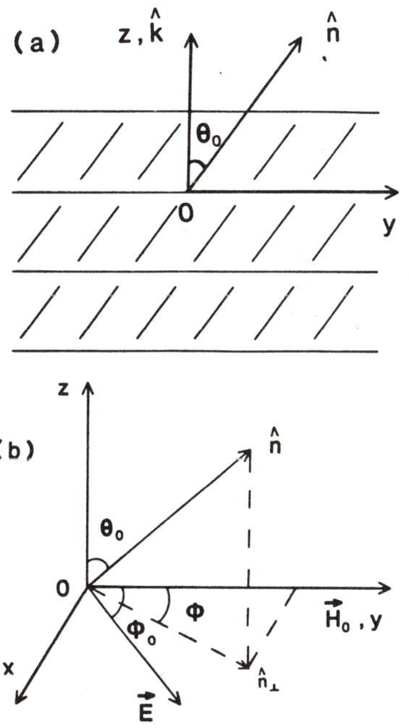

FIG. 1. Assumed structure of a smectic-C liquid crystal. (a) A homogeneous dc magnetic field H_0 is directed along the y direction so that the unperturbed state of the SmC sample has $\hat{k}$ along the z axis and $\hat{n}=\hat{n}_0$ in the y-z plane making an angle θ_0 with the z axis. (b) An orienting optical beam is linearly polarized in the x-y plane with the electric field directed at an angle ϕ_0 to the y axis and is normally incident on the SmC sample. In equilibrium, the director $\hat{n}$ of the SmC is oriented at an azimulthal angle ϕ with the y axis with the polar angle θ_0 being fixed.

the following discussion, an orienting uniform dc magnetic field $\vec{H}_0=(0,H_0,0)$ directed along the y direction is applied to the sample so that in the absence of an optical field, the preferred azimuthal direction is along the y axis. The unperturbed state then has $\hat{k}$ along the z axis, and $\hat{n}=\hat{n}_0$ in the z-y plane making an angle θ_0 with $\hat{k}$. It is convenient to introduce a vector $\hat{n}_\perp\equiv\hat{n}-(\hat{n}\cdot\hat{k})\hat{k}$ which lies in the plane normal to $\hat{k}$ and which satisfies $\hat{n}_\perp^2=\text{const}$.

When a light beam linearly polarized in the x-y plane with the elastic field directed at an angle ϕ_0 to the y axis is normally incident on the SmC layer, there is an additional term in the total free energy density of the system due to the optical field. For the discussion of the optical-field-induced Freedericksz transition, the total free energy density F (erg/cm^3) can be written as

$$F=\tfrac{1}{2}\alpha_{11}(\text{div}\hat{n}_\perp)^2+\tfrac{1}{2}\alpha'_{22}(\hat{n}_\perp\cdot\text{curl}\hat{n}_\perp)^2+\tfrac{1}{2}\alpha_{33}(\hat{k}\cdot\text{curl}\hat{n}_\perp)^2-\alpha'_{23}(\hat{n}_\perp\cdot\text{curl}\hat{n}_\perp)(\hat{k}\cdot\text{curl}\hat{n}_\perp)-\frac{1}{8\pi}(\vec{E}\cdot\vec{D}+\vec{B}\cdot\vec{H})-\tfrac{1}{2}\chi_a(\hat{n}\cdot\vec{H}_0)^2\,, \tag{2.1}$$

where χ_a is the magnetic susceptibility anisotropy, $\alpha'_{ij}=\alpha_{ij}\sin^2\theta_0$, and α_{ij} are the smectic curvature elastic moduli.[12] The smectic curvature elastic moduli have the same dimensions and the same magnitudes as the Frank elastic constants for NLC.[16,20] The first three terms of Eq. (2.1) are analogous to splay, twist, and bend in a NLC. These types of deformations were first discussed by Saupe.[21] The term $-(\vec{E}\cdot\vec{D}+\vec{B}\cdot\vec{H})/8\pi$ is the electromagnetic energy density of the light beam and the last term is the dc magnetic energy density. The Euler equations for the director $\hat{n}(\vec{r})$ have the form

$$\frac{\delta F}{\delta n_i}-\frac{\partial}{\partial x_k}\frac{\delta F}{\delta(\partial n_i/\partial x_k)}=\lambda n_i\,, \tag{2.2}$$

where $\lambda(\vec{r})$ is an undetermined Lagrange multiplier which ensures that the condition $|\hat{n}|=1$ is satisfied. The orientation of the SmC sample is then completely described by the solution to Eq. (2.2) subject to initial and boundary conditions.

From the results of Rapini[12] on the dc-field-induced Freedericksz transition, which show that one needs an extremely high field to vary the polar angle of the SmC sample, it is reasonable for us to assume that the polar angle θ_0 remains fixed. We denote the angle between the director and the y axis by ϕ [Fig. 1(b)], then the director is given by

$$\hat{n}=(\sin\theta_0\sin\phi,\sin\theta_0\cos\phi,\cos\theta_0)\,,$$

and (2.3)

$$\hat{n}_\perp=(\sin\theta_0\sin\phi,\sin\theta_0\cos\phi,0)\,.$$

The elastic deformation energy density F_d is then given by

$$F_d=\tfrac{1}{2}\sin^2\theta_0(\alpha_{11}\cos^2\phi+\alpha_{33}\sin^2\phi)\left[\frac{\partial\phi}{\partial x}\right]^2+\tfrac{1}{2}\sin^2\theta_0(\alpha_{11}\sin^2\phi+\alpha_{33}\cos^2\phi)\left[\frac{\partial\phi}{\partial y}\right]^2+\tfrac{1}{2}\sin^2\theta_0\sin(2\phi)(\alpha_{33}-\alpha_{11})\frac{\partial\phi}{\partial x}\frac{\partial\phi}{\partial y}\,. \tag{2.4}$$

Experimentally, the exact values of the elastic constants of SmC have not yet been measured but are of the same order of magnitudes as NLC's,[16,20] and since the equations that arise from unequal elastic constants are very complicated,[22] we shall use a single-elastic-constant approximation, $\alpha_{11}=\alpha_{33}=\alpha$, to simplify the discussion. Then the deformation energy is reduced to the form

$$F_d=\tfrac{1}{2}\alpha\sin^2\theta_0(\text{grad}\phi)^2\,. \tag{2.5}$$

For the optical field, the electric and magnetic energy densities are equal[23] so that the total electromagnetic energy density can be written as $F_{\text{opt}}=-\vec{E}\cdot\vec{D}/4\pi$. It has been shown that SmC is nearly uniaxial with the optical axis along $\hat{n}$.[16,24] Using the uniaxial dielectric tensor[16]

$$\epsilon_{ij}=\epsilon_\perp\delta_{ij}+\epsilon_a n_i n_j \tag{2.6}$$

for the SmC, $\hat{\epsilon}$ can be written as

$$\hat{\epsilon}=\begin{bmatrix}\epsilon_\perp+\epsilon_a\sin^2\theta_0\sin^2\phi & \epsilon_a\sin^2\theta_0\sin\phi\cos\phi & \epsilon_a\sin\theta_0\cos\theta_0\sin\phi\\ \epsilon_a\sin^2\theta_0\sin\phi\cos\phi & \epsilon_\perp+\epsilon_a\sin^2\theta_0\cos\phi & \epsilon_a\sin\theta_0\cos\theta_0\cos\phi\\ \epsilon_a\sin\theta_0\cos\theta_0\sin\phi & \epsilon_a\sin\theta_0\cos\theta_0\cos\phi & \epsilon_\perp+\epsilon_a\cos^2\theta_0\end{bmatrix} \tag{2.7}$$

where $\epsilon_a=\epsilon_\parallel-\epsilon_\perp$, and $\epsilon_\perp$ $(=n_0^2)$ and $\epsilon_\parallel$ $(=n_e^2)$ are the dielectric constants perpendicular and parallel to the local director, respectively, at the incident optical-field frequency. Thus, the total electromagnetic energy density of the optical field F_{opt} can be written as

$$F_{\text{opt}}=-\frac{1}{8\pi}[\epsilon_\perp|\vec{E}|^2+\epsilon_a\sin^2\theta_0|\vec{E}|^2\cos^2(\phi_0-\phi)]\,. \tag{2.8}$$

Similarly, the magnetic energy density of the dc magnetic field can be written as

$$F_H=-\tfrac{1}{2}\chi_a\sin^2\theta_0 H_0^2\cos^2\phi \,. \tag{2.9}$$

Consequently, the total free energy density of the SmC can be written as

$$F=\tfrac{1}{2}\alpha\sin^2\theta_0(\mathrm{grad}\phi)^2 -\frac{1}{8\pi}[\epsilon_\perp|\vec{E}|^2+\epsilon_a|\vec{E}|^2\sin^2\theta_0\cos^2(\phi_0-\phi)] -\tfrac{1}{2}\chi_a H_0^2\sin^2\theta_0\cos^2\phi \,. \tag{2.10}$$

In general, the fields produced by a monochromatic wave incident upon a slab of SmC consists of four partial waves.[25] The fields as well as the polarization of the optical beam in the SmC clearly depends on the orientation of the SmC. The complete determination of the electromagnetic fields in the SmC is complicated and will be left for further investigation. To simplify the discussion, we shall fix the amplitude and polarization of the electric field. Then the term $-\epsilon_\perp|\vec{E}|^2/8\pi$ in the optical energy makes no contribution to the Euler equation. By the cylindrical symmetry of the problem, the relevant total free energy of the sample can be written as

$$\mathcal{F}=d\sin^2\theta_0\int_0^\infty \tilde{F}\,dr$$

with the final planar total free energy density per area in the x-y plane $\tilde{F}$ (erg/cm^2) given by

$$\tilde{F}=\begin{cases}\dfrac{1}{2}\left[\alpha\left[\dfrac{d\phi}{dr}\right]^2+\dfrac{\epsilon_a|\vec{E}|^2}{4\pi}\sin^2(\phi-\phi_0)+\chi_a H_0^2\sin^2\phi\right]r, & r\le r_0\\[2ex] \dfrac{1}{2}\left[\alpha\left[\dfrac{d\phi}{dr}\right]^2+\chi_a H_0^2\sin^2\phi\right]r, & r>r_0\end{cases} \tag{2.11}$$

where d is the thickness of the sample and r_0 is the cutoff radius defined by the profile of the intensity of light beam I:

$$I=\begin{cases}\text{const}, & r\le r_0\\ 0, & r>r_0\,.\end{cases} \tag{2.12}$$

The resulting Euler equations take the forms

$$\frac{\partial^2\phi}{\partial r^2}+\frac{1}{r}\frac{\partial\phi}{\partial r}-\tfrac{1}{2}[u^2\cos(2\phi_0)+q^2]\sin(2\phi) +\tfrac{1}{2}u^2\sin(2\phi_0)\cos(2\phi)=0 \quad \text{for } r\le r_0 \tag{2.13a}$$

and

$$\frac{\partial^2\phi}{\partial r^2}+\frac{1}{r}\frac{\partial\phi}{\partial r}-\tfrac{1}{2}q^2\sin(2\phi)=0 \quad \text{for } r>r_0 \tag{2.13b}$$

where $u\equiv|\vec{E}|\sqrt{\epsilon_a/4\pi\alpha}$ and $q\equiv H_0\sqrt{\chi_a/\alpha}$ are the inverse optical and magnetic coherence lengths. Since the intensity of the incident field in related to the electric field through $I=cn_0|\vec{E}|^2/8\pi$, we have $u^2=2\epsilon_a I/cn_0\alpha$. For a typical SmC liquid crystal, $\chi_a\sim10^{-7}$ in cgs units, $\alpha\sim0.5\times10^{-6}$ dyn, and $\epsilon_a\sim0.6$, we have $q^2\sim0.2H_0^2$ and $u^2\sim0.5I$, where I is expressed in mW/cm^2. For a spot size of 100 μm in radius and a magnetic field of about 1 kG, $q\sim450$ cm^{-1} and $qr_0\sim4.5$, $u\sim400$ cm^{-1} for an optical field of about 100 mW in total power $P=\pi r_0^2 I$. In the following, we consider solutions which minimize the total free energy. Consequently, ϕ must be a continuous function of r and must tend to zero (i.e., aligned in the magnetic field direction) as $r\to\infty$. At the boundary $r=r_0$, the solution and its first derivative must be continuous. Since $\cos[2(2\pi-\phi_0)]=\cos(\pm2\phi_0)$ and $\sin[2(2\pi-\phi_0)]=\sin(-2\phi_0)=-\sin(2\phi_0)$, the orienting effects of an optical field with polarization ϕ_0 is the same as that of an optical field with polarization $-\phi_0$ or $2\pi-\phi_0$, except that ϕ changes sign, i.e., if $\phi_0\to-\phi_0$ or $\phi_0\to2\pi-\phi_0$, then $\phi(r)\to-\phi(r)$.

Equations (2.13) are nonlinear in ϕ and general solutions cannot be found except by numerical means.[22] However, by considering small distortions, the equations can be linearized and solved. Physically, there are now two competing alignment fields: the external magnetic field and the optical polarization field. It is these two fields together with the elastic energy that determines the spatial orientation of the SmC director.

III. GENERAL PROPERTIES AND SMALL-DISTORTION-REGIME SOLUTION

A. General properties of the solution at the origin and at infinity

We first formulate the general conditions for which the solution must satisfy. From the form of Eq. (2.13a), we can establish the general behavior of ϕ at the origin. Equation (2.13a) is exactly satisfied by the constant deformation angle determined from the following equation

$$\tan(2\phi)=u^2\sin(2\phi_0)/[u^2\cos(2\phi_0)+q^2] \tag{3.1}$$

and so we may have $\partial\phi/\partial r=0$ at $r=0$. Since ϕ is bounded, we deduce that

$$\lim_{r\to0}\left[r\frac{\partial^2\phi}{\partial r^2}+\frac{\partial\phi}{\partial r}\right]=0\,. \tag{3.2}$$

Suppose that in the neighborhood of $r=0$ there exists a series

$$\frac{\partial\phi}{\partial r}=\sum_{n=0}^{\infty}C_n r^n\,. \tag{3.3}$$

Then

$$\lim_{r\to 0}\sum_{n=0}^{\infty} C_n r^n = -\lim_{r\to 0}\sum_{n=1}^{\infty} nC_n r^n , \quad (3.4)$$

from which follows that $C_0=0$, so that the first derivative of the solution with respect to r must vanish at the origin:

$$\frac{\partial\phi}{\partial r}=0 \text{ at } r=0 . \quad (3.5)$$

Thus, there is no point defect at the origin. Consequently, the exponential functions $\exp(ar)$, the normal and modified Bessel functions of the second kind, $Y_n(ar)$ and $K_n(ar)$, cannot be the solutions of the deformation angle for $r<r_0$, where a is a characteristic wave vector and n is any integer.

We now consider the asymptotic behavior of the deformation angle. The constribution to $\mathcal{F}$ for $r>r_0$ is

$$\mathcal{F}_> = \tfrac{1}{2} ad \sin^2\theta_0 \int_{r_0}^{\infty} \left[\left[\frac{\partial\phi}{\partial r}\right]^2 + q^2\sin^2\phi \right] r\,dr . \quad (3.6)$$

Therefore, the solution must tend to $e^{-qr}/\sqrt{qr}$ so that the integral converges at infinity. Since the solution approaches zero for large r, the optical alignment effect is localized.

B. Linearized regime

We now consider the equilibrium orientation state in the linearized regime which can be described by linearizing Eqs. (2.13a) and (2.13b):

$$\frac{\partial^2\phi}{\partial r^2}+\frac{1}{r}\frac{\partial\phi}{\partial r}-g^2\phi+\tfrac{1}{2}u^2\sin(2\phi_0)=0 \text{ for } r\le r_0 \quad (3.7a)$$

and

$$\frac{\partial^2\phi}{\partial r^2}+\frac{1}{r}\frac{\partial\phi}{\partial r}-q^2\phi=0 \text{ for } r>r_0 , \quad (3.7b)$$

where $g^2\equiv u^2\cos(2\phi_0)+q^2$. g^2 is always positive for $q>u$. However, if $q\le u$, g^2 can be positive, zero, or negative depending on the polarization angle ϕ_0. We let ϕ_c be the smaller positively valued angle satisfying $\phi_c=\frac{1}{2}\cos^{-1}(-q^2/u^2)$. Then $0<\phi_c\le\pi/2$ and $g^2>0$ if $n\pi-\phi_c<|\phi_0|<n\pi+\phi_c$, $g=0$ if $|\phi_0|=n\pi\pm\phi_c$, and $g^2<0$ if $n\pi+\phi_c<|\phi_0|<(n+1)\pi-\phi_c$, where $n=0,1,2,\ldots$. In the following discussion, we always set $g=|g^2|^{1/2}$.

For $r>r_0$, ϕ satisfies the modified Bessel equation for which the solutions are the zeroth-order modified Bessel functions of the first and second kinds, $I_0(qr)$ and $K_0(qr)$. Since as $r\to\infty$, the solution must tend to $e^{-qr}/\sqrt{qr}$, the physically acceptable solution is the zeroth-order modified Bessel function $K_0(qr)$ of the second kind:

$$\phi=C^s_> K_0(qr) \text{ for } r>r_0 . \quad (3.8)$$

The amplitude of the deformation angle for $r>r_0$, $C^s_>$, will be determined later by matching the solution for $r\le r_0$ at $r=r_0$.

For $r\le r_0$, the nature of the solution depends on the azimuthal polarization angle ϕ_0. We shall consider the following two cases separately.

Case 1. The incident wave is polarized at an oblique angle to the dc magnetic field: $\phi_0\neq\pi/2$.

Case 2. The incident wave is polarized normal to the dc magnetic field: $\phi_0=\pi/2$.

C. Oblique polarization

For oblique polarization ($\phi_0\neq\pi/2$), the differential equation is inhomogeneous for $r\le r_0$ with a particular solution given by

$$\phi=C^s_0=\tfrac{1}{2}\tan^{-1}[u^2\sin(2\phi_0)/g^2] . \quad (3.9)$$

Notice that C^s_0 is positive for $g^2>0$, negative for $g^2<0$, $\pi/4$ for $g=0$. The solution to the homogeneous part can be modified Bessel functions or normal Bessel functions of order zero, depending on the sign of g^2 being positive or negative. Since the deformation angle is finite at $r=0$, the solution for the homogeneous part is $I_0(gr)$ for $g^2>0$ and $J_0(gr)$ for $g^2<0$. Together with the particular solution, by imposing the boundary condition that the solutions and their first derivatives must be continuous at the boundary $r=r_0$, the amplitudes of the deformation angles can also be determined. For $g^2>0$, the solution which is finite at $r=0$ and which satisfies the boundary condition at $r=r_0$ can be expressed in terms of the zeroth-order modified Bessel function of the first kind:

$$\phi=C^s_< I_0(gr)+C^s_0 \text{ for } r\le r_0 \quad (3.10a)$$

with

$$C^s_< = \frac{qK_1(qr_0)}{qI_0(gr_0)K_1(qr_0)+gI_1(gr_0)K_0(qr_0)}C^s_0 \quad (3.10b)$$

and

$$C^s_> = \frac{gI_1(gr_0)}{qI_0(gr_0)K_1(qr_0)+gI_1(gr_0)K_0(qr_0)}C^s_0 \quad (3.10c)$$

for Eq. (3.8). The solution for $g^2<0$ is the zeroth-order Bessel function of the first kind and has the form

$$\phi=C^s_< J_0(gr)+C^s_0 \text{ for } r\le r_0 \quad (3.11a)$$

with

$$C^s_< = -\frac{qK_1(qr_0)}{qJ_0(gr_0)K_1(qr_0)-gJ_1(gr_0)K_0(qr_0)}C^s_0 \quad (3.11b)$$

and

$$C^s_> = \frac{gJ_1(gr_0)}{qJ_0(gr_0)K_1(qr_0)-gJ_1(gr_0)K_0(qr_0)}C^s_0 \quad (3.11c)$$

for Eq. (3.8).[26] Since $\sin[2(2\pi-\phi_0)]=\sin(-2\phi_0)=-\sin(2\phi_0)$ we have, as $\phi_0\to-\phi_0$ or $\phi_0\to 2\pi-\phi_0$, then $C^s_<\to -C^s_<$ and $\phi(r)\to-\phi(r)$, which show that the orienting effects of an optical field with polarization ϕ_0 is the same as that of an optical field with polarization $-\phi_0$ or $2\pi-\phi_0$, except that ϕ changes sign $[\phi(r)\to-\phi(r)]$ as shown earlier in Sec. II.

For the special case where $g=0$, i.e., $\cos(2\phi_0)=-q^2/u^2$, the linearized equation for $r\le r_0$ takes the form

$$\frac{1}{r}\frac{\partial}{\partial r}\left[r\frac{\partial\phi}{\partial r}\right]+\tfrac{1}{2}u^2\sin(2\phi_0)=0\,. \qquad (3.12)$$

Again, by using the boundary conditions at $r=0$ and $r=r_0$, the complete solution for $g=0$ is

$$\phi=\tfrac{1}{8}u^2\sin(2\phi_0)\left[r_0^2-r^2+\frac{2r_0K_0(qr_0)}{qK_1(qr_0)}\right] \quad \text{for } r\le r_0 \qquad (3.13a)$$

with

$$C_>^s=r_0u^2\sin(2\phi_0)/[4qK_1(qr_0)] \qquad (3.13b)$$

for Eq. (3.8).

The spatial orientation of the azimuthal angle of the director as a function of the optical-field intensity is shown in Fig. 2. For $qr_0=5$ and $\phi_0=40°$, substantial alignment is already present for $u=q/2$ or incident total power of 30 mW. Almost complete alignment at the center of the spot can be achieved with $u=3q$ or $P\sim1.1$ W. (For $r_0\sim100$ μm, $qr_0\sim5$ corresponds to a magnetic field of 1.1 kG.) Figure 3 shows the spatial orientation of the azimuthal angle of the director as a function of the radius of the laser illumination spot for $q=500$ cm^{-1}, $ur_0=2.5$, and $\phi_0=40°$. With magnetic field and incident optical power fixed, further increases the alignment can be achieved by using a smaller laser spot at low laser power. The alignment at the center of the spot can increase dramatically by a reduction of the illumination-spot radius from 100 to 10 μm. The dependence of the deformation on the polarization of the optical field is shown in Fig. 4 for a fixed magnetic field and optical beam power. The results indicate that there is a polarization angle ϕ_0 which gives the maximum alignment effect.

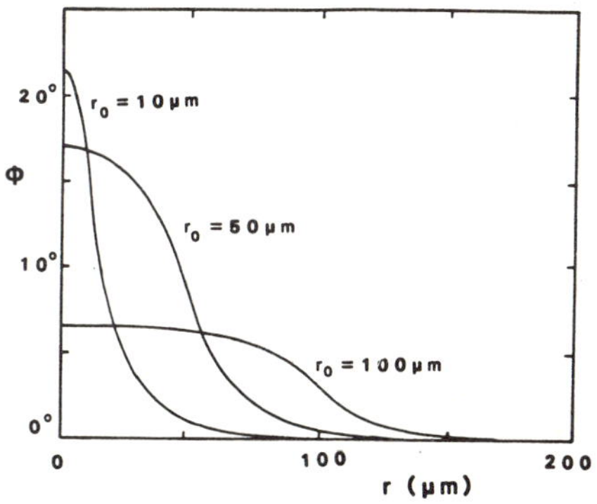

FIG. 3. Spatial orientation of the azimuthal angle of the director as a function of the radius of laser illumination spot for $q=500$ cm^{-1}, $ur_0=2.5$, and $\phi_0=40°$.

D. Normal polarization

For a normally polarized incident light beam ($\phi_0=\pi/2$), there exists a threshold intensity below which no molecular reorientation can be induced. To investigate the threshold behavior, we again linearize Eq. (2.13a) at small distortion. The resulting equation takes the form of the zeroth-order Bessel equation

$$\frac{\partial^2\phi}{\partial r^2}+\frac{1}{r}\frac{\partial\phi}{\partial r}+(u^2-q^2)\phi=0 \quad \text{for } r\le r_0\,. \qquad (3.14)$$

We let $w^2\equiv|u^2-q^2|$. We first consider the case where $u<q$. Then the possible solutions are 0, $K_0(wr)$, and $I_0(wr)$. Neither $K_0(wr)$ nor $I_0(wr)$ can be the solution because the first derivative of $K_0(wr)$ diverges at the origin and $I_0(wr)$ does not satisfy the condition of continuity of the logarithmic derivative to a $K_0(qr)$ function at the boundary $r=r_0$. Therefore, the solution zero is the only acceptable solution for $u<q$.

When $u\ge q$, the solutions of Eq. (3.14) are 0, $J_0(wr)$, and $Y_0(wr)$. $Y_0(wr)$ is not an acceptable solution since it diverges at $r=0$. Consequently, the only possible nonzero solution is the zeroth-order Bessel function of the first kind $J_0(wr)$. The continuity of the logarithmic derivative of the solutions at the boundary $r=r_0$ yields the following equation for determining the threshold intensity[27,28]:

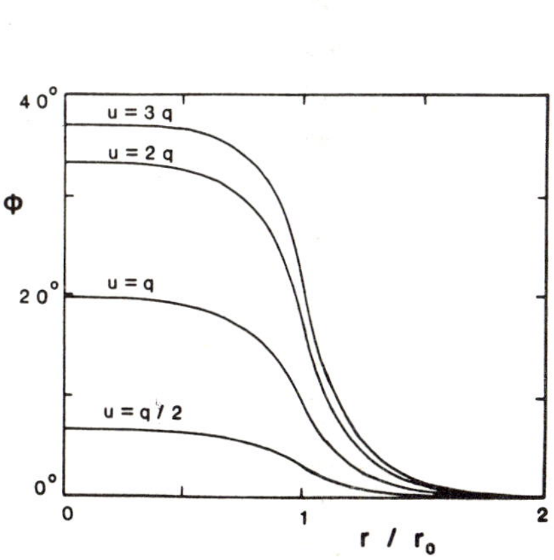

FIG. 2. Spatial orientation of the azimuthal angle of the director as a function of the reduced optical-field intensity $u=[2\epsilon_a I/cn_0\alpha]^{1/2}$ for $qr_0=5$ and $\phi_0=40°$.

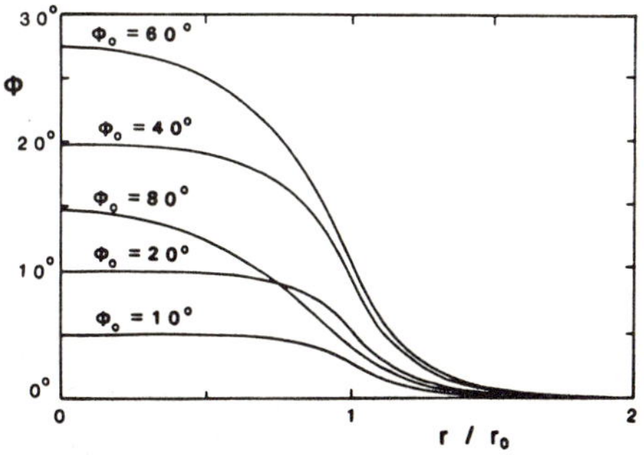

FIG. 4. Spatial orientation of the azimuthal angle of the director as a function of the polarization of the optical field ϕ_0 for $ur_0=qr_0=5$.

$$\frac{w_{th}J_1(w_{th}r_0)}{J_0(w_{th}r_0)}=\frac{qK_1(qr_0)}{K_0(qr_0)} . \quad (3.15)$$

In the case $qr_0 \ll 1$ we obtain, from Eq. (3.15),

$$w_{th}r_0 \sim -2/[0.5772+\ln(qr_0/2)] . \quad (3.16)$$

For $qr_0 \sim 1$ we have

$$w_{th}r_0 \sim 1.4 . \quad (3.17)$$

For $qr_0 > \pi$,

$$w_{th}r_0 \sim J_{01}[1-J_{01}K_0(qr_0)/qr_0K_1(qr_0)] , \quad (3.18)$$

where $J_{01} \approx 2.405$ is the first zero of the Bessel function J_0. In the limit of $qr_0 \gg 1$, Eq. (3.18) shows that $w_{th}r_0 \sim J_{01}$. The threshold intensity is then given by

$$I_{th}=cn_0\alpha u_{th}^2/2\epsilon_a , \quad (3.19)$$

where $u_{th}^2=w_{th}^2+q^2$. The threshold power of the light beam is

$$P_{th}=\pi r_0^2 I_{th}=\pi c n_0 \alpha u_{th}^2 r_0^2/2\epsilon_a .$$

For $qr_0 \sim 5$, $u_{th}r_0 \sim 5.4$ and P_{th} is of the order of 145 mW. In the limit $qr_0 \ll 1$, $w_{th}r_0$ is much greater than qr_0 so that $u_{th} \sim w_{th} \sim -2/r_0[0.5772+\ln(qr_0/2)]$. Thus the threshold optical energy is essentially independent of the external magnetic field and works against the elastic deformation energy. For $qr_0 \gg 1$, $w_{th}r_0 \sim J_{01} \ll qr_0$ so that $u_{th} \sim q$ which shows that $|\vec{E}_{th}| \sim 2\sqrt{\chi_a/\epsilon_a}H_0$ and the threshold optical field is essentially competing against the magnetic field. The reduced threshold power as a function of the spot size and reduced magnetic field strength qr_0 is shown in Fig. 5.

The amplitude of the azimuthal angle in the region above the threshold can not be determined from the linear approximation, but can be determined with allowance for the nonlinear terms in the Euler equation. By expanding Eq. (2.13a) up to and including terms $\sim\phi^3$ we obtain

$$\frac{\partial^2\phi}{\partial r^2}+\frac{1}{r}\frac{\partial\phi}{\partial r}+w^2(1-\tfrac{2}{3}\phi^2)\phi=0 \text{ for } r\le r_0 . \quad (3.20)$$

We look for an approximate solution of the form

$$\phi=C_<^s J_0(w_{th}r) , \quad (3.21)$$

where $C_<^s=\phi(r=0)$ is the deformation angle at the origin. By putting Eq. (3.21) into Eq. (3.20) and evaluating at $r=0$ we obtain

$$C_<^s=[3(1-w_{th}^2/w^2)/2]^{1/2} . \quad (3.22)$$

Therefore, when the incident intensity is above threshold, the approximate solutions are given by Eqs. (3.8) and (3.21) with the respective amplitudes given by Eq. (3.22) and

$$C_>^s=C_<^s J_0(w_{th}r_0)/K_0(qr_0) \quad (3.23)$$

outside the illuminated region.

E. Dynamic response

We now consider the time dependence of the molecular reorientation. For the total free energy density defined by Eq. (2.1) we include a dissipative term $\eta(\partial\hat{n}/\partial t)/2$ which will contribute a viscous torque opposing any rapid change of the director, where η is the viscosity of the SmC. The dynamic behavior is described by the resulting Euler equations of the forms

$$\frac{\partial^2\phi}{\partial r^2}+\frac{1}{r}\frac{\partial\phi}{\partial r}-\tfrac{1}{2}g^2\sin(2\phi)+\tfrac{1}{2}u^2\sin(2\phi_0)\cos(2\phi)=h^2\frac{\partial\phi}{\partial t} \text{ for } r\le r_0 \quad (3.24a)$$

and

$$\frac{\partial^2\phi}{\partial r^2}+\frac{1}{r}\frac{\partial\phi}{\partial r}-\tfrac{1}{2}q^2\sin(2\phi)=h^2\frac{\partial\phi}{\partial t} \text{ for } r>r_0 \quad (3.24b)$$

where $h^2\equiv\eta/\alpha$.

Although backflow effects are completely ignored in the dynamic equations (3.24a) and (3.24b), these equations are still complicated even in the linearized regime. In the following, we consider only the case of normal polarization in the small-distortion regime which can be described by

$$\frac{\partial^2\phi}{\partial r^2}+\frac{1}{r}\frac{\partial\phi}{\partial r}+w^2(1-\tfrac{2}{3}\phi^2)\phi=h^2\frac{\partial\phi}{\partial t} \text{ for } r\le r_0 . \quad (3.25)$$

By assuming a weak time dependence of the spatial distribution of the deformation angle, we look for an approximate solution for $I \gtrsim I_{th}$ of the form

$$\phi(r,t)=\begin{cases} C_<^d(t)J_0(w_{th}r), & r\le r_0 \\ C_>^d(t)K_0(qr), & r>r_0 \end{cases} \quad (3.26)$$

where $C_<^d$ and $C_>^d$ are the amplitudes of the distortion for $r\le r_0$ and $r>r_0$, respectively. Putting Eq. (3.26) into Eq. (3.25) and evaluating at $r=0$, the amplitude for the distortion within the optical-field illuminating region is described by

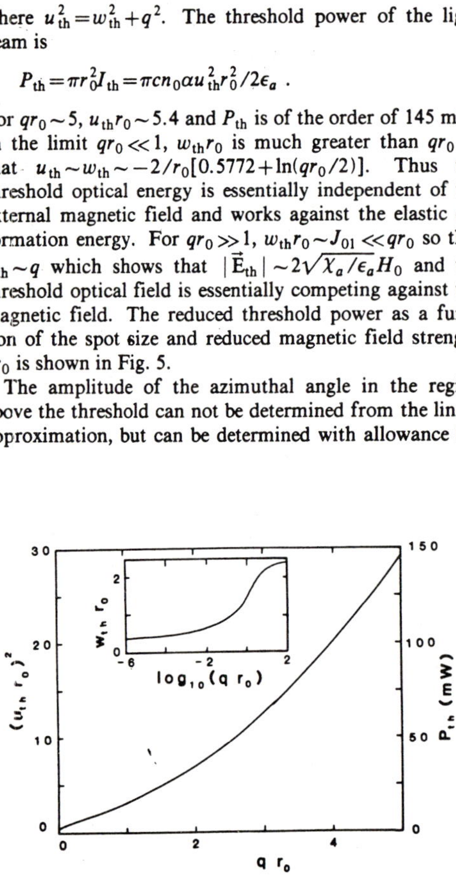

FIG. 5. Threshold power as a function of the spot size and reduced magnetic field strength $qr_0=\sqrt{\chi_a/\alpha}H_0r_0$. Threshold power P_{th} is related to $(u_{th}r_0)^2$ by $P_{th}=\pi cn_0\alpha(u_{th}r_0)^2/2\epsilon_a$. For a typical SmC sample, $P\sim5(ur_0)^2$ and the thresold power for $P_{th}=5(u_{th}r_0)^2$ is shown in the figure where the power is expressed in mW. Insert shows the variation of the function $w_{th}r_0=(u_{th}^2-q^2)^{1/2}r_0$ with respect to $\log_{10}(qr_0)$.

$$\frac{\partial}{\partial t}C^d_< = C^d_<[a - b(C^d_<)^2] , \quad (3.27)$$

where $a \equiv (w^2 - w_{th}^2)/h^2$ and $b \equiv 2w^2/3h^2$. Equation (3.27) is of the same form for the optically induced Freedericksz transition in a NLC.[10] Since there is no constant term on the right-hand side of the equation, $C^d_< = 0$ can be a solution. Therefore, we need a small fluctuation C_i at $t=0$ to get the distortion started. Equation (3.27) can then be solved to give

$$C^d_< = \left[\frac{1}{1 + C_r^2 e^{-2at}}\right]^{1/2} C^s_< , \quad (3.28)$$

where $C_r = [(C^s_< / C_i)^2 - 1]^{1/2}$ and

$$C^s_< = \sqrt{a/b} = [3(1 - w_{th}^2/w^2)/2]^{1/2} = \lim_{t \to \infty} C^d_<$$

which is the static solution [Eq. (3.22)]. By matching the solution at $r = r_0$, we have

$$C^d_>(t) = C^d_<(t) J_0(w_{th} r_0)/K_0(q r_0) . \quad (3.29)$$

At $t \sim 0$, Eq. (3.28) describes the exponential growth of a small fluctuation C_i with a time constant $1/a$:

$$C^d_<(t \sim 0) \sim C_i e^{at} . \quad (3.30)$$

For $t \gg 1$, the deformation amplitude $C^d_<$ reaches its final value $C^s_<$ exponentially with a smaller time constant $1/2a$:

$$C^d_<(t \gg 1) \sim (1 - \tfrac{1}{2} C_r^2 e^{-2at}) C^s_< . \quad (3.31)$$

Note that the time constant $1/a$ is proportional to $(w^2 - w_{th}^2)^{-1}$ and is dependent on the incident laser power. For a 10-μm-radius laser beam incident on a sample in an external 1-kG magnetic field, the threshold power is ~ 8 mW. Using $\eta \sim 1$ cp and $\alpha \sim 0.5 \times 10^{-6}$ dyn for a typical Sm*C* sample, the time constant is ~ 16 msec for incident power twice threshold, decreasing to 2 msec for a power ten times threshold. Hence, in general, the response time is of the order of milliseconds, which is faster than typical response time[7,8,10] in the nematic case.

IV. OPTICAL REFLECTIVITY AND TRANSMISSIVITY

Experimentally, the molecular reorientation can be qualitatively measured by the reflectivity or transmissivity of a normally incident probe beam. In the following, we consider a probe beam incident normally onto the Sm*C* polarized along *OP* which is selected by a polarizer making an angle ϕ_p with the y axis (Fig. 6). On entering the Sm*C*, each ray is divided into two rays with different effective refractive indices, and with their electric displacement vectors D_p and D_v vibrating in two mutually orthogonal directions at right angles to the Sm*C* normal. D_p lies in the plane containing the optical axis of the Sm*C*. The ray vibrating along D_p acts like the extraordinary ray having an effective refractive index

$$n_p = n_{op} n_{ep} / (n_{op}^2 \sin^2\theta_0 + n_{ep}^2 \cos^2\theta_0)^{1/2} ,$$

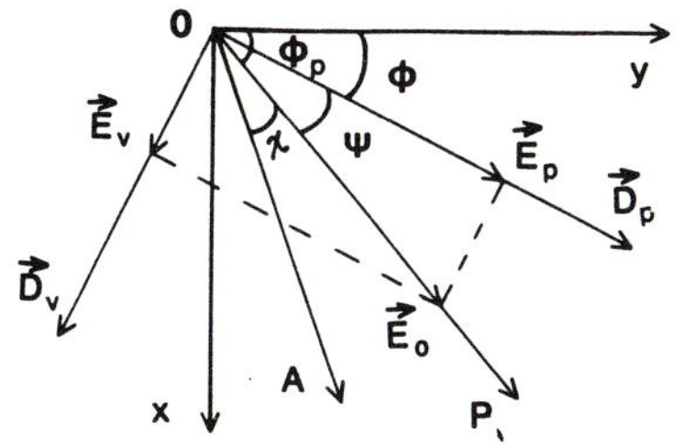

FIG. 6. Construction of the vibration components transmitted by a polarizer and analyzer for the reflectivity and transmissivity measurements. The director of the Sm*C* is oriented at an azimuthal angle ϕ with the y axis. A probe beam is incident normally onto the Sm*C* with polarization selected by a polarizer which is denoted by *OP* making an angle ϕ_p with the y axis. On entering the Sm*C*, each ray is divided into two rays with different effective refractive indices, and with their electric displacement vectors D_p and D_v vibrating in two mutually orthogonal directions at right angles to the Sm*C* normal. D_p lies in the plane containing the optical axis of the Sm*C*. The ray vibrating along D_p acts like the extraordinary ray. The ray vibrating along D_v acts like the ordinary ray. $\vec{E}_0$ is the electric field of the probe beam. $\Psi = \phi_p - \phi$ is the angle that the polarizer makes with the plane containing the optical axis OD_p. The measured quantity can be either the reflectivity or the transmissivity. The direction of vibration of the measured field is selected by an analyzer which is represented by *OA* making an angle χ with the polarizer.

where n_{op} and n_{ep} are the ordinary and the extraordinary refractive indices at the probe-beam wavelength λ_p. The ray vibrating along D_v acts like the ordinary ray having an effective refractive index $n_v = n_{op}$. The direction of vibration of the measured field is selected by an analyzer which is denoted by *OA* and the measured quantity can be either the reflectivity or the transmissivity.

In the following, we denote by subscripts p and v quantities referring to the ray vibrating along D_p and D_v. We let R_j and T_j be the reflection and transmission coefficients of the ray vibrating along D_j ($j = p$ or v). Since the azimuthal angle of the director depends only on r, the Sm*C* medium can be considered as an anisotropic dielectric medium with an refractive index depending on r but is independent of z. Then the refractive index N of the system is

$$N = \begin{cases} 1, & z < 0 \\ n_j(r), & 0 \le z \le d \\ 1, & z > d . \end{cases} \quad (4.1)$$

The amplitudes of the reflected (r) and transmitted (t) waves at the interfaces $z=0$ (subscript 1) and $z=d$ (subscript 2) are, respectively,[29]

$$(r_1)_j=(n_j-1)/(1+n_j)\ ,$$
$$(r_2)_j=(1-n_j)/(1+n_j)=-(r_1)_j\ ,$$

and (4.2)

$$(t_1)_j=2/(1+n_j)\ ,$$
$$(t_2)_j=2n_j/(1+n_j)=n_j(t_1)_j\ .$$

By taking into account multiple reflections at the two interfaces $z=0$ and $z=d$, the reflection and transmission coefficents are

$$R_j=\frac{(r_1)_j+(r_2)_j e^{i2\beta_j}}{1+(r_1)_j(r_2)_j e^{i2\beta_j}}$$

and (4.3)

$$T_j=\frac{(t_1)_j(t_2)_j e^{i\beta_j}}{1+(r_1)_j(r_2)_j e^{i2\beta_j}}\ ,$$

where $\beta_j\equiv 2\pi n_j d/\lambda_p$. With the use of Eq. (4.2), R_j is reduced to

$$R_j=|R_j|e^{i\delta_j^r} \tag{4.4a}$$

with

$$|R_j|=\frac{(n_j^2-1)\sin\beta_j}{[4n_j^2+(n_j^2-1)^2\sin^2\beta_j]^{1/2}} \tag{4.4b}$$

and

$$\tan\delta_j^r=-\frac{2n_j}{1+n_j^2}\cot\beta_j\ . \tag{4.4c}$$

Consequently, the phase difference between the two reflected rays $\delta^r=\delta_p^r-\delta_v^r$ is given by

$$\tan\delta^r=2\frac{n_v(1+n_p^2)\tan\beta_p-n_p(1+n_v^2)\tan\beta_v}{4n_p n_v+(1+n_p^2)(1+n_v^2)\tan\beta_p\tan\beta_v}\ . \tag{4.5}$$

Similarly, the transmission coefficient can be reduced to

$$T_j=|T_j|e^{i\delta_j^t} \tag{4.6a}$$

with

$$|T_j|=\frac{2n_j}{[4n_j^2+(n_j^2-1)^2\sin^2\beta_j]^{1/2}} \tag{4.6b}$$

and

$$\tan\delta_j^t=\frac{1+n_j^2}{2n_j}\tan\beta_j=\tan\left[\frac{\pi}{2}+\delta_j^r\right]\ . \tag{4.6c}$$

The phase difference between the two transmitted rays $\delta^t=\delta_p^t-\delta_v^t$ is the same as δ^r, i.e., $\delta^r=\delta^t\equiv\delta$. By Eqs. (4.4)–(4.6), the reflection and transmission coefficients satisfy $|R_j|^2+|T_j|^2=1$ which is in agreement with the law of conservation of energy.

Since the analyzer transmits only the components parallel to OA, the resultant fields of two rays after passing through the analyzer are given by

$$E_{op}=E_p Q_p\cos(\Psi-\chi)e^{i\delta_j}$$
$$=E_0 Q_p\cos\Psi\cos(\Psi-\chi)e^{i\delta_j}$$

and (4.7)

$$E_{ov}=E_v Q_v\sin(\Psi-\chi)e^{i\delta_j}$$
$$=E_0 Q_v\sin\Psi\sin(\Psi-\chi)e^{i\delta_j}\ ,$$

where $\Psi\equiv\phi_p-\phi$ is the angle that the polarizer makes with OD_p, χ is the angle between the analyzer and the polarizer, $Q_j=|R_j|$ and $\delta_j=\delta_j^r$ for the reflectivity measurement, $Q_j=|T_j|$ and $\delta_j=\delta_j^t$ for the transmissivity measurement, and E_0 is the electric field of the probe beam. The two rays E_{op} and E_{ov} are superposed at the observation point with the field given by $\vec{E}_{ob}=\vec{E}_{op}+\vec{E}_{ov}$:

$$I_{ob}=\frac{c}{4\pi}\langle\vec{E}_{ob}^{\,2}\rangle$$
$$=I_0\left[[Q_p\cos\chi+(Q_v-Q_p)\sin\Psi\sin(\Psi-\chi)]^2-Q_pQ_v\sin(2\Psi)\sin[2(\Psi-\chi)]\sin^2\frac{\delta}{2}\right]\ , \tag{4.8}$$

where $I_0=c\langle\vec{E}_0^{\,2}\rangle/4\pi$ is the probe-beam intensity. Equation (4.8) is the general expression for the reflected and transmitted intensities for the probe beam at a point where the local orientation of the director is known. In our calculations, the molecular orientation is cylindrically symmetric. Hence, the total reflected or transmitted power of the probe beam covering an area of radius r_{ob} is given by

$$P_t=2\pi\int_0^{r_{ob}}I_{ob}(r)r\,dr\ . \tag{4.9}$$

Consequently, the reflectivity ($\mathscr{R}$) and transmissivity ($\mathscr{T}$) are given by P_t/P_0 where $P_0=\pi r_{ob}^2 I_0$ is the power of the probe beam:

$$\mathscr{R}=\frac{2}{r_{ob}^2}\int_0^{r_{ob}}\left[[|R_p|\cos\chi+(|R_v|-|R_p|)\sin\Psi\sin(\Psi-\chi)]^2-|R_p||R_v|\sin(2\Psi)\sin2[(\Psi-\chi)]\sin^2\frac{\delta}{2}\right]r\,dr$$

and (4.10)

$$\mathscr{T}=\frac{2}{r_{ob}^2}\int_0^{r_{ob}}\left[[|T_p|\cos\chi+(|T_v|-|T_p|)\sin\Psi\sin(\Psi-\chi)]^2-|T_p||T_v|\sin(2\Psi)\sin[2(\Psi-\chi)]\sin^2\frac{\delta}{2}\right]r\,dr\ .$$

We now consider two special cases corresponding to polarized and depolarized reflection and transmission, respectively.

Case 1. Polarizer and analyzer are parallel. Then $\chi=0$ and

$$I_{\rm ob}=I_0\left[[Q_p+(Q_v-Q_p)\sin^2\Psi]^2-Q_pQ_v\sin^2 2\Psi\sin^2\frac{\delta}{2}\right]. \tag{4.11}$$

Case 2. Polarizer and analyzer are perpendicular. Then $\chi=\pi/2$ and

$$I_{\rm ob}=I_0\sin^2 2\psi\left[\tfrac{1}{4}(Q_v-Q_p)^2-Q_pQ_v\sin^2\frac{\delta}{2}\right]. \tag{4.12}$$

Figure 7 shows the depolarized reflectivity and transmissivity for a 3000-Å-thick Sm*C* film as a function of reduced reorientating intensity $u/q=[2\epsilon_a I/cn_0\chi_a H^2]^{1/2}$. The probe beam has a wavelength of 6328 Å, polarized along the direction of the unperturbed director (y axis) and has the same spot size as the orienting beam which is polarized at 40° to the y axis (same case as presented in Fig. 2). The depolarized intensities increase strongly and approach limiting values as the power of the orienting laser beam is increased. Therefore, these molecular reorientation effects are measurable and can be quantitatively compared with the calculations presented in this paper.

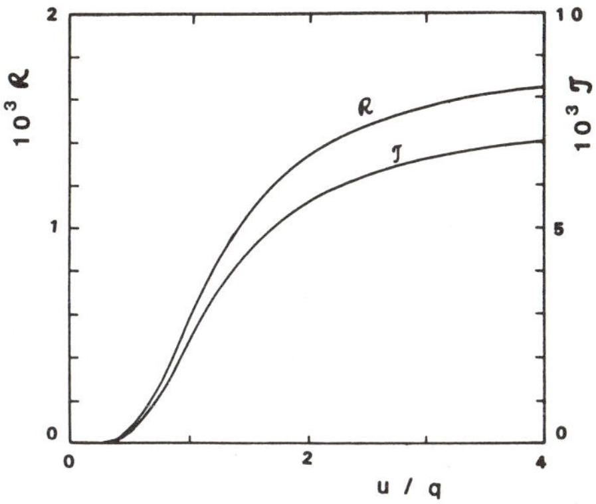

FIG. 7. Optical reflectivity and transmissivity as a function of the reduced intensity $u/q=[2\epsilon_a I/cn_0\chi_a H^2]^{1/2}$ for a cell of 0.3 μm thick with $qr_0=5$ and $\phi_0=40°$. For the probe beam, we set $\lambda_p=6328$ Å, $n_{op}=1.55$, $n_{ep}=1.75$, $\theta_0=30°$, $\chi=90°$, and $\phi_p=0°$.

ACKNOWLEDGMENTS

We gratefully acknowledge Alan J. Hurd and Robert B. Meyer for useful discussions. H. L. O. was supported by the U.S. Army Research Office under Contract No. DAAG-29-80-K-0050. This work was also supported by the National Science Foundation, Solid State Chemistry Program, under Grant No. DMR-80-06492.

*Present address: Hewlett Packard Laboratories, 1501 Page Mill Road, Building 1U, Palo Alto, CA 94304.

[1]Y. R. Shen, Rev. Mod. Phys. 48, 1 (1976); S. M. Arakelian, G. A. Lyakhov, and Yu. S. Chilingarian, Usp. Fiz. Nauk 131, 3 (1980) [Sov. Phys.—Usp. 23, 245 (1980)].

[2]J. W. Shelton and Y. R. Shen, Phys. Rev. Lett. 25, 23 (1970); 26, 538 (1971); Phys. Rev. A 5, 1867 (1972).

[3]V. A. Belyakov and N. V. Shipov, Zh. Eksp. Teor. Fiz. 82, 1159 (1982) [Sov. Phys.—JETP 55, 674 (1982)].

[4]N. V. Tabiryan and B. Ya. Zel'dovich, Mol. Cryst. Liq. Cryst. 69, 19 (1981); B. Ya. Zel'dovich and N. V. Tabiryan, Zh. Eksp. Teor. Fiz. 82, 167 (1982) [Sov. Phys.—JETP 55, 99 (1982)].

[5]H. G. Winful, Phys. Rev. Lett. 49, 1179 (1982).

[6]A. S. Zolot'ko, V. F. Kitaeva, N. Kroo, N. N. Sobolev, and L. Chilag, Zh. Eksp. Teor. Fiz. Pis'ma Red 81, 72 (1980) [JETP Lett. 32, 158 (1980)].

[7]B. Ya. Zel'dovich, N. V. Tabiryan, and Yu. S. Chilingaryan, Zh. Eksp. Teor. Fiz. 81, 71 (1981) [Sov. Phys.—JETP 54, 32 (1981)]; B. Ya. Zel'dovich and N. V. Tabiryan, Zh. Eksp. Teor. Fiz. 82, 1126 (1982) [Sov. Phys.—JETP 55, 656 (1982)].

[8]S. D. Durbin, S. M. Arakelian, and Y. R. Shen, Phys. Rev. Lett. 47, 1411 (1981); S. D. Durbin, S. M. Arakelian, M. M. Cheung, and Y. R. Shen, J. Phys. (Paris) Colloq. 44, C2-161 (1983).

[9]I. C. Khoo, Phys. Rev. A 25, 1040 (1982); 25, 1636 (1982); 26, 1131(E) (1982).

[10]H. L Ong, Phys. Rev. A 28, 2393 (1983); H. L. Ong and R. B. Meyer, in Optical Bistability II, proceedings of the Topical Meeting on Optical Bistability, 1983, University of Rochester, edited by C. M. Bowdon, H. M. Gibbs, and S. L. McCall (Plenum, New York, in press).

[11]J. W. Helfrich, J. Chem. Phys. 55, 839 (1971).

[12]A. Rapini, J. Phys. (Paris) 33, 237 (1972); Prog. Solid State Chem. 8, 337 (1973).

[13]J. P. Hurault, J. Chem. Phys. 59, 2068 (1973).

[14]E. Meirovitch, Z. Luz, and A. Alexander, Phys. Rev. A 15, 408 (1977); Mol. Phys. 37, 1489 (1979); E. Meirovitch and J. H. Freed, J. Phys. Chem. 84, 2459 (1980).

[15]See also, M. J. Stephen and J. P. Straley, Rev. Mod. Phys. 46, 617 (1974); P. I. Ktorides and D. L. Uhrich, Mol. Cryst. Liq. Cryst. 87, 69 (1982), and Ref. 16.

[16]P. G. de Gennes, *The Physics of Liquid Crystals* (Oxford University Press, Oxford, 1974); S. Chandrasekhar, *Liquid Crystals* (Cambridge University Press, Cambridge, England, 1977).

[17]P. H. Lippel and C. Y. Young, Appl. Phys. Lett. (in press).

[18]N. V. Tabiryan and B. Ya. Zel'dovich, Mol. Cryst. Liq. Cryst. 69, 31 (1981); B. Ya. Zel'dovich and N. V. Tabiryan, Zh.

Eksp. Teor. Fiz. Pis'ma Red. 34, 72 (1981) [JETP Lett. 34, 67 (1981)].

[19]Orsay Liquid Crystal Group, Solid State Commun. 9, 653 (1969).

[20]Y. Galerne, J. L. Martinand, G. Durand, and H. Veyssie, Phys. Rev. Lett. 29, 562 (1972).

[21]A. Saupe, Mol. Cryst. Liq. Cryst. 7, 59 (1969); J. Nehring and A. Saupe, J. Chem. Soc. Faraday Trans. 2 68, 1 (1972).

[22]Even within the single-elastic-constant approximation, the Euler equations (2.13a) and (2.13b) are still very complicated. For $r > r_0$, the Euler equation (2.13b) is relatively simple but has a form of the Sine-Gordon equation for which no general solution exists except by numerical means. For a review on the Sine-Gordon equation, see, for example, A. Barone, F. Esposito, C. J. Magef, and A. C. Scott, Nuovo Cimento 1, 227 (1971).

[23]M. Born and E. Wolf, *Principle of Optics*, 5th ed. (Pergamon, Oxford, 1975), Chap. III.

[24]T. R. Taylor, S. L. Arova, and J. L. Fergason, Phys. Rev. Lett. 25, 722 (1970).

[25]See, for examples, J. Schesser and G. Eichmann, J. Opt. Soc. Am. 62, 786 (1972); E. G. Njoku, J. Appl. Phys. 54, 524 (1983); R. F. Alvarez-Estrada and M. L. Calvo, Opt. Acta 30, 481 (1983).

[26]The linearized differential equation (3.7a) for $r \leq r_0$ involves two series expansions of two different trigonometric functions with different coefficients. We have taken the linear term in the sine expansion and drop the quadratic term in the cosine expansion in Eq. (2.13a). For this truncation to be valid, the linear term must be larger than the quadratic term and hence the solutions (3.10) and (3.11) are valid only if $|g^2/[u^2\sin(2\phi_0)]| \gg |\phi|$.

[27]It is interesting to note that in a study of the influence of the finite size of the light spot on the laser-induced reorientation of a planar nematic liquid crystal, it has been shown by Zel'dovich *et al.* and by Csillag *et al.* that the threshold intensity of the transition is also of the same form as Eq. (3.15). See Ref. 7 and L. Csillag, I. Janossy, V. F. Kitaeva, N. Kroo, and N. N. Csobolev, Mol. Cryst. Liq. Cryst. 84, 125 (1982).

[28]It can be shown that the dispersion relation of an electromagnetic wave propagating in a cylindrical dielectric wave guide also satisfies an equation similar to Eq. (3.15). See J. D. Jackson, *Classical Electrodynamics*, 2nd ed. (Wiley, New York, 1975), Chap. VIII.

[29]See Ref. 23, Chap. I.

Transient laser-induced molecular reorientation and laser heating in a nematic liquid crystal

H. Hsiung, L. P. Shi,* and Y. R. Shen
Department of Physics, University of California, Berkeley, California 94720
(Received 6 March 1984)

Transient molecular reorientation induced in a liquid crystal by a giant laser pulse was observed for the first time in a nematic substance 4′-*n*-pentyl-4-cyanobiphenyl (5CB). Even though the response time of molecular reorientation was several orders of magnitude longer than the laser pulsewidth, the effect on the refractive indices was still large enough to be easily detectable by a Mach-Zehnder interferometer. Transient laser heating of the medium also affected the refractive indices, but with proper geometric and polarization arrangement, it could be decoupled from molecular reorientation and measured separately. Theoretical calculations using the Ericksen-Leslie theory for molecular reorientation in liquid crystals and the heat-diffusion theory for the laser heating effect were able to explain the experimental observations quantitatively.

I. INTRODUCTION

Liquid crystals are generally characterized by the strong correlation between molecules, which respond cooperatively to external perturbations. That strong molecular reorientation (or director reorientation) can be easily induced by a static electric or magnetic field is a well-known phenomenon.[1] The same effect induced by optical fields was, however, only studied recently.[2–6] Unusually large nonlinear optical effects based on the optical-field-induced molecular reorientation have been observed in nematic liquid-crystal films under the illumination of one or more cw laser beams.[3,4,6,7] In these cases, both the static and dynamical properties of this field-induced molecular motion are found to obey the Ericksen-Leslie continuum theory,[8] which describes the collective molecular reorientation by the rotation of a "director" (average molecular orientation).

The superiority of using lasers for material studies often lies in its spatial and temporal flexibilities, that is, the material can be selectively excited and probed in space and time. These qualities may allow us to elucidate fundamental material properties not accessible to conventional techniques. The location, dimension, direction, and duration of the material excitation can be readily controlled through adjustment of the beam spot, direction, polarization, and pulse width of the exciting laser field. The flexibilities can be further enhanced when two or more light waves are used to induce excitations. Such a technique, however, has not yet been fully explored in liquid-crystal research. Although the recent studies of optical-field-induced molecular reorientation in nematic liquid-crystal films have demonstrated the ability of the technique to resolve spatial variation of excitations, corresponding transient phenomena induced by pulsed optical fields have not yet been reported in the literature.[9] Because of the possibility of using lasers to induce excitations on a very short time scale, such studies could provide rare opportunities to test the applicability of the continuum theory in the extreme cases.

We report here the first observation of transient molecular reorientation induced in nematic 4′-*n*-pentyl-4-cyanobiphenyl (5CB) thin films by a 6-nsec laser pulse. Other than inducing the molecular reorientation, the intense laser pulse also heated the sample and thermally induced transient birefringence change. On the relatively short-time scale, this thermal effect was quite appreciable, and hence could prevent clean observation of the molecular reorientation. In our experiment, therefore, a static magnetic field was also applied to induce a stationary molecular reorientation in the nematic films: First, to enhance the transient optical-field-induced molecular reorientation, and second, to provide a condition for temporal discrimination against the thermal effect. The laser-induced changes in the 5CB sample were then probed by a weak cw laser beam with the help of a Mach-Zehnder interferometer. The detailed experimental arrangement is described in Sec. II.

The experimental results are given in Sec. III. We found that the magnitude of the transient molecular reorientation was proportional to the total energy of the laser pulse, and the relaxation time of such excitation was 9 orders of magnitude longer than the duration of the exciting laser pulse. The laser-induced thermal birefringence was also present. Its time dependence could be well described by the sum of two exponentials. The theory of these laser-induced effects is presented in Sec. IV. It is seen that the Ericksen-Leslie continuum theory, taking into account the finite beam size, agrees well with the experiment, and a simple heat-diffusion model can explain the temporal behavior of the thermal effect. More general discussion of the problem is given in Sec. V.

II. EXPERIMENTAL ARRANGEMENT

The 5CB used in this experiment was purchased from British Drug Houses, Inc. Its nematic phase occurs between 22.5 and 35.2 °C. Single-domain, homeotropic films were formed between two glass substrates which had been treated with *n*,*n*-dimethyl-*n*-octadecyl-3-aminopropyltrimethoxysilyl chloride (DMOAP).[10] The sample temperature was stabilized at 24.4±0.02 °C. Two samples, 130 and 190 μm thick, were studied. The thickness was controlled by Mylar spacers and its actual value was deduced from the observed threshold magnetic field (H_c) for Fréedericksz transition.[1]

The experimental arrangement is shown in Fig. 1(a). A 6-nsec laser pulse at 1.06 μm from a Q-switched Nd:YAG laser was focused with nearly normal incidence to an e^{-2} diameter of 500 μm on the sample to induce transient molecular reorientation in the sample. The resulting change of refractive indices was probed by a cw He-Ne laser beam at 632.8 nm, which was normally incident and closely overlapped with the pump laser beam at the sample. Spatial filtering was used at the output to ensure that only the central portion of the cylindrical region illuminated by the pump pulse was probed. The sample was placed in one arm of a Mach-Zehnder interferometer.[11] The induced refractive-index change in the sample resulted in a phase shift in the output interferogram which could be accurately measured. By displacing one mirror with a piezoelectric transducer, we could adjust the quiescent optical path difference between the two arms of the interferometer to an odd multiple of $\lambda/4$ to achieve the best signal-to-noise ratio. Vibrational and acoustic isolations were needed to ensure stability of the interferometer. The accuracy of our phase-shift measurements was better than 5×10^{-3} rad, corresponding to an average refractive-index change of $\lesssim 4\times10^{-6}$ in the samples. This is the best method we could think of that allows us to measure separately the refractive-index changes resulting from the laser-induced molecular reorientation and the laser-induced thermal effect mentioned earlier. The measurements were carried out in the following way.

A dc magnetic field H was applied along the direction parallel to the plane surfaces of the sample. With H

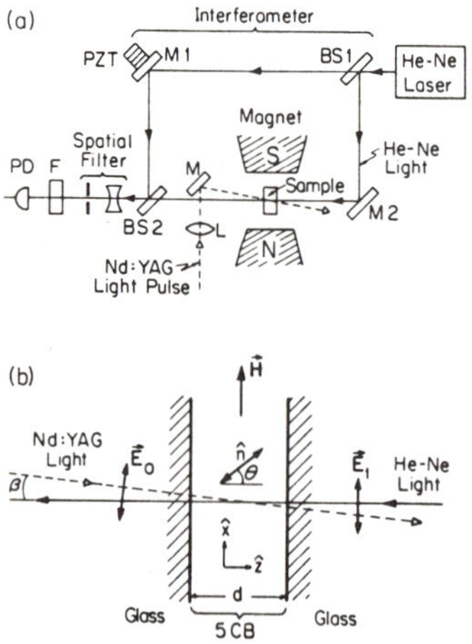

FIG. 1. (a) Experimental arrangement. BS1, BS2—50%-50% beam splitters; M, $M1$, $M2$—mirrors; L—lens; F—narrow-band filter (632.8 nm); PD—fast photodiode. (b) Experimental geometry. $\vec{H}=H\hat{x}$, $\vec{E}_1 \| \vec{H}$. $\beta \simeq 1°$.

larger than the critical field H_c for the Freedericksz transition, it could induce a molecular reorientation in the $\hat{H}$-$\hat{z}$ plane with a spatial variation $\theta_0(z)$ [see Fig. 1(b)]. A linearly polarized pump beam could, in addition, modify the refractive indices of the sample through both molecular reorientation and induced thermal effect; the latter was the result of laser heating of the medium. However, if the pump beam was polarized perpendicular to the $\hat{H}$-$\hat{z}$ plane, the molecular reorientation by the pump beam was negligibly small; only laser heating contributed to the refractive-index change, which could be probed by the probe beam polarized either parallel or perpendicular to the $\hat{H}$-$\hat{z}$ plane.

If the pump beam was polarized in the $\hat{H}$-$\hat{z}$ plane, even a relatively weak pump intensity could induce a molecular reorientation in the $\hat{H}$-$\hat{z}$ plane, in addition to the laser heating effect. The probe beam polarized perpendicular to $\hat{H}$-$\hat{z}$ again monitored only the thermally induced refractive-index change, but when polarized parallel to $\hat{H}$, it should feel both the thermal effect and the molecular reorientation effect. The refractive-index changes due to both effects were actually fairly small; we can therefore write the laser-induced phase shift experienced by the probe beam in traversing the sample cell of thickness d as

$$\Delta\Phi(t)=\Delta\Phi_{MR}+\Delta\Phi_{th}$$

$$\Delta\Phi_{MR}=\frac{2\pi}{\lambda}\int_{-d/2}^{d/2}\left[\frac{\partial n_{eff}}{\partial\theta}\right]_{\theta_0}\Delta\theta(z,t)dz \tag{1}$$

$$\Delta\Phi_{th}=\frac{2\pi}{\lambda}\int_{-d/2}^{d/2}\left[\frac{\partial n_{eff}}{\partial T}\right]_{\theta_0}\Delta T(z,t)dz ,$$

where $\Delta\Phi_{MR}$ and $\Delta\Phi_{th}$ are the induced phase shifts due to molecular reorientation $\Delta\theta$ and temperature rise ΔT, respectively, and $n_{eff}(z,t)$ is the refractive index seen by the probe beam at position z and time t.

In the above case, the temperature rise across the sample should be uniform initially. We then have, at $t\sim0$,

$$\Delta\Phi_{th}(t\sim0)\simeq\frac{2\pi}{\lambda}\Delta T_0\int_{-d/2}^{d/2}\left[\frac{\partial n_{eff}}{\partial T}\right]_{\theta_0}dz . \tag{2}$$

The effective refractive index n_{eff} can be written as[12]

$$n_{eff}=\left[\frac{\cos^2\theta}{n_0^2}+\frac{\sin^2\theta}{n_e^2}\right]^{-1/2} , \tag{3}$$

where n_0 and n_e are ordinary and extraordinary refractive indices, respectively, and hence

$$\frac{\partial n_{eff}}{\partial T}=\frac{\partial n_{eff}}{\partial n_0}\frac{\partial n_0}{\partial T}+\frac{\partial n_{eff}}{\partial n_e}\frac{\partial n_e}{\partial T} . \tag{4}$$

Since for 5CB, $\partial n_0/\partial T>0$ and $\partial n_e/\partial T<0$ (also, $\partial n_{eff}/\partial n_0>0$ and $\partial n_{eff}/\partial n_e>0$),[13] it was possible to choose a certain θ_0 such that $\Delta\Phi_{th}(t\sim0)=0$. This then allowed us to obtain $\Delta\Phi_{MR}(t\sim0)$ directly from the measured $\Delta\Phi(t\sim0)$.

As time went on, heat diffusion in the cell became significant, and the temperature across the cell was no longer uniform. The result was $\Delta\Phi_{th}(t>0)\neq0$. However, in

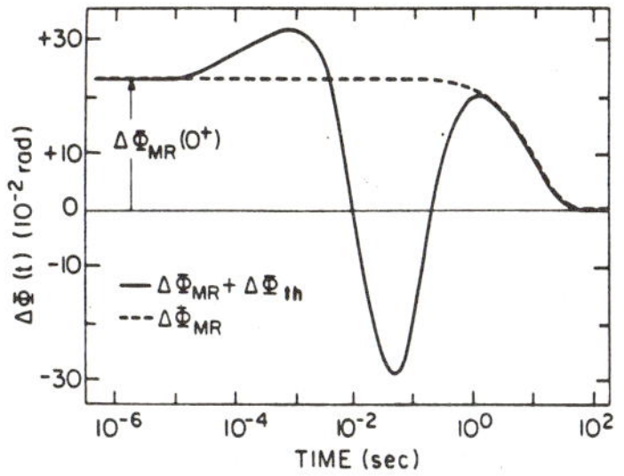

FIG. 2. Phase shift of the probe beam induced by a 27-mJ pulsed laser excitation at $t=0$. The dashed curve represents the molecular reorientation contribution obtained by subtracting the thermal contribution from the total phase shift (solid curve). The sample is a nematic 5CB film of 190 μm thickness.

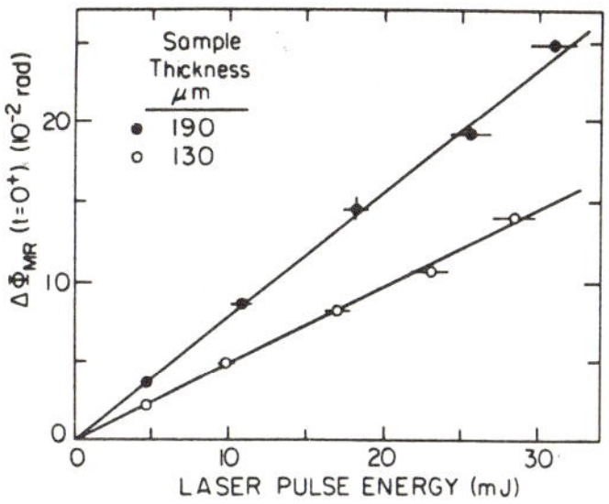

FIG. 4. Initial phase shift associated with molecular reorientation in 5CB films induced by pulsed laser excitation as a function of pulse energy.

liquid crystals, the relaxation time of the thermal effect is generally much shorter than the orientational relaxation time. Therefore, $\Delta\Phi_{MR}(t)$ could be obtained simply by subtracting $\Delta\Phi_{th}(t)$ from $\Delta\Phi(t)$.

III. EXPERIMENTAL RESULTS

Figure 2 displays a typical trace of the total phase shift, $\Delta\Phi(t)=\Delta\Phi_{MR}(t)+\Delta\Phi_{th}(t)$, on a logarithmic time scale detected with both the pump beam and the probe beam polarized in the $\hat{H}$-$\hat{z}$ plane. During the first 10 μsec after the pump laser excitation (which is essentially a δ function at $t=0$ on the graph), the thermal signal $\Delta\Phi_{th}$ is practically zero with our discrimination technique. A purely thermal signal $\Delta\Phi_{th}(t)$ was obtained using a pump pulse with polarization perpendicular to the $\hat{H}$-$\hat{z}$ plane. A typical trace is shown in Fig. 3. The molecular reorientation part, $\Delta\Phi_{MR}(t)$, of the total signal could then be singled out, as indicated by the dashed curve in Fig. 2.

From the measurements we found that after the pump pulse was over, $\Delta\Phi_{MR}(t)$ decayed exponentially with a time constant 9.2 ± 0.9 sec for the 130-μm sample, or 23 ± 4.8 sec for the 190-μm scale. The magnitude of $\Delta\Phi_{MR}$ at $t=0^+$ immediately after the pump pulse excitation was linearly proportional to the pump pulse energy, as shown in Fig. 4. The proportional constant was 4.84×10^{-3} rad/mJ for the 130-μm sample, or 7.88×10^{-3} rad/mJ for the 190-μm sample.

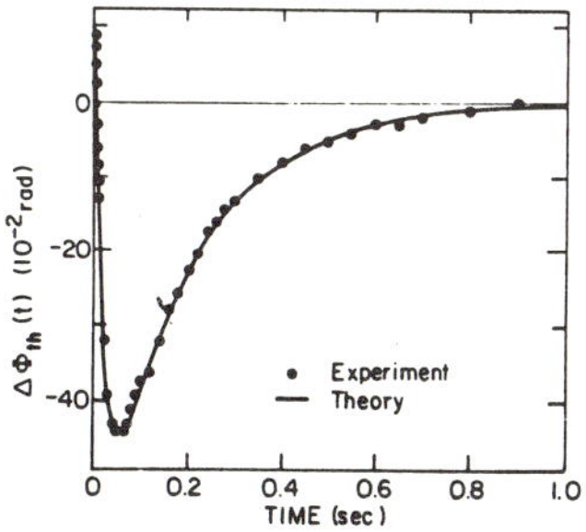

FIG. 3. An example of the observed phase shift as a function of time induced by pulsed laser heating in a 5CB film. The data were obtained for a 27-mJ laser pulse propagating in a sample 190 μm thick under the condition $\vec{E}_0\perp\vec{H}$. The solid curve is described by the expression $\Delta\Phi_{th}=-ae^{-t/t_0}+be^{-t/t_1}$ with $t_0=190$ ms, $t_1=19$ ms, $a=0.67$, and $b=0.83$.

The thermal signal $\Delta\Phi_{th}(t)$, in general, could be fit empirically by a function of the form $-a\exp(-t/t_0)+b\exp(-t/t_1)$ with $a,b>0$. An example is given in Fig. 3 where the solid curve is the empirical fit. While the coefficients a and b depend on the pump pulse energy, the time constants t_0 and t_1 are only functions of the sample thickness and characteristics of 5CB. We found for the 130-μm sample, $t_0=135$ msec and $t_1=13.5$ msec, and for the 190-μm sample, $t_0=190$ msec and $t_1=19$ msec. In both cases, the ratio of the two time constants was $t_0/t_1=10$.

IV. THEORY AND COMPARISON WITH EXPERIMENT

We neglect in the following theoretical derivation the coupling between laser heating and molecular reorientation. In general, with a finite pump beam traversing through a liquid-crystal cell, the induced temperature gradient in the medium can lead to a fluid flow which in turn orients the molecules through the flow-alignment mechanism.[14] In our case, because of the pumping geometry and the relatively large beam size, the thermally induced flow alignment is not significant. We can therefore treat the laser-induced molecular reorientation and the laser-induced thermal effect separately as two independent parts contributing to the observed phase shift $\Delta\Phi$.

A. Transient optical-field induced molecular reorientation

We use here the Ericksen-Leslie continuum theory[8] to describe the effect. The rotational motion of the director (i.e., molecular reorientation) is driven by the pump laser pulse, but it is also coupled with the translation motion (flow) of the fluid through viscosity. Thus, with a finite pump beam, a rigorous theoretical calculation would require the solution of a set of coupled three-dimensional nonlinear partial differential equations for the angle of director rotation θ and the fluid velocity $\vec{v}$. The mathematical complexity, however, can be greatly simplified through appropriate approximations.

If the laser-beam diameter is much larger than the sample thickness, and if only the core region is probed, then the pump beam can be approximated as an infinite plane wave. The rotation of the director can then be described by a one-dimensional equation of motion following the theory of Ericksen and Leslie. For the system in Fig. 1(b), the equation takes the form[15]

$$\rho_1\frac{\partial^2\theta}{\partial t^2}+\gamma_1\frac{\partial\theta}{\partial t}=[K_{33}+(K_{11}-K_{33})\sin^2\theta]\frac{\partial^2\theta}{\partial z^2}+\tfrac{1}{2}(K_{11}-K_{33})\sin(2\theta)\left[\frac{\partial\theta}{\partial z}\right]^2+\Gamma_H+\Gamma_{opt}+\tfrac{1}{2}[\gamma_1-\gamma_2\cos(2\theta)]\frac{\partial v_x}{\partial z}\,. \tag{5}$$

Here, $\theta(z,t)$ defines the orientation of the director $\hat{n}=(\cos\theta, 0, \sin\theta)$, and $v_x(z,t)$ is the x component of the flow velocity $\vec{v}=(v_x, 0, 0)$; ρ_1 is the "moment of inertia" per unit volume, which will be taken as zero in our calculation because its effect can be significant only during a subnanosecond pulse excitation; K_{11} and K_{33} are the Frank elastic constants, and γ_1 and γ_2 are the Leslie viscosity coefficients; finally, Γ_H and Γ_{opt} are the torques on the medium produced by the static magnetic field $\vec{H}$ and the pulsed optical field $\vec{E}$, respectively:[1]

$$\Gamma_H=\Delta\chi(\hat{n}\cdot\vec{H})\,|\hat{n}\times\vec{H}\,|=\tfrac{1}{2}\Delta\chi H^2\sin(2\theta)\,,\qquad \Gamma_{opt}=\frac{\Delta\epsilon}{4\pi}\langle(\hat{n}\cdot\vec{E})\,|\hat{n}\times\vec{E}\,|\rangle\,, \tag{6}$$

where $\Delta\chi=\chi_{\parallel}-\chi_{\perp}$ and $\Delta\epsilon=\epsilon_{\parallel}-\epsilon_{\perp}$, with $\chi_{\parallel}$ and $\chi_{\perp}$ being the diamagnetic susceptibilities parallel and perpendicular to the director, respectively, and $\epsilon_{\parallel}$ and $\epsilon_{\perp}$ being the corresponding optical dielectric constants.

The pump field at the input end is represented approximately by

$$\vec{E}_0=\hat{x}\tfrac{1}{2}\mathscr{E}_0(t)\exp(ik_0z-i\omega t)+\text{c.c.} \tag{7}$$

Inside the liquid crystal, the optical axis of the medium, which is along $\hat{n}$, varies with z because of molecular reorientation. Accordingly, the $\hat{z}$ component of the $\vec{E}$ field becomes nonvanishing in order to satisfy the requirement $\hat{z}\cdot\overleftrightarrow{\epsilon}\cdot\vec{E}=0$; we have[12]

$$E_z/E_x=-\tfrac{1}{2}\Delta\epsilon\sin(2\theta)/(\epsilon_{\parallel}-\Delta\epsilon\sin^2\theta) \tag{8}$$

with $|E_x|=\mathscr{E}_0$. Absorption and scattering loss in the medium are negligible in our case.

The appropriate boundary conditions for Eq. (5) are $\theta(z=\pm d/2)=0$ and $v_x(z=\pm d/2)=0$, assuming strong anchoring of the homeotropically aligned molecules at the windows. If the optical-field-induced molecular reorientation is much weaker than the dc magnetic-field-induced reorientation, θ and v_x in Eq. (5) can be written approximately in the form

$$\theta(z,t)=\theta_0(z)+\Delta\theta(z,t)\,,\qquad v_x(z,t)=v_m(t)[\sin(qz)-2z/d]\,, \tag{9}$$

with

$$\theta_0(z)=\theta_m\cos(qz)\,,$$

$$\Delta\theta(z,t)=\phi_m(t)\cos(qz)\,,$$

where $q=\pi/d$ and $|\phi_m|\ll|\theta_m|$. The initial conditions for $\phi_m(t)$ and $v_m(t)$ are $\phi_m(0)=v_m(0)=0$. The dc field-induced director reorientation θ_0 can be determined from the static equation

$$[K_{33}+(K_{11}-K_{33})\sin^2\theta_0]\frac{\partial^2\theta_0}{\partial z^2}+\tfrac{1}{2}(K_{11}-K_{33})\sin(2\theta_0)\left[\frac{\partial\theta_0}{\partial z}\right]^2+\tfrac{1}{2}\Delta\chi H^2\sin(2\theta_0)=0\,, \tag{10}$$

which shows $\theta_0\neq0$ only if H is larger than the critical field $H_c=q(K_{33}/\Delta\chi)^{1/2}$ for the Fréedericksz transition.[1] With the known values[16] of $\Delta\chi$ and K's for 5CB, we find $\theta_m=1.14$ rad in the $d=130$-μm sample with $H=900$ Oe, and $\theta_m=1.12$ rad in the $d=190$-μm sample with $H=605$ Oe.

Equation (5) can now be transformed into an equation for the optical-field-induced director orientation $\phi_m(t)$. Since $|\phi_m|\ll|\theta_m|$, the equation can be linearized by keeping only terms up to the first order in ϕ_m and v_m. In reducing the equation further, the quantity Γ_{opt} is expanded into a power series of $\Delta\epsilon/\epsilon_{\parallel}$ ($=0.187$ for 5CB at 24.4 °C) and terms of orders higher than $(\Delta\epsilon/\epsilon_{\parallel})^2$ are neglected. The coefficients of the resultant equation are further expanded into power series of θ_m; with $\theta_m\sim1.1$ rad, these series all converge rapidly. Finally, by multiplying the equation by $\cos qz$ and integrating from $z=-d/2$ to $d/2$, it reduces to the form

$$\gamma_1\frac{\partial\phi_m}{\partial t}+\gamma_1\tau^{-1}\phi_m=f_{opt}+\gamma q v_m\,, \tag{11}$$

where

$$\gamma_1\tau^{-1}(t)=q^2[K_{33}+(K_{11}-K_{33})G_1]+\Delta\chi H^2G_2+\frac{\Delta\epsilon}{8\pi}(\mathscr{E}_0(t))^2G_3\,,$$

$$f_{opt}(t)=\frac{\Delta\epsilon}{8\pi}(\mathscr{E}_0(t))^2G_4\,, \tag{12}$$

$$\gamma=\frac{1}{2}\left[1-\frac{8}{\pi^2}\right](\gamma_1-\gamma_2)+\gamma_2G_5\,.$$

The G's are defined as

$$G_1=\sum_{n=0}^{\infty}(-1)^n\frac{2n+3}{4(n+1)}\frac{P_n}{(2n+1)!}(2\theta_m)^{2n+2}\,,$$

$$G_2=\sum_{n=0}^{\infty}(-1)^{n+1}\frac{2P_n}{(2n)!}(2\theta_m)^{2n}\,,$$

$$G_3=\sum_{n=0}^{\infty}(-1)^{n+1}\frac{2P_nQ_n}{(2n)!}(2\theta_m)^{2n}\ ,\qquad(13)$$

$$G_4=\sum_{n=0}^{\infty}(-1)^{n}\frac{P_nQ_n}{(2n+1)!}(2\theta_m)^{2n+1}\ ,$$

$$G_5=\sum_{n=0}^{\infty}(-1)^{n+1}\left[P_n-\frac{2}{\pi^2}\frac{P_n^{-1}}{(n+1)}\right]\frac{(2\theta_m)^{2n}}{(2n)!}\ ,$$

where

$$P_n=\frac{1}{2}\times\frac{3}{4}\times\frac{5}{6}\times\cdots\times\frac{2n+1}{2n+2}\ ,$$

$$Q_n=1-2^{2n}(\Delta\epsilon/\epsilon_{\parallel})-(1+2^{2n+3}-3^{2n+2})(\Delta\epsilon/\epsilon_{\parallel})^2/16\ .$$

In Eq. (11) the γ term arises from the flow-orientation coupling, the magnitude of which depends on the static molecular orientation through the coefficient G_5. To find ϕ_m we need to know v_m. In general, the solution can only be obtained by solving Eq. (11) coupled with the equations of motion for the fluid flow. Physically, we expect the effect of the flow-orientation coupling to be as follows. First, the pump pulse rotates the director, and through the flow-orientation coupling, induces a flow. Then, even after the pulse is over, the inertia of the flow should drag the director reorientation further before it turns around and relaxes. Such behavior was, however, not observed in our experiment. It therefore suggests that in our case, the flow-orientation coupling may be rather weak. Indeed, with the previously mentioned values of θ_m in G_5, and using $\gamma_1=0.85$ P and $\gamma_2=-0.93$ P (Ref. 17) for 5CB (at 24.4 °C), we find $\gamma\simeq10^{-3}$ P for both the 130-μm and the 190-μm samples. This indicates a negligibly small flow-orientation coupling which agrees with the experimental observation. Thus, we can safely neglect the γ term in Eq. (11) and reduce it to the simpler form

$$\left[\frac{\partial}{\partial t}+\frac{1}{\tau}\right]\phi_m=f_{\rm opt}/\gamma_1\ .\qquad(14)$$

The response time τ given by Eq. (12) is a function of the pump intensity, but in our case, τ (of the order of 10–100 μsec during the pump pulse) is always much longer than the pump pulse width. The solution of Eq. (14) is then trivial. Immediately after the pump pulse is over, the optical-field-induced reorientation is simply

$$\phi_m(t=0^+)\cong\int_{-\infty}^{\infty}(f_{\rm opt}/\gamma_1)dt\ ,\qquad(15)$$

and from Eqs. (1), (3), and (9), we have the corresponding phase shift

$$\Delta\Phi_{\rm MR}(t=0^+)=\left[\frac{2\pi}{\lambda}\int_{-d/2}^{d/2}\left[\frac{\partial n_{\rm eff}}{\partial\theta}\right]_{\theta_0}\cos(qz)dz\right]\times\int_{-\infty}^{\infty}(f_{\rm opt}/\gamma_1)dt=AU_0\ ,\qquad(16)$$

where $U_0=\pi W^2(c/8\pi)\int_{-\infty}^{\infty}\mathscr{E}^2dt$ is the incoming pump pulse energy in air, W is the e^{-2} beam radius, and A is the proportional constant given by

$$A=\left[\frac{2\pi}{\lambda}\int_{-d/2}^{d/2}\left[\frac{\partial n_{\rm eff}}{\partial\theta}\right]_{\theta_0}\cos(qz)dz\right]\times\frac{G_4\Delta\epsilon}{\pi W^2c\epsilon^{1/2}\gamma_1}\left[\frac{4\epsilon^{1/2}}{(1+\epsilon^{1/2})^2}\right]\qquad(17)$$

with $4\epsilon^{1/2}/(1+\epsilon^{1/2})^2$ being the transmission coefficient at the air-sample boundary. The above result shows explicitly that $\Delta\Phi_{\rm MR}(t=0^+)$ is directly proportional to the pump pulse energy, in agreement with the experimental observation depicted in Fig. 4. Using the measured value of $W=250$ μm and the known quantities[13,17] for 5CB, we obtain from Eq. (17) $A=5.0\times10^{-3}$ rad/mJ for the 130-μm sample and $A=7.6\times10^{-3}$ rad/mJ for the 190-μm sample. These values are in excellent agreement with those deduced from the experimental data in Fig. 4. Note that if $\theta_m=0$, then $f_{\rm opt}=0$ and $\phi_m(0^+)=0$, and we would see no optical-field-induced reorientation effect. This is why the application of a dc magnetic field to induce a finite θ_m is necessary in our experiment.

According to Eq. (14), after the pump pulse is over, ϕ_m should decay exponentially to zero,

$$\phi_m(t)=\phi_m(0^+)\exp(-t/\tau_0)$$

with a time constant $\tau_0=\tau$ ($\mathscr{E}_0^2=0$). The exponential decay is indeed what was observed, as shown in Fig. 2. The values of τ_0 calculated from Eq. (12) are 13 sec for the 130-μm sample and 30 sec for the 190-μm sample. Both, however, are significantly larger than the experimental values given in Sec. III.

A better prediction for τ_0 can be obtained if the finite beam size is taken into account. The equation of motion for ϕ_m should now include a term proportional to the second derivative of ϕ_m in the transverse direction.[8] For $t>0^+$ (after the pump pulse is over), it has the form

$$\gamma_1\frac{\partial\phi_m}{\partial t}+(\gamma_1/\tau_0)\phi_m=K\nabla_\rho^2\phi_m\ ,\qquad(18)$$

where $\nabla_\rho^2=\partial^2/\partial\rho^2+(1/\rho)(\partial/\partial\rho)$, ρ is the polar coordinate in the transverse plane, and for simplicity we have assumed $K_{11}=K_{22}=K_{33}=K$ for the $\nabla_\rho^2\phi_m$ term. The general solution of Eq. (18) with the simplified boundary condition $\phi_m(\rho\geq W)=0$ should have the expression

$$\phi_m(\rho,t)=\sum_{n=1}^{\infty}a_nJ_0(S_n\rho/W)\exp(-t/\tau_n)\qquad(19)$$

with

$$\tau_n=[\tau_0^{-1}+(K/\gamma_1)(S_n/W)^2]^{-1}\ ,\qquad(20)$$

where J_0 is the Bessel function of order zero, S_n is the nth root of J_0, and a_n's are coefficients. We assume that the $\exp(-t/\tau_1)$ term dominates and the solution takes the simple form

$$\phi_m(\rho,t)=\begin{cases}\phi_m(0,0^+)J_0(S_1\rho/W)\exp(-t/\tau_1) & \text{for } \rho\leq W\\ 0 \quad \text{for } \rho>W\end{cases}\qquad(21)$$

with $S_1=2.405$. The relaxation time τ_1 is now given by Eq. (20). As expected, the relaxation time decreases with the beam size. Taking $K\simeq K_{33}=0.8\times10^{-6}$ dyn,[16]

$W=250$ μm and the value of τ_0 obtained earlier, we find $\tau_1=12$ sec for the 130-μm sample, and $\tau_1=24$ sec for the 190-μm sample, in fair agreement with the measured values.

B. Laser-induced thermal effect

Residual absorption at the pump laser frequency can lead to heating of the liquid-crystal medium. Although the absorption in 5CB at 1.06 μm is very weak, it can still result in a detectable temperature change. Right after the pump pulse is over, the temperature rise ΔT has a distribution nearly constant along z, but proportional to the pulse energy distribution transversely. Then, through heat diffusion, the temperature variation should relax back to the uniform distribution at equilibrium. At any given instant, a weighted integration of the temperature change $\Delta T(\rho,z)$ along z is reflected in the phase shift $\Delta\Phi_{\rm th}$ described in Eq. (1).

A general analysis of the variation of $\Delta T(\rho,z)$ with time is quite complicated. However, since the pump beam diameter was much larger than the sample thickness in our experiment, we can use the approximation of neglecting heat diffusion in the transverse plane. The temperature change ΔT now obeys the one-dimensional heat-diffusion equation

$$\frac{\partial}{\partial t}\Delta T=\alpha\frac{\partial^2}{\partial z^2}\Delta T . \tag{22}$$

We have assumed here that the thermal diffusivity α is independent of z, that is, we have neglected the dependence of α on the director orientation in the medium. With the initial condition $\Delta T(t=0^+)=\Delta T_0$ throughout the sample and the boundary condition $\Delta T(z=\pm d/2,\ t)=0$, we readily find the solution of Eq. (22) as[18]

$$\Delta T(z,t)=\frac{4\Delta T_0}{\pi}\sum_{n=0}^{\infty}\frac{(-1)^n}{2n+1}\cos[(2n+1)qz]e^{-t/t_n} , \tag{23}$$

where $t_n=1/\alpha(2n+1)^2q^2$ and $q=\pi/d$. The corresponding phase shift $\Delta\Phi_{\rm th}$ due to ΔT can then be obtained from Eq. (1):

$$\Delta\Phi_{\rm th}(t)=\frac{8\Delta T_0}{\lambda}\sum_{n=0}^{\infty}\frac{(-1)^n}{2n+1}e^{-t/t_n}\times\int_{-d/2}^{d/2}\left[\frac{\partial n_{\rm eff}}{\partial T}\right]_{\theta_0}\times\cos[(2n+1)qz]dz \tag{24}$$

with $\theta_0=\theta_m\cos qz$.

As we mentioned in Sec. II, it is possible to choose a value of θ_m, by adjusting the magnetic field H, to make $\Delta\Phi_{\rm th}(0^+)=0$. In principle, this value of θ_m can be estimated from Eq. (2) using Eq. (3) and is independent of the sample thickness. However, the large uncertainty in the known ratio $(\partial n_e/\partial T)/(\partial n_0/\partial T)$ allows only a rather crude estimate. Our experimental result yielded $\theta_m\simeq1.1$ (corresponding to $H/H_c\simeq1.35$) for the two samples of different thickness. Even with $\Delta\Phi(0^+)=0$, Eq. (24) shows that at later times, $\Delta\Phi_{\rm th}(t)$ would still become finite. For $\Delta\Phi_{\rm th}(t)$ in the relatively long-time regime, we need to keep only the two terms with the longest relaxation times in Eq. (24). By expanding $(\partial n_{\rm eff}/\partial T)_{\theta_0}$ into a series up to the θ_0^2 term and evaluating the integral in Eq. (24), we find

$$\Delta\Phi_{\rm th}(t)\cong{\rm const}\times[-\exp(-t/t_0)+\tfrac{3}{5}\exp(-t/t_1)] . \tag{25}$$

The result here shows that $t_0=9t_1=d^2/\pi^2\alpha$, and the ratio of the amplitudes of the two terms is $-\frac{3}{5}$, independent of the pump pulse energy and the sample thickness.

Our experimental data on $\Delta\Phi_{\rm th}(t)$ can indeed be fit approximately by the linear combination of two exponentials with proper signs on their amplitudes. The ratio of the two relaxation times was found to be $t_0/t_1\simeq10$ for the two samples of different thickness, in fair agreement with the theoretical prediction. By using $t_0=d^2/\pi^2\alpha$, the value of α deduced from the measured t_0 is, however, different for the two samples of different thickness. It is 1.26×10^{-4} cm^2/sec for the 130-μm sample, or 1.93×10^{-4} cm^2/sec for the 190-μm sample. These values of α are in the correct range of thermal diffusivities of liquid crystals. The dependence of α on sample thickness can be qualitatively attributed to the effect of transverse heat diffusion in the sample cell which we have neglected in our analysis. In a thicker cell, the transverse heat diffusion is more important, and therefore, the effective one-dimensional thermal diffusivity α should be larger. The values of α deduced above roughly scale with the sample thickness. A more serious discrepancy between theory and experiment is in the relative amplitude of the two exponentials of $\Delta\Phi_{\rm th}(t)$. The theoretical value is $-\frac{3}{5}$, while the experimental value is -1.2 in Fig. 3. This discrepancy presumably also arises from the approximations we have made in the analysis.

V. DISCUSSION

We have studied in this work transient refractive-index change in a nematic liquid crystal 5CB arising from the pulsed-laser-induced molecular reorientation and thermal effect. The same mechanisms actually also lead to a change in optical birefringence. This is because the ordinary and extraordinary refractive indices of the medium have different functional dependences on the director orientation angle and on the temperature. In the steady state, the induced refractive-index change from molecular reorientation in 5CB is larger than that from the thermal effect by about 1 order of magnitude.[7] The former, however, has a much slower response time than the latter. Consequently, when pulsed laser excitation is used, the thermal effect can appear dominating the molecular reorientation effect. The two effects often can be separated by measurements with different pump polarizations; this is because the thermal effect is generally independent of the pump polarization.

The change in refractive index induced in a nematic liquid crystal by a steady-state optical field is known to be extremely large, with $\Delta n\sim0.05$ for a pump field of ~100 W/cm^2.[5] However, because of the very slow time response of the correlated molecular orientational motion, the amount of Δn that can be induced by a $\sim$10-nsec laser pulse through molecular reorientation is rather small. It is $\sim(T_p/\tau)$ times smaller than the steady-state Δn, assuming equal cw and pulsed beam intensity, where T_p is the pulse width and τ is the response time of molecular reorientation ($\tau \sim 10$ sec for 100-W/cm^2 pump intensity). Thus, to attain an average Δn of 10^{-4} in a 100-μm cell (corresponding to a phase shift of $\Delta\Phi_{MR} \sim 0.1$ rad) by a 10-nsec pulsed laser excitation, we would need a laser intensity of $\sim 2\times10^8$ W/cm^2. This is indeed roughly the same order of magnitude as the one we had to use in our experiment. Since $\Delta\Phi_{MR} \sim 0.1$ rad is readily detectable, we can then use the time variation of $\Delta\Phi_{MR}$ to study the dynamics of molecular reorientation. We found from our measurement that it obeys the continuum theory of Ericksen and Leslie very well.

The Ericksen-Leslie theory has been widely used in discussing the dynamic phenomena of the nematic state. However, the inertia term in the equation is always neglected. While this approximation should be valid for long-pulse excitation, it may break down if the exciting pulse is sufficiently short. In our experiment with 6-nsec exciting laser pulses, there was no evidence of the inertia term being important. This sets a limit on the magnitude of the moment of inertia ρ_1 in Eq. (5) to be less than $\sim 10^{-8}$ g cm^{-1}. If we assume that ρ_1 corresponds to the moment of inertia of a bunch of oriented molecules within a volume of the size of the correlation length ($\sim$100 Å) of the liquid crystal, then we expect $\rho_1 \sim 10^{-12}$ g cm^{-1}. The effect of such a "moment of inertia" on molecular reorientation could be measured by picosecond laser pulses.

In our measurements, the situation was greatly simplified by using a geometric arrangement such that the flow-orientation coupling was negligible. As a result, the effect of fluid flow could be disregarded, and the molecular reorientation part and the thermal part could be decoupled. In general, however, the flow-orientation coupling can be important, and it can lead to interesting but complicated dynamic phenomena in liquid crystals. For example, laser heating resulting in a strong local temperature gradient can give rise to a convection flow in the fluid, which then orients the molecules. The back reaction of the molecular reorientation should in turn affect the flow. That laser heating can be well localized and switched on or off very rapidly should allow us to excite and investigate certain dynamic phenomena in a fluid that cannot be studied with other techniques.

More generally, acoustic waves can also be generated in a liquid crystal by pulsed laser excitation. They arise from both the direct electrostrictive coupling between the laser field and the acoustic waves and the indirect excitation via thermal expansion caused by laser heating. Both mechanisms rely on the existence of an intensity gradient in the laser-beam profile. For a larger beam cross section, the acoustic waves generated are less intense. In our experiment, the pump beam size was sufficiently large so that the effect of acoustic-wave generation was negligible. It is, however, possible to generate acoustic waves in a liquid crystal more effectively by crossing two laser pulses at an angle and inducing a transient density grating in the sample. The resulting acoustic waves and their properties can be probed and studied by a third laser pulse.

Generally speaking, excitation of a medium by short laser pulses can be used to study dynamic properties of the medium over a very wide time range. Here, we have shown that nanosecond-pulse excitation can yield information about the dynamics of molecular reorientation on the $\sim$10-sec time scale, and thermal effect on the 10–100-msec time scale. The power of this technique lies in the fact that a single δ-function-like laser pulse may induce a number of fundamental excitation modes of vastly different time constants. Consider, for example, molecular reorientation coupled with flow induced by a picosecond laser pulse in a liquid crystal. It can be shown that, aside from the thermal effect, the transient behavior will manifest itself with three characteristic time constants:

$$\tau_i \simeq \rho_1/\gamma_1 ,$$

$$\tau_f \sim \rho/\bar{\gamma} q^2 ,$$

$$\tau_s \sim \bar{\gamma}/\bar{K} q^2 ,$$

where ρ_1, γ_1, and q have the meaning defined earlier, and $\bar{\gamma}$ and $\bar{K}$ are the effective viscosity coefficient and elastic constant, respectively, and ρ is the density of the fluid. The time constants τ_f and τ_s are associated with the so-called "fast" and "slow" modes[8] in a liquid crystal and dominated by the damping of flow and the relaxation of molecular reorientation, respectively. The molecular reorientation observed in our experiment corresponds to the slow mode. The time constant τ_i, which is independent of pumping power and geometry, is associated with a cooperative librational mode. If we take $\rho_1 \sim 10^{-12}$ g cm^{-1}, $\bar{\gamma} \sim 1$ P, $\bar{K} \sim 10^{-6}$ dyn, $\rho \simeq 1$ g cm^{-3}, and $q^{-2} \sim 10^{-5}$ cm^2 (for $d \sim 100$ μm), then $\tau_i \sim 1$ psec, $\tau_f \sim 10$ μsec, and $\tau_s \sim 10$ sec. These three modes excited by the same picosecond laser pulse will therefore have very different time responses, and can, in principle, be studied by a probe laser in different time regimes.

VI. CONCLUSION

We have presented here the first observation of transient molecular reorientation induced in a liquid crystal by a Q-switched laser pulse. The response time of molecular reorientation in the nematic phase is of the order of 10–100 μsec. Although this is 10^3–10^4 times longer than the duration of the laser pulse, transient molecular reorientation is still strong enough to yield an easily detectable phase shift in the probe beam. Residual absorption and subsequent very rapid radiationless conversion into heat can result in a temperature rise in the medium which decays via heat diffusion with relaxation times in the 10–200 msec range. The temperature rise also induces a refractive-index change in the medium and hence a phase shift in the probe beam. This thermal effect and the molecular reorientation are initiated simultaneously by the pulsed laser excitation. They are in general coupled via fluid flow. In our experiment, an initial orientation of molecules by a dc magnetic field allowed us to decouple the two. Then, by using different pump polarizations, the two effects could be separately measured. Theoretical calculation following the continuum theory of Ericksen and Leslie for liquid crystals showed excellent agreement with the experimental results on molecular reorientation. On the other hand, a simple theory of heat diffusion explained the thermal effect fairly well.

The experiment reported in this work demonstrates the possibility of using a short laser pulse (or pulses) to simultaneously excite various dynamic modes in a medium and subsequently probing the responses of these modes which could vary on vastly different time scales. The technique allows the studies of dynamic properties of several material excitations by a single measurement in the time domain. This is in contrast to the usual spectroscopic probing technique. Aside from being indirect, the latter has the disadvantage of requiring several different analyzing systems in order to study material excitations with very different time responses.

ACKNOWLEDGMENTS

This work was supported by the National Science Foundation under Grant No. DMR 81-17366.

---

*Permanent address: Department of Physics, Zhongshan University, Guangzhou, Guangdong, People's Republic of China.

[1]See, for example, P. G. de Gennes, *The Physics of Liquid Crystals* (Clarendon, Oxford, 1974), Chap. 3.

[2]R. M. Herman and R. J. Serinko, Phys. Rev. A **19**, 1757 (1979).

[3]B. Ya. Zel'dovich and N. V. Tabiryan, Pis'ma Zh. Eksp. Teor. Fiz. **30**, 510 (1979) [JETP Lett. **30**, 478 (1979)]; B. Ya. Zel'dovich, N. V. Pilipetskii, A. V. Sukhov, and N. V. Tabiryan, *ibid.* **31**, 287 (1980) [*ibid.* **31**, 263 (1980)]; A. S. Zolot'ko, V. F. Kitaeva, N. Kroo, N. N. Sobolev, and L. Chillag, *ibid.* **32**, 170 (1980) [*ibid.* **32**, 158 (1980)].

[4]I. C. Khoo and S. L. Zhuang, Appl. Phys. Lett. **37**, 3 (1980); IEEE J. Quantum Electron. **QE-18**, 246 (1982); I. C. Khoo, Phys. Rev. A **25**, 1636 (1982); **27**, 2747 (1983).

[5]S. D. Durbin, S. M. Arakelian, and Y. R. Shen, Phys. Rev. Lett. **47**, 1411 (1981).

[6]S. D. Durbin, S. M. Arakelian, and Y. R. Shen, Opt. Lett. **6**, 411 (1981); **7**, 145 (1982).

[7]Mi-Mee Cheung, S. D. Durbin, and Y. R. Shen, Opt. Lett. **8**, 39 (1983).

[8]See, for example, S. Chandrasekhar, *Liquid Crystals* (Cambridge University Press, Cambridge, 1977), Chap. 3, and references therein.

[9]Transient optical Kerr effect and the associated nonlinear optical effects induced in the *isotropic phase* of liquid crystals by laser pulses have already been studied extensively. See, for example, Y. R. Shen, Rev. Mod. Phys. **48**, 1 (1976), and references therein; J. R. Lalanne, B. Martin, B. Pouligny, and S. Kielich, Mol. Cryst. Liq. Cryst. **42**, 153 (1977), and references therein.

[10]F. J. Kahn, Appl. Phys. Lett. **22**, 386 (1973).

[11]M. Born and E. Wolf, *Principles of Optics* (Pergamon, Oxford, 1975), p. 312.

[12]M. Born and E. Wolf, *Principles of Optics*, Ref. 11, Chap. 14.

[13]P. P. Karat and N. V. Madhusudana, Mol. Cryst. Liq. Cryst. **36**, 51 (1976); K. C. Chu, C. K. Chen, and Y. R. Shen, *ibid.* **59**, 97 (1980).

[14]P. Pieranski, E. Dubois-Violette, and E. Guyon, Phys. Rev. Lett. **30**, 736 (1973); E. Dubois-Violette, E. Guyon, and P. Pieranski, Mol. Cryst. Liq. Cryst. **26**, 193 (1974).

[15]C. Z. Van Doorn, J. Phys. (Paris) Colloq. **36**, C1-261 (1975).

[16]J. D. Bunning, T. E. Faber, and P. L. Sherrell, J. Phys. (Paris) **42**, 1175 (1981).

[17]K. Skarp, S. T. Lagerwall, and B. Stebler, Mol. Cryst. Liq. Cryst. **60**, 215 (1980).

[18]See, for example, F. J. Bayley, J. M. Owen, and A. B. Turner, *Heat Transfer* (Barnes & Noble, New York, 1972), p. 54.

Mol. Cryst. Liq. Cryst., 1985, Vol. 122, pp. 59–75
0026-8941/85/1224-0059/$20.00/0

Nonlinear Optical Effects in the Nematic Phase

D. ARMITAGE and S. M. DELWART

Lockheed Palo Alto Research Laboratory, Palo Alto, California 94304

(*Received July 30, 1984*)

Large nonlinear optical effects are associated with nematic order. This is due to reorientation of the nematic by the optic field and or reduction in nematic order by absorption of optical energy. These effects were analyzed taking into account viscosity and turbidity. Experiments were performed using an absorption cell containing dye doped pentylcyanobiphenyl. A thermal grating is induced by absorbing interfering 10-ns YAG laser pulses. The grating is continuously monitored by a HeNe laser readout. The amplitude and response time of the grating increase as the isotropic phase transition temperature is approached. A diffraction efficiency of 1 percent with 200-ns response is observed for a write energy of 1 mJ/cm^2, 1°C below the transition temperature. Modulating the thermal grating across the phase transition produces an index grating with a nonthermal decay characteristic ~10 ms, typical of the reorientation time of a disordered nematic.

INTRODUCTION

Nonlinear optics is concerned with effects such as harmonic generation and wave mixing which follow from a nonlinear dependence of polarization on electric field.[1] The response at optical frequencies in the liquid crystal phase is discussed elsewhere.[2] Here we are interested in nonlinearities based on director reorientation and thermal effects in the nematic phase.

The control of director orientation by a static magnetic or electric field has been established for many years. It follows as a direct consequence that the director will be influenced by an optical field. This has been demonstrated in recent years; however, the effect was observed much earlier.[3–10] The current activity stems from potential application of liquid crystals in nonlinear optical devices.

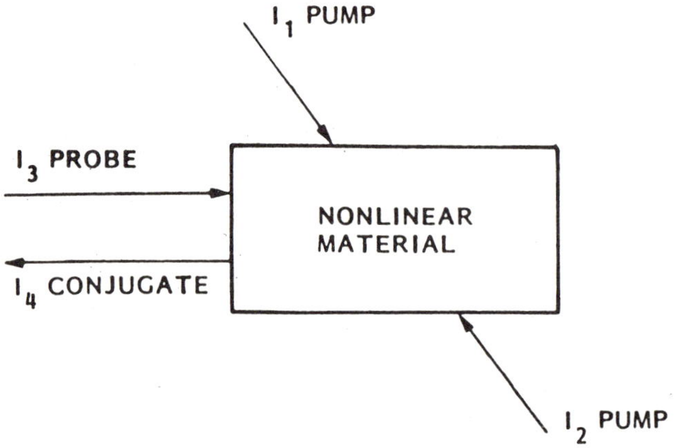

FIGURE 1 Degenerate four-wave mixing.

For example, the phase conjugate operation is of particular interest in optical systems.[11, 12] In identifying the wavefront phase distribution $\phi(x, y)$, a phase conjugate operation produces a counterpropagating wave with opposite phase $-\phi(x, y)$. Therefore, it is possible to construct systems where the phase aberrations are automatically cancelled by a double pass through the distorting media via a phase conjugate mirror.

Figure 1 shows phase conjugation via degenerate four-wave mixing. This can be interpreted as the interaction of the probe and pump beams to form a hologram in the nonlinear medium, which is also read by the pump beam. The relation between the four waves is described analytically in terms of the nonlinear third-order susceptibility tensor. The advantage of liquid crystals in such applications is a very large effect for small input power. The disadvantage is a slow viscous response and optical path length limited by turbidity.

The response can be accelerated by increased optical power. Eventually the retarding force is dominated by viscosity. We calculate the response in this regime.

Recently it has been observed that the thermal grating response of a nematic is unusually large.[13–15] The property exploited is the temperature dependence of the refractive index. This effect has considerable history, where prior experiments used a solvent containing a dye, e.g., rhodamine 6G in ethanol.[16, 17] The optical energy is absorbed by the dye and then transferred to the solvent, resulting in a thermal grating and consequent index grating. The time constant of this process in ordinary liquids is of nanosecond order.

The temperature sensitivity of the nematic refractive index, particularly close to the isotropic phase transition, is determined by the temperature dependence of the order parameter. However, the greater sensitivity is associated with a larger response time of 100-ns order. We show that when submillisecond response is demanded, the thermal effect is more efficient than the orientation effect.

TURBIDITY

Turbidity is a distinctive feature of the nematic phase. This is a consequence of large birefringence and weak elasticity. These properties favor low power devices. The optical path length in a nematic device is limited ~100 μm by turbidity. The following expression for the optical scattering cross section can be written in the limit $\Delta n/n \to 0$[18]

$$\sigma = V(\epsilon_a \pi/\lambda^2)^2 (kT/Kq^2)P$$

where σ = scattering cross section/solid angle, V = volume = Ad, ϵ_a = optical dielectric anisotropy = $2n\Delta n$, λ = vacuum wavelength = 5×10^{-5} cm, kT = thermal energy, K = elastic constant $\simeq 10^{-6}$ dyne, q = scattering vector = $(4\pi n \sin 1/2\phi)/\lambda$, ϕ = scattering angle, n = refractive index = 1.5, Δn = birefringence = 0.2, P = polarization and orientation factor $\simeq 1$; for the case shown in Fig. 2, we calculate $P = 1.5$.

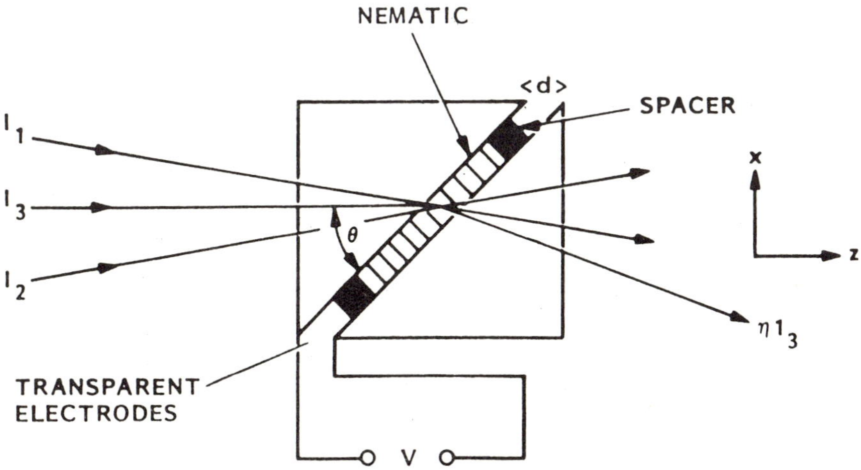

FIGURE 2 Configuration of liquid crystal, laser beams, and electric fields.

The integrated scattering is written

$$2\pi \int_{\phi_0}^{\pi} \sigma \sin\phi \, d\phi = 1.6 \times 10^{-4}\ \mathrm{K}^{-1} \Delta n^2 A d \ln(2/\phi_0) \qquad (2)$$

where ϕ is a minimum angle determined by the minimum energy of fluctuations. The nematic is contained in an optical cell with spacing $d < 200\ \mu$m. Assuming a strong anchoring condition, the minimum distortion energy corresponds to taking $q = \pi/d$ or $\phi_0 = \lambda/2nd = 3.4 \times 10^{-3}$.

Hence, the scattering attenuation of optical intensity I is written

$$I \propto \exp(-ad) = \exp(1.6 \times 10^{-3}\ \mathrm{K}^{-1} \Delta n^2 d) \qquad (3)$$

The result is insensitive to the choice of ϕ_0. For typical values, the $1/e$ point $\simeq 0.2$ mm.

ORIENTATIONAL RESPONSE

A beam propagating at θ to the nematic axis as shown in Fig. 2 experiences a local birefringence $\Delta n \sin^2\theta$, in the limit $\Delta n \ll n$. This exerts a torque given by[7]

$$\text{Torque} = 2I\Delta n c^{-1} \sin\theta \cos\theta \qquad (4)$$

where $c = 3 \times 10^{10}$ cm/s and $\theta = 45°$ for maximum torque.

Coherent input beams, as shown in Fig. 2, give an intensity distribution in the nematic.

$$I = I_1 + I_2 + 2(I_1 I_2)^{1/2} \cos qx \qquad (5)$$

This modulates the nematic orientation at spatial frequency $\Lambda = 2\pi/q$, producing a phase grating. Using the thin phase grating approximation ($J_1^2(\psi) = \psi^2/4$), the equilibrium diffraction efficiency has the form.[7]

$$\eta_0 = I_1 I_2 \left[\frac{\Delta n^2 d}{\lambda c K (4\Lambda^{-2} + d^{-2})} \right]^2 \qquad (6)$$

substituting typical values gives $\eta_0 \simeq 1$ percent for $(I_1 I_2)^{1/2} \simeq 1$ W/cm^2, when $d = \Lambda = 100\ \mu$m.

In the high speed limit the response is viscous limited. Therefore, ignoring the elasticity gives the torque balance equation.

$$I\Delta n/c = \gamma\, d\theta/dt \tag{7}$$

where γ = effective viscosity ~0.1 poise.

This implies a diffraction efficiency[19]

$$\eta_0 = I_1 I_2 (2\pi d\Delta n^2 \Delta t \gamma^{-1} \lambda^{-1} c^{-1})^2 \tag{8}$$

where Δt is the integration time.

Taking into account turbidity, I should be averaged according to $\exp(-\alpha z)$ and η_0 attenuated $\exp(-\alpha d)$. Hence, Eq. (8) becomes

$$\eta = \eta_0 (\alpha d)^{-2} [1 - \exp(-\alpha d)]^2 [\exp(-\alpha d)] \tag{9}$$

substituting Eqs. (3) and (8) gives a maximum

$$\max \eta = 2 \times 10^6 I_1 I_2 (\Delta t\, K \gamma^{-1} \lambda^{-1} c^{-1})^2 \tag{10}$$

when

$$\exp(1.6 \times 10^{-3}\, K^{-1} \Delta n^2 d) = 3 \tag{11}$$

If d is optimized according to Eq. (11), η favors materials with high $(K/\gamma)^2$. Substituting the given typical values and requiring $\eta = 1$ percent for $\Delta t = 10^{-3}$ s gives $(I_1 I_2)^{1/2} = 1000$ W/cm^2 and optimum $d = 160$ μm.

Equation (6) can be similarly optimized for turbidity. Making the assumption $\Lambda < d$, optimum $d \simeq 160$ μm and $\eta \simeq 0.1\, \eta_0$. However, the averaging process used in deriving Eq. (9) introduces error in this case. The η described here characterizes the performance in a 4-wave mixing configuration.

SUPPRESSION OF SCATTERING BY APPLIED ELECTRIC FIELD

Applying an electric field reduces the thermal fluctuations and hence light scattering.[18,20] Therefore, η could be raised by increasing d.[19]

The applied electric field introduces an energy term $E^2\Delta\epsilon/2$, where $\Delta\epsilon$ is the low-frequency permittivity anisotropy. This term is added to the elastic term Kq^2, and the scattering cross section becomes[18, 20]

$$\sigma_E = \sigma_0 Kq^2/(Kq^2 + E^2\Delta\epsilon/2) \tag{12}$$

Integrating as before, results in a field-dependent beam attenuation

$$I = \exp\left[-7.84 \times 10^{-5}\Delta n^2\, \mathrm{K}^{-1}\ln(1 + 2.8 \times 10^{11}K/E^2\Delta\epsilon)\right] \tag{13}$$

Taking an upper limit $E = 10^5$ V/cm and $\Delta\epsilon = 10\epsilon_0$, the $1/e$ point becomes 2.2 mm, which is a factor 15 over the zero field case.

The applied field also suppresses the optically written phase grating. However, since we are committed to short optical pulses, we can take advantage of the slow viscous relaxation when the applied field is removed.

When the field is switched off, the fluctuations grow at rate

$$\tau = \gamma/Kq^2 \tag{14}$$

for

$$\tau\langle\Delta t, q^2\rangle\gamma/K\Delta t \tag{15}$$

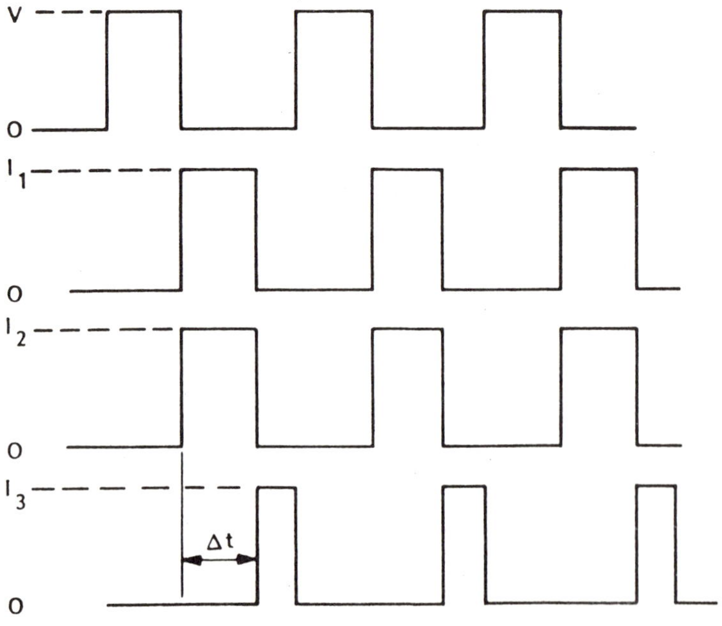

FIGURE 3 Sequence of laser and voltage pulses.

This implies a higher cutoff angle ϕ_0 in the scattering integral. Approximating ϕ_0 according to $q^2 = \gamma / K\Delta t$, the beam attenuation becomes

$$I = \exp\{-79\Delta n^2 d[26 + \ln(K\Delta t/\gamma)]\}$$

for $\Delta t = 10^{-3}$ s the $1/e$ point is 0.44 mm, which is 2.9 over the zero field case, implying $\eta \times (2.9)^2$. Figure 3 indicates the pulse sequence.

THERMAL GRATINGS

The absorption of optical energy is associated with a temperature rise and hence a change in refractive index. This effect is exploited in the dye cell, where the absorption is controlled by a dye which is solubilized in a solvent chosen for high dn/dT.[16, 17, 21, 22]

In the nematic phase, dn/dT is exceptionally large. A wide range of soluble dyes are available. The potential of thermal gratings in nematics has already been recognized.[13] We analyze the performance of the nematic dye cell as follows.

Assume intersecting laser beams of equal intensity I, which give an interference pattern

$$\text{intensity} = 2I(1 + \sin qx) \qquad (16)$$

The absorbed energy spatially modulates the refractive index, which in the adiabatic approximation is written

$$\delta n = 2C^{-1}\, dn/dT\, It\alpha \exp(\ \alpha z) \sin qx \qquad (17)$$

In the thin phase approximation ($J_1^2(\psi) = \psi^2/4$)

$$\eta = (2\pi/\lambda)^2 I^2 t^2 C^{-2} (dn/dT)^2 [1 - \exp(-\alpha d)]^2 \exp(-\alpha d) \qquad (18)$$

where t = integration time, C = heat capacity/volume, T = temperature, α = absorption coefficient.

This is maximum at $\alpha = 1.1/d$ when

$$\eta = 5.85 I^2 t^2 \lambda^{-2} C^{-2} (dn/dT)^2 \qquad (19)$$

Clearly, the material property $C^{-2}(dn/dT)^2$ should be maximized.

The refractive index can be approximated.[23]

$$dn_e/dT = 1/2n_i\, d\rho/dT + 2/3\Delta n'\, dS/dT \tag{20}$$

$$dn_0/dT = 1/2n_i\, d\rho/dT - 1/3\Delta n'\, dS/dT \tag{21}$$

where n_e = extraordinary index, n_0 = ordinary index, n_i = index in isotropic phase, ρ = density, S = order parameter, $\Delta n' = \Delta n/S$.

Clearly $|dn_e/dT| > |dn_0/dT|$. As the isotropic point is approached, dS/dT dominates and $|dn_e/dT| \rightarrow |2dn_0/dT|$. The normalized birefringence $\Delta n/S$ should be used in comparing nematogens.

The mean field approximation[23]

$$S = \left(1 - 0.98\, TV^2/T_n V_n^2\right)^{0.22} \tag{22}$$

gives some indication of the behavior over a wide temperature range.

Experimental data for methoxybenzylidenebutyaniline (MBBA) and pentylcyanobiphenyl (5CB) are available. Two sets of refractive index data were compared for each material and found consistent when expressed in terms of reduced temperature $T_n - T$.[24–26] The heat capacity as a function of temperature is available for MBBA.[27, 28] For 5CB the value at $T_n - T = 10°K$ is known,[29] and assuming the same reduced temperature dependence as MBBA should introduce negligible error in this calculation.

The experimental values of $C^{-2}(dn/dT)^2$ are plotted against reduced temperature in Fig. 4. The values derived from the mean field expression (22) are also plotted. As expected, the mean field approximation is poor close to the transition point.

There is about a factor 100 variation with temperature. At $T_n - T = 1°K$, the predicted 5CB, $\eta = 1$ percent for input energy 0.8 mJ/cm^2.

A further advantage of the nematic follows from the incorporation of pleochroic dyes.[30] The dye is oriented by the nematic and results in a polarization-dependent absorption coefficient. This can be exploited by orthogonally polarized write and read beams. The ratio of optical densities can be written.[30]

$$L \text{ dye } \alpha_e/\alpha_0 = r_L = (1 + 2S)/(1 - S) \tag{23}$$

$$T \text{ dye } \alpha_0/\alpha_e = r_T = (2 + S)/(1 - S) \tag{24}$$

where L and T refer to dyes absorbing light polarized along or

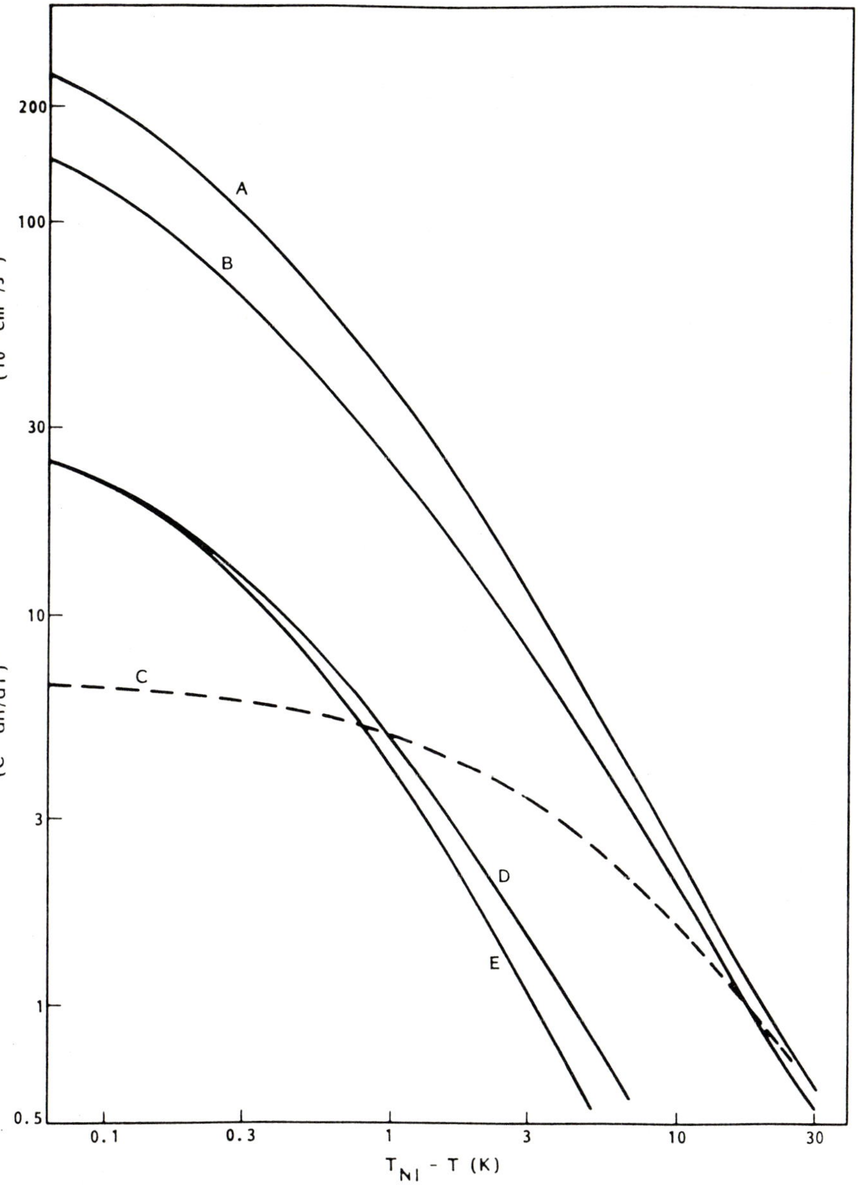

FIGURE 4 Nematic materials parameter $(C^{-1}\, dn/dT)^2$ against reduced temperature $T_n - T$. (A) MBBA extraordinary axis, (B) 5CB extraordinary axis, (C) mean-field prediction for 5CB extraordinary axis, (D) MBBA ordinary axis, (E) 5CB ordinary axis.

perpendicular to the optic axis. The order parameter of the dye in the liquid is approximately the same as the nematic order parameter.

Substituting the α ratios in Eq. (18) and recalculating the maximum gives the result

$$\eta_L = 39.5\left[1 - (1 + 2r_L)^{-1}\right]^2 (1 + 2r_L)^{-1/r_L} I^2 t^2 \lambda^{-2} C^{-2} (dn_0/dT)^2 \tag{25}$$

$$\eta_T = 39.5\left[1 - (1 + 2r_T)^{-1}\right]^2 (1 + 2r_T)^{-1/r_T} I^2 t^2 \lambda^{-2} C^{-2} (dn_e/dT)^2 \tag{26}$$

Note the association of L with n_0 and T with n_e. The dyes are more effective for increasing S, but high dS/dT favors low S. Substituting $S = 0.5$ in Eqs. (23)–(26) gives $\eta_L = 3.08\ \eta_0 I$ and $\eta_T = 2.29\ \eta_{eI}$, where η_I = the efficiency for a polarization independent dye. This is about the level of improvement that can be attained in pursuit of high η, which is dominated by dS/dT hence $S < 0.5$.

In some applications, low attentuation of the readout beam is important. Therefore, expanding the exponentials in Eq. (18)

$$\alpha d \ll 1, \qquad \eta_T = r_T^2 \eta_{eI} \tag{27}$$

with a similar expression for η_L. It should be noted that for $T_n - T > 20°K$, $r_T^2 > 10$ and $r_L^2 > 30$

EXPERIMENTAL

The nematogen used in the experiments was 5CB containing a small quantity of L-dye D81 (EM Chemicals). The glass sample cell spacing was 10 or 100 μm, (Hellma Cells). Uniform parallel alignment of the nematic was achieved by the rubbed PVA process. The glass sample cell was inserted into a temperature controlled aluminum block.

Figure 5 shows the optical arrangement. A 10-μm period thermal grating is written via M1 and M2. Continuous readout with a HeNe laser records the grating evolution. Insertion of M3 allows readout with the YAG pulse.

Pure 5CB, without dye, fails to show diffraction for write energy as high as 0.2 J/cm^2. Inclusion of a small quantity of dye provides optical absorption, and diffraction is easily seen.

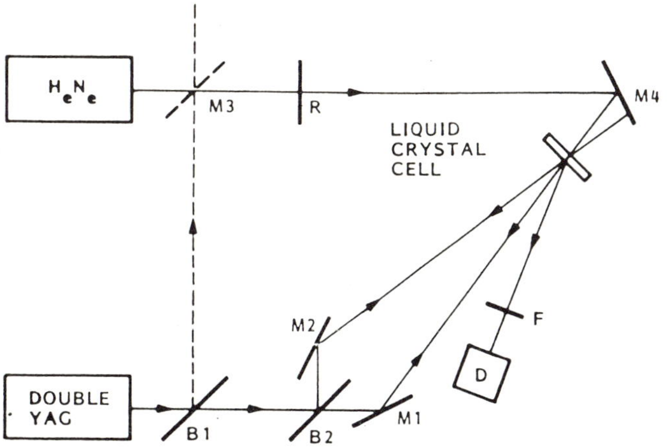

FIGURE 5 Optical arrangement, M1, M2, M3, M4 mirrors, B1, B2 beamsplitter, R polarization control, F red filter, D diode detector.

The first-order diffraction intensity is recorded on a storage oscilloscope. The YAG pulse is 10 ns long, and the detector time constant is 10 ns. This must be kept in mind when interpreting the scope trace.

The response time of the nematic to an absorbed energy pulse is determined by the order parameter response. Here a connection can be made to experiments on ultrasonic propagation.[31,32] The order parameter cannot change faster than a molecular vibration period. Since S is determined by the interactions of many molecules, the characteristic time is much longer than the molecular time scale. Moreover, as the phase transition point is approached, there is a slowing of order parameter response typical of critical phenomena.[18] This has been verified by ultrasonic experiments. Therefore, the thermal grating response time should be comparable with the ultrasonic data.[31,32]

Figure 6 shows the time evolution of the first-order diffraction intensity in the isotropic phase. An experimental time constant $\tau \simeq 30$ ns is indicated at $T - T_n = 6°K$. The time constant is not sensitive to temperature $T - T_n$. The magnitude of the time constant is comparable with ultrasonic result, $\tau = 25$ ns.[32] However in the ultrasonic case there is a strong temperature dependence $\tau \propto (T_n^* - T)^{-1}$.

Figure 7 shows similar recordings for the nematic phase at three temperatures. The time constant $\tau = 60$, 140, and 220 ns at $T_n - T = 14$, 2, and 0.4°K. The magnitude is of the same order, but greater than observed in the ultrasonic experiment. However, the temperature dependence is weaker.[32]

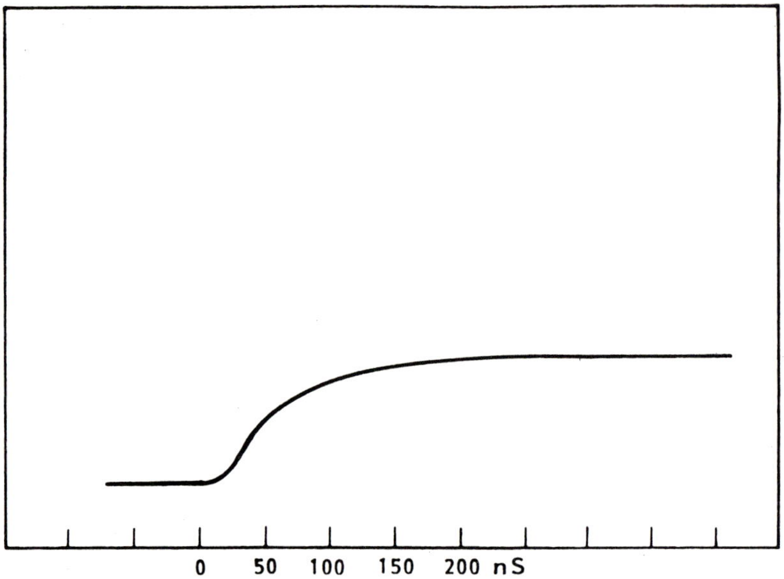

FIGURE 6 Time response of first-order diffraction intensity in isotropic phase of 5CB. $T - T_n = 6°K$. Cell thickness = 100 μm.

The decay of the thermal grating is determined by normal heat transfer processes. It can be shown[33] that the decay rate $\tau_d = (D_q^2)^{-1}$, where $D_e = 1.25 \times 10^{-3}$ and $D_0 = 7.9 \times 10^{-4}$ cm^2 s^{-1} in the extraordinary and ordinary directions respectively, and q is the grating wave vector. This is consistent with the decay trace shown in Fig. 8 for a 10-μm grating period.

The grating can be modulated into the isotropic phase, forming alternating layers of nematic and isotropic liquid. The discontinuity in

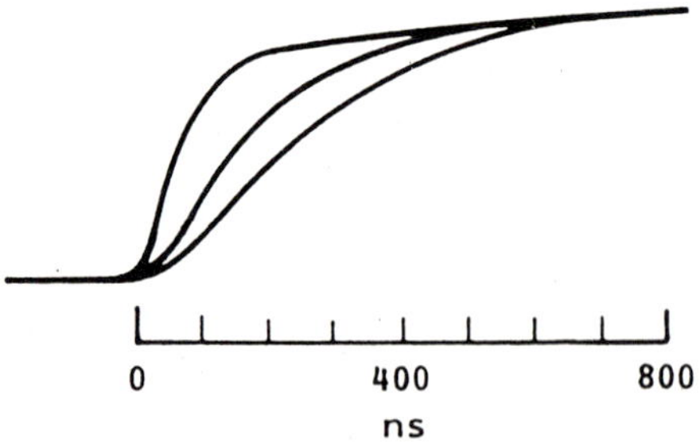

FIGURE 7 Rise time of first-order diffraction intensity in nematic phase of 5CB at $T_n - T = 14$, 2, and 0.4°K. Rise time increases at T approaches T_n. Cell thickness = 100 μm.

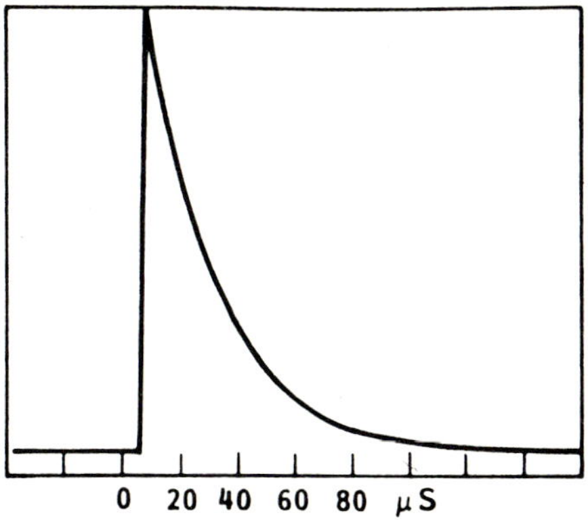

FIGURE 8 Decay of first-order diffraction intensity in nematic phase of 5CB at $T - T_n = 6°K$. Cell thickness = 100 μm.

refractive index ~0.1 at the nematic-isotropic interface implies a large amplitude, approximately rectangular phase grating. In this case 8 diffraction orders are visible. The rise time ~100 ns is consistent with Fig. 7.

The decay time as shown in Figs. 9 and 10 is of order 10 ms. This is much longer than the thermal decay. When modulated into the

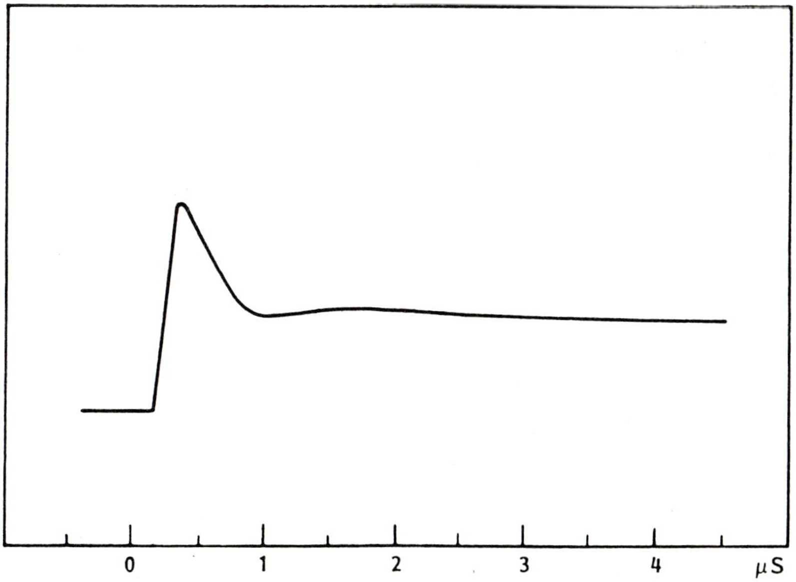

FIGURE 9 First-order diffraction intensity for modulation across nematic-isotropic transition cell thickness = 10 μm.

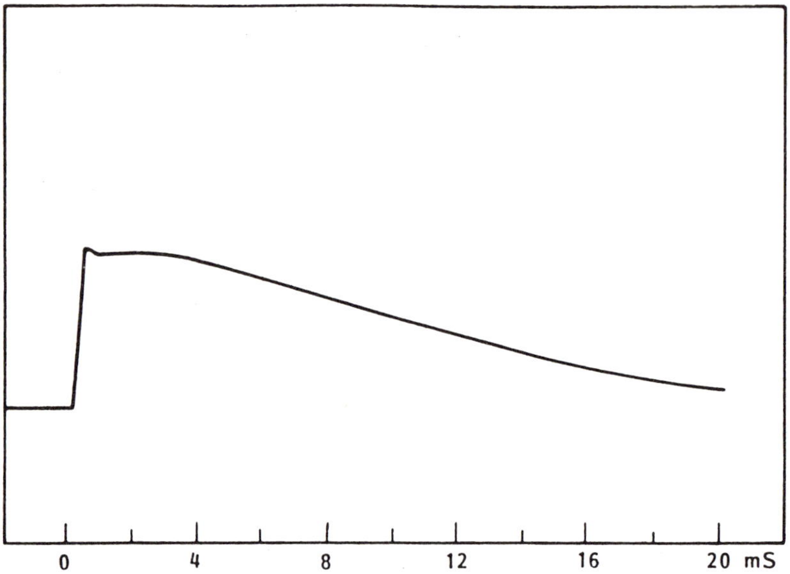

FIGURE 10 Decay of first-order diffraction intensity for modulation across nematic-isotropic transition. Cell thickness = 10 μm.

isotropic phase, the consequent rapid quench results in a disoriented nematic. The following reorientation of the nematic is viscous limited, which is consistent with a 10-ms time constant.

An initial peak or oscillatory response is shown in Fig. 9 and 10. The detail varies considerably from pulse to pulse. In this regime, a sinusoidal form is a poor approximation to the grating structure. Also, the transition to the isotropic is associated with large phase modulations $> 2\pi$. Therefore, the behavior should be sensitive to write pulse conditions. Detailed interpretation of these effects requires an improvement in experimental control.

The visibility of high-order diffractions is enhanced by the long integration time ~10 ms. Readout with the YAG pulse restricts visibility to second-order.

The measured diffraction efficiencies η_e or η_0 as a function of temperature for $T_n - T > 1°C$ are consistent with data presented in Fig. 4. Discrepancies as large as a factor 2 can be attributed to fluctuations in the YAG output. For $T_n - T < 1°C$, the observed η does not achieve the predicted increase. This is explained by a number of practical restrictions. Fluctuations in the write pulse energy and hence average heating limit the temperature uniformity and stability. The light scattering increases close to the transition.

Acoustic waves can be generated by optical pulses. Laser induced phonon spectroscopy is a technique applied to the study of molecular processes.[34] In our case the 10-μm grating implies an acoustic period less than the 10-ns YAG pulse; therefore, the acoustic response is weak and can be neglected.

RESPONSE TIME

For an energy pulse input the order parameter response can be approximated

$$S = S_0 + \Delta S \exp(-t/\tau) \tag{28}$$

where S_0 is the new equilibrium value for the increase in energy or temperature. A single relaxation time τ is less applicable close to the transition point. The change in order ΔS controls the diffraction efficiency via Eq. (18):

$$\eta \ \alpha \ (dn/dT)^2 \ \alpha \ (\Delta S)^2[1 - \exp(t/\tau)]^2 \tag{29}$$

The analysis of ultrasonic data is complicated by coupling to both the magnitude and local orientation of the order parameter.[31,32] There does not appear to be a simple connection between the optical and ultrasonic relaxation times.

In the isotropic phase the ultrasonic wave couples to scalar order parameter fluctuations.[18] An analytic treatment is consistent with experimental results.[32]

The order parameter fluctuations in the isotropic phase are a weak optical effect which is not directly significant in the thermal grating diffraction. Therefore, again there is no direct relation between the optical and ultrasonic relaxation. In the isotropic phase the refractive index change is due to density change $\Delta\rho$; therefore $\eta \ \alpha \ \Delta\rho^2$. Nematic correlation in the isotropic phase will influence the relaxation time observed in the optical experiment. However the weak temperature dependence of τ implies that this is not substantial.

The optical energy absorbed by the dye is transferred to the liquid crystal over a finite time known as the thermalization time. For ordinary solvents this is typically ~nanosecond, but can be as high as ~10 ns.[21,22] The observed response time in the isotropic phase could be attributed to dye thermalization and viscous limited response.

The order parameter response is now the difference between the nematic and isotropic phase responses. Because of the square law response in Eq. (29), the optical response should be about a factor 2 longer than the ultrasonic relaxation time. When these adjustments are made, $1/2(\tau - 30/2 \text{ ns})$, becomes 22, 62, and 102 ns at $T_n - T = 14$, 2, and 0.4°K. This approaches the ultrasonic data, e.g., 16 and 50 ns at 14 and 2°K.[32]

CONCLUSION

We have derived a simple expression for the diffraction efficiency of the orientational response in the limit of viscous retarding force. This allows a simple incorporation of optical scattering effects to predict an optimum cell thickness ~100 μm. A similar optimum follows when the elastic forces are included.

The thermal grating response of a nematic is analyzed. The large effects are related to the temperature dependence of the order parameter. The magnitude of the response is offset by an increasing response time as the transition temperature to the isotropic is approached.

The observed optical response time is related to ultrasonic relaxation rates. It is noted that the optical response time can be identified with the order parameter, allowing simple interpretation of the result.

At the nematic-isotropic transition, the response time ~300 ns. Modulating the thermal grating into the isotropic phase results in a viscous limited decay associated with a quenched disordered nematic phase. These results could be of interest in the analysis of thermally written smectic display devices, where the picture element is pulsed into the isotropic phase.[29]

Acknowledgements

Discussion with G. Eyring, M. D. Fayer, H. J. Hoffman, and T. J. Karr and assistance from M. J. Murphy and R. E. Stone are acknowledged. The work is funded by the Lockheed Independent Research program.

References

1. A. Yariv, *Quantum Electronics*, (Wiley, 1975).
2. M. I. Barnik, L. M. Blinov, A. M. Dorozhkin, and N. M. Shtykov, *Mol. Cryst. Liq. Cryst.* **V**, 1-12 (1983).

3. R. M. Herman and R. J. Serinko, *Phys. Rev.* **A19** 1757–69 (1979).
4. I. C. Khoo, *Phys. Rev.* **A25**, 1636–1644, and **A26**, 1131 (1982).
5. B. Ya. Zel' Dovich and N. V. Tabiryan, *Zh. Eksp. Teor. Fiz.* **82**, 1126–1146 (1982) [*Sov. Phys. JETP* **55**, 656–666 (1982)].
6. S. M. Arakelyan, A. S. Karayan, and Yu. S. Chilingaryan, *Kvant. Elek.* **9**, 2481–2490 (1982) [*Sov. J. Q. E.* **12**, 1619–25 (1982)].
7. S. D. Durbin, S. M. Arakelian, and Y. R. Shen, *Opt. Lett.* **7**, 145–47 (1982).
8. H. L. Ong, *Phys. Rev.* **A28**, 2392–2407 (1983).
9. H. L. Ong and C. Y. Young, *Phys. Rev.* **A29**, 297–307 (1984).
10. A. Saupe, Private Communciation.
11. R. A. Fisher, *Optical Phase Conjugation* (Academic, 1983).
12. I. C. Khoo and S. L. Zauang, *IEEE J. Q. E.* QE 18, 246–49 (1982).
13. I. C. Khoo and S. Shepard, *J. Appl. Phys.* **54**, 5491–5493 (1983).
14. I. C. Khoo and R. Normandin, *J. Appl. Phys.* **55**, 1416–1418 (1984).
15. I. C. Khoo and R. Normandin *Opt. Lett.* **9**, 285–286 (1984).
16. G. Martin and R. N. Hellwarth, *App. Phys. Lett.* **34**, 371–373 (1979).
17. M. H. Garret and H. J. Hoffman, *J. Opt. Soc. Am.* **73**, 617–623 (1983).
18. P. G. deGennes, *The Physics of Liquid Crystals*, (Clarendon, 1974).
19. D. Armitage, *Applied Optics* (in press).
20. E. Wiener-Avnear, *App. Phys. Lett.* **29**, 635 (1976).
21. H. J. Hoffman and P. E. Perkins, CLEO (1983).
22. H. J. Hoffman, IEEE J. Q. E. (in press).
23. W. H. De Jeu, *Physical Properties of Liquid Crystalline Materials* (Gordon and Breach, 1980).
24. R. G. Horn, *J. De Phy* **39**, 167–172 (1978).
25. H. A. Tarry, RSRE Baldock, unpublished data.
26. M. Brunet-Germain, *C. R. Acad. SC.* **B271**, 1075–1077 (1970).
27. D. Armitage and F. P. Price, *Phys. Rev.* **A15**, 2496–2500 (1977).
28. T. Shinoda, Y. Maeda, and H. Endkido, *J. Chem. Them.* **6**, 921 (1974).
29. D. Armitage, *J. Appl. Phys.* **52**, 1294–1300 (1981).
30. M. Schadt, *J. Chem. Phys.* **71**, 2336–2334 (1979).
31. S. Candau and S. V. Letcher, *Advances in Liquid Crystals*, G. H. Brown, ed. (Academic, 1978).
32. S. Nakai, P. Martinoty, and S. Candau, *J. De Phys.* **37**, 769–780 (1976).
33. F. Rondelez, W. Urbach, and H. Hervert, *Phys. Rev. Lett.* **61**, 1058–1061 (1978).
34. G. Eyring and M. D. Fayer, *J. Chem. Phys.* (in press).

Dynamic Gratings and the Associated Self Diffractions and Wavefront Conjugation Processes in Nematic Liquid Crystals

IAM-CHOON KHOO, MEMBER, IEEE

(Invited Paper)

Abstract—The origins and the dynamics of optical nonlinearities in nematic liquid crystal films, namely, laser-induced molecular reorientational and thermal refractive index changes, are analyzed in the context of optical wave mixings. Theoretical expressions for the basic nonlinearities, the rise and decay time, diffraction efficiencies, and other pertinent parameters involved in the dynamic grating formation are derived. Experimental results obtained with visible and infrared laser pulses are analyzed. Some newly observed novel nonlinear processes are also reported.

I. Introduction

LINEAR and nonlinear optical properties of liquid crystals in their mesophases have been studied in several contexts, in both fundamental and application-oriented pursuits. In the context of nonlinear optical processes, they have recently received considerable renewed interests as a result of the newly discovered extraordinarily large optical nonlinearity due to the laser-induced molecular reorientation, and a renewed effort explicitly at the large thermal index effect in liquid crystals. In the last few years, several groups [2]–[10] have looked at the optical nonlinearity in the mesophases of liquid crystals and the associated nonlinear processes. A brief review of some of these nonlinear optical processes and the fundamental mechanisms in both the liquid crystal and the isotropic phases has recently appeared [1]. In this paper, therefore, we will concentrate only on optical wave mixing processes that are relevant to this Special Issue.

Specific effects include optical self diffractions, generation of high-frequency acoustic waves, optical wave front conjugation (with gain) and self oscillations, and infrared-to-visible image conversion. The origin of the optical nonlinearities responsible for these processes is either the optical-induced molecular reorientation or the laser-induced thermal refractive index change. In the next section, we will review these two types of nonlinearities, and especially their dynamical dependences. This is followed by a summary discussion of the aforementioned nonlinear processes.

II. Liquid Crystal Nonlinearity and Dynamics

A. Orientational

In theory, all three mesophases (nematics, smectics, and cholesterics) of liquid crystals posses extraordinarily large optical nonlinearity [1]–[3]. To date, the most extensive and conclusive experimental studies have been performed in the nematic phase [1]–[10] where the molecules are directionally correlated but positionally random. The direction is characterized by the director axis n, and the correlation by an order parameter S^{11}. Two of the most commonly employed cell alignments are depicted in Fig. 1(a) and (b), termed homeotropic and planar cells, respectively. Homeotropic samples are obtained by coating the glass slides with a surfactant, while planar samples are often made using glass slides that have been rubbed in a unidirectional direction [11]. Nematics are highly birefringent, with $\Delta n = n_1 - n_2 = 0.2$ or larger where n_1 and n_2 are the refractive indexes for the optical field parallel and perpendicular to the director axis, respectively. In analogy to well-known dc field-induced electrooptical effects, the primary effect of an optical field is the dielectric torque exerted by the optical electric field in the liquid crystal. Under suitable conditions, (which we will presently elaborate) the optical torque creates a distortion or reorientation of the director axis and an accompanying self (laser)-induced refractive index change.

Nematic director axis reorientation by an optical field and the associated wave mixing and self-focusing effect were first theoretically quantitatively studied by Tabiryan and Zel'dovich [4], who also discussed similar processes in smectic and cholesteric liquid crystals. Herman and Serinko [5] presented a theory of nematic liquid crystal axis reorientation by an optical field in the presence of a strong bias dc magnetic field and the associated wave mixings and optical diffraction effects. Others [2], [3], [5]–[10] have also considered a similar reorientation process under steady-state or time-dependent optical illumination. A quantitative analysis and experimental study of transient nanosecond laser-induced molecular reorientation and heating have recently been presented by Hsiung

Manuscript received October 7, 1985; revised March 3, 1986. This work was supported by the National Science Foundation under Grant ECS 8415387 and the U.S. Air Force Office of Scientific Research under Grant AFOSR840375.

The author is with the Department of Electrical Engineering, Pennsylvania State University, University Park, PA 16802.

IEEE Log Number 8609096.

0018-9197/86/0800-1268$01.00

et al. [12] and by Khoo and Normandin [13]. In this paper, we shall therefore provide a brief simplified version of the theories for orientation and thermal effect whereby the physics of the nonlinearity and the nonlinear optical processes in the transient regime can be clearly appreciated. The interested readers could also refer to the lengthy detailed treatments by Tabiryan and Zel'dovich [4] and Ong [7].

Consider two linearly polarized lasers incident on a homeotropic nematic film as shown in Fig. 1(a). The two beams' interference gives rise to an intensity sinusoidal on the film, i.e., the total electric field amplitude E on the film is given by

$$E^2 = E_1^2 + E_2^2 + 2E_1E_2 \cos q \cdot y \tag{1}$$

where E_1 and E_2 are the amplitudes of the two optical fields

$$q = |\vec{q}| = |\vec{K}_1 - \vec{K}_2|.$$

The free-energy density of the system in the one-elastic constant approximation is given by [11]

$$F = \frac{K}{2}\left[\left(\frac{\partial\theta}{\partial y}\right)^2 + \left(\frac{\partial\theta}{\partial z}\right)^2\right] - \frac{\Delta\epsilon}{8\pi}(E_1^2 + E_2^2 + 2E_1E_2 \cos qy)\sin^2(\theta + \beta) \tag{2}$$

where $\Delta\epsilon$ is the optical dielectric constant anisotropy. $\Delta\epsilon = \epsilon_1 - \epsilon_2$ and θ is the reorientation angle of the director axis $\hat{n}$ from the initial alignment.

The time dependence of the molecular reorientation following the usual Ericksen–Leslie approach becomes [11]

$$\gamma\frac{\partial\theta}{\partial t} = K\left(\frac{\partial^2\theta}{\partial y^2} + \frac{\partial^2\theta}{\partial z^2}\right) + \frac{\Delta\epsilon}{8\pi}E^2 \sin 2(\theta + \beta) \tag{3}$$

where γ is the effective viscosity coefficient. In writing (1)–(3), we have neglected several terms associated with flow and inertia (see Hsiung *et al.* [12]) in the medium which have been shown to be negligible in all of the wave-mixing experiments so far.

In the small-angle (θ) limit (which is often the case in actual wave-mixing experiments), (3) becomes

$$\gamma\frac{\partial\theta}{\partial t} = K\left(\frac{\partial^2\theta}{\partial y^2} + \frac{\partial^2\theta}{\partial z^2}\right) + \frac{\Delta\epsilon E^2}{4\pi}\left[(\cos 2\beta)\,\theta + \frac{\sin 2\beta}{2}\right]. \tag{4}$$

An approximate solution of this equation is

$$\theta = \theta(t)\cos qy \sin\frac{\pi z}{d} \tag{5}$$

which satisfies the so-called hard-boundary conditions ($\theta = 0$ at $z = 0$ and at $z = d$).

Substituting (5) into (4) gives

$$\gamma\frac{\partial\theta}{\partial t} = \left[\frac{\Delta\epsilon E^2 \cos 2\beta}{4\pi} - K\left(\frac{\pi^2}{d^2} + q^2\right)\right]\theta + \frac{\Delta\epsilon E_{op}^2}{8\pi}\sin 2\beta \tag{6}$$

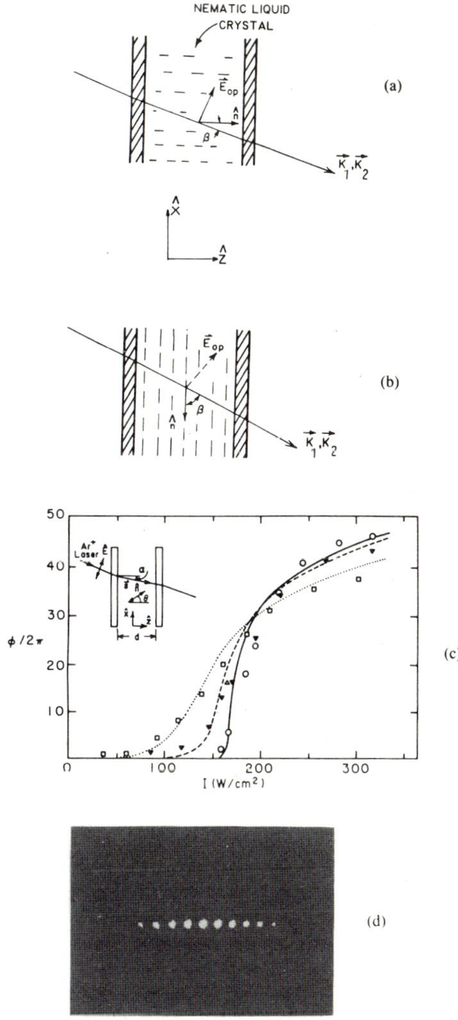

Fig. 1. Schematic of two optical waves propagating in (a) a homeotropic and (b) a planar nematic liquid crystal film. $\vec{K}$'s are the propagation directions, E_{op} is the optical electric field, and $\hat{n}$ is the director axis direction. The two optical beams are crossed at a wave-mixing angle in a plane perpendicular to the plane of the paper. (c) Experimental data and theoretical curves for the phase shift $\Delta\phi$ induced in a 250 μm, homeotropically aligned, 5CB film by an Ar^+ laser beam at different angles a: circles and solid curve, $\alpha = 0°$; solid triangles and dashed curve, $\alpha = 3°$; squares and dotted curve, $\alpha = 11°$. Inset shows the experimental geometry (after [3]). (d) Photograph of the multiorder diffractions via two wave mixings in a nematic liquid crystal film using CW lasers with intensities on the order of a few watts/cm^2. Similar multiorder diffractions are observed via transient wave mixings with nanosecond laser pulses (I. C. Khoo, unpublished).

or

$$\frac{\partial\theta}{\partial t} + \frac{1}{\tau}\theta = \frac{\Delta\epsilon E^2}{8\pi\gamma}\sin 2\beta \tag{7}$$

where

$$\tau = \left[\frac{-\gamma}{\frac{\Delta\epsilon}{4\pi}E^2\cos 2\beta - K(\pi^2/d^2 + q^2)}\right]. \tag{8}$$

τ is the time constant characterizing the buildup of the reorientation process.

Depending on the magnitude of the optical field, the growth of the reorientation process can be slow or fast. If we define an optical Freedericks transition field E_F^1,

$$E_F = \frac{4\pi^3 K}{\Delta\epsilon d^2},$$

in analogy to dc field effect, (8) gives

$$\frac{1}{\tau} = \frac{4\pi}{\Delta\epsilon} d^2 \frac{\gamma}{\left|E^2\cos 2\beta - E_F^2\left(1 + \frac{q^2}{\pi^2}d^2\right)\right|}. \tag{9}$$

Most previous studies of molecular reorientation employed an optical field below or just above the Freedericks field E_F, and thus τ is large. It is obvious that if the optical field strength increases, e.g., to 10^4 statvolt/cm^2 or more, τ can be very short (microseconds or tens of nanoseconds), using standard values for d, $\Delta\epsilon$, K, γ, and q ($q = 2\pi/\Lambda$ where Λ is the grating constant) (e.g., $d \approx 50$ μm $\approx \Lambda$, $\Delta\epsilon = 0.6$, $K = 10^{-7}$ dynes, $\gamma = 0.1$ poise). The complication arising from optical reorientation with a high-power pulsed laser is that the liquid crystal will also be heated [12], [13], via some finite absorption. This heating effect and the associated refractive index changes have been characterized by different time scales from the orientational process, as will be discussed in the next section.

The decay time of the process is simply

$$\frac{1}{\tau_{\text{decay}}} = \frac{\gamma}{K(\pi^2/d^2 + q^2)}. \tag{10}$$

For a typical thickness $d \sim 50$ μm and $\Lambda \sim 50$ μm, the decay time is on the order of 1 s. Millisecond decay time is achieved for d or Λ on the order of a few micrometers.

The magnitude of the reorientation angle θ depends obviously on the magnitude as well as the duration of the pulsed optical field, and will saturate when all the molecules are aligned in the direction of the optical field. We note here, however, that for large reorientation angles, (6) is no longer valid and one has to use (3), the solution of which has been discussed in various contexts by several workers [3], [4], [7], [12]. On the other hand, for small θ, the stationary maximum value of θ is roughly given by

$$\theta \sim \frac{\Delta\epsilon E^2 \sin 2\beta \sin \pi Z/d \cos qy}{8\pi\left[K(\pi^2/d^2 + q^2) - \frac{\Delta\epsilon E^2\cos 2\beta}{4\pi}\right]}. \tag{11}$$

For a given reorientation θ, the change in the refractive index Δn associated with the two extraordinary rays (e.g., Fig. 1 where the laser beams propagate as an extraordinary ray) is given by

$$\Delta n = n_e(\beta + \theta) - n_e(\beta)$$

where n_e is given by

$$n_e(\beta + \theta) = \frac{n_1 n_2}{\sqrt{n_1^2\cos^2(\beta + \theta) + n_2^2\sin^2(\beta + \theta)}}. \tag{12}$$

Equation (12) gives, in the small θ limit, an index grating

$$\Delta n(\theta) \sim \theta \frac{n\Delta\epsilon}{2\epsilon_1}\sin 2\beta. \tag{13}$$

In the small θ limit, and for $E_{op}^2 << E_F^2$, (11) reduces to a form analogous to those obtained in previous publications [1]–[4]. The significance of $\Delta n(\theta)$ is its magnitude. For $d \sim 100$ μm, $\beta \sim 0.5$ rad, an optical intensity on the order of 100 W/cm^2 will induce a Δn of 10^{-3} ($\theta \sim 10^{-2}$). For example, Fig. 1(c) shows a detailed measurement by Durbin *et al.* [3] of the laser-induced phase shift [due to $\Delta n(\theta)$] in a nematic liquid crystal film for various values of the angle β (denoted as α in [3]). In the context of wave mixings involving two laser beams E_1 and E_2 as described earlier, the theory and experiments were first quantitatively studied by this author [2]. The dependence of the self-diffraction effect on the angle β, the grating spatial frequency, and also the temperature *independence* of the nonlinear diffraction were experimentally verified. In the presence of a strong dc bias field, the optical nonlinearity of a nematic liquid crystal film is enhanced; as may be seen in the photograph of Fig. 1(d). This shows the multiorder diffractions from a 75 μm thick nematic film where an applied bias magnetic field is used to first reorient the director axis. Diffraction efficiencies on the order of more than 10 percent are observed for laser intensities of only a few watts/cm^2. The effect was first confirmed by an observation by Khoo [2] following the prediction of Herman and Serinko [5]. Subsequently, Durbin *et al.* [10] have also studied the problem involving the multiple diffraction orders, and have presented a detailed analysis of the multiple diffraction effects.

II. Thermal Nonlinearity

The dependence of the refractive index of nematic liquid crystal on the temperature has occupied central importance in the study of the fundamental and applied properties of liquid crystals [11]. In this discussion, we will follow the literature [11], [14] and choose as the starting point of our analysis of the thermal nonlinearities the dielectric constants $\epsilon_1 = n_1^2$ and $\epsilon_2 = n_2^2$. Depending on the levels of sophistication one desires, there are several possible forms of ϵ_1 and ϵ_2 in terms of molecular parameters (see, for example, De Jeu, [11]). In the simplest case, they are given by

$$\epsilon_1 = 1 + (N\rho/3\epsilon_0 M)\,[\alpha_l K_l(2S + 1) + \alpha_t K_t(1 - 2S)] \tag{14}$$

and

$$\epsilon_2 = 1 + (N\rho/3\epsilon_0 M)\,[\alpha_l K_l(1 - S) + \alpha_t K_t(2 + S)] \tag{15}$$

where $\alpha_{e,t}$ and $K_{e,t}$ are the longitudinal and transverse components of the molecular electronic polarization α and the internal field tensor K; N is the Avagadro number, ρ is the density of the liquid crystal, M is the molecular weight, and S is the order parameter. The degree of sophistication depends on the various models used for evaluating K. To understand the key factor affecting the thermal refractive index change, we shall use a slightly simplified version where the local field correction factors are ignored. In that case, ϵ_1 and ϵ_2 from (14) and (15) can be expressed in the form

$$\epsilon_1 = \epsilon_l + \tfrac{2}{3}\Delta\epsilon \qquad (16)$$

$$\epsilon_2 = \epsilon_l - \tfrac{1}{3}\Delta\epsilon \qquad (17)$$

where

$$\epsilon_l = 1 + \frac{N\rho}{3\epsilon_0 M}(\alpha_l k_l + 2\alpha_t k_t) \sim 1 + \text{const. } \rho \qquad (18)$$

and

$$\Delta\epsilon \sim \frac{N\rho S}{\epsilon_0 M}(\alpha_l k_l - \alpha_t k_t) \sim \rho S. \qquad (19)$$

On the other hand, the order parameter S (defined in many standard texts) can be shown to be well approximated by the expression

$$S = [1 - 0.98\, TV^2/T_{ni} V_{ni}^2]^{0.22} \qquad (20)$$

where T_{ni} is the nematic → isotropic phase transition temperature and V_{ni} is the corresponding volume. The most significant contributions to the refractive index dependence on temperature are from ρ and S. Other parameters are also highly temperature dependent, such as the volume V and the specific heat C_V that is involved in the laser heating of the liquid crystals. One can think of the contributions from ρ and S as being primary, while others are only secondary and may be neglected.

The most significant feature of laser-induced thermal index change is the magnitude of dn_1/dT and dn_2/dT. Fig. 2 is a plot of the thermal index gradients (dn_1/dT and dn_2/dT) of PCB from experimental values of the refractive indexes [14].

Typically, (dn_1/dT) is on the order of 4×10^{-4} K^{-1} while (dne/dT) is about 1.5×10^{-3} K^{-1}, which are already much larger than most other high thermal index materials (e.g., cyclohexane). Near T_c, both dn/dT's increase by more than an order of magnitude.

For transient wave mixings, the detailed calculations for the three-dimensional thermal grating buildup and temperature distribution and dissipation are obviously very complex, and are further complicated by the anisotropic thermal diffusion constants of the liquid crystals, as well as the enclosing glass slides. In the simplest case where the thermal grating is reducible to a one-dimensional problem [14] (e.g., the case of very small grating constant Λ compared to the cell thickness d), the thermal decay time constants for heat dissipation along and per-

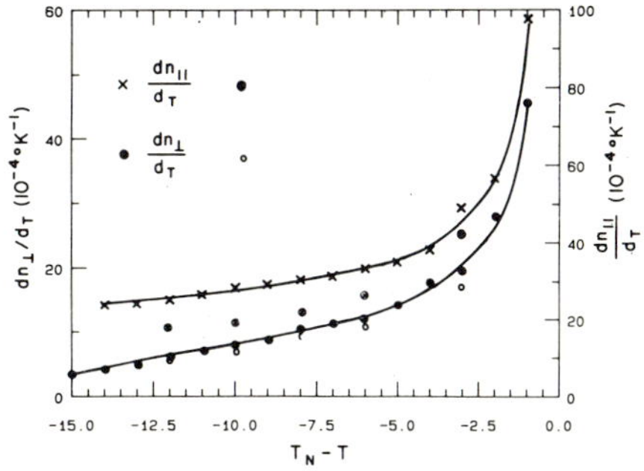

Fig. 2. The dependences of ($dn_{\parallel}/dT$) ($n_{\parallel} = n_1$) and ($dn_{\perp}/dT$) ($n_{\perp} = n_2$) on temperature deduced from data in [14]. T_n is the nematic → isotropic transition temperature. Some theoretical values obtained using (16)–(20) (cf. [13]) are also indicated (θ and 0).

pendicular to the nematic axis are, respectively $\tau_1 = (D_1 q_1^2)^{-1}$ and $\tau_2 = (D_2 q_2^2)^{-1}$. For typical liquid crystals, $D_2 \sim 7.9 \times 10^{-4}$ cm^{2}s^{-1} and $D_1 \sim 1.25 \times 10^{-3}$ cm^{2}s^{-1}. For a grating constant $\Lambda_2 = 2\pi/q_2 \sim 17$ μm, one gets $\tau_2 \sim 110$ μm. On the other hand, $\tau = 11\,(n_2/n_1)^2\tau_2 \approx 50$ μs. These estimated values are in good agreement with the experimental data obtained in a recent study [13].

III. Experimentally Observed Dynamic Grating Effects

A. Nonlinear Diffraction and High-Frequency Acoustic Waves

From (15), (16), and (20), it is obvious that the change in the refractive index as a function of a temperature rise is due to a change in the order parameter S and the density ρ. In the transient regime involving very short laser pulses (short compared to the acoustic phonon lifetime), the interference of two laser beams will give rise to an index grating comprising a nonpropagating component associated with S and a propagating component from ρ. These two components, under appropriate conditions (temperature, angle, etc., as detailed in [13]), will interfere and produce modulation in the diffraction of a probe beam from the grating produced by the two incident beams [15], [16]. The frequency of modulation is given by C_s/Λ where C_s is the sound velocity in the bulk liquid crystalline film. Fig. 3(a) shows the dynamics of the defraction of a CW He–Ne probe laser from a grating produced by two linearly polarized nanosecond Nd:YAG second harmonic laser pulses in a nematic film [17]. The initial portion clearly shows the modulation caused by the acoustic interference. A principal requirement for generating these gigahertz acoustic waves is that the bandwidth of the two exciting lasers be large enough to accommodate the acoustic frequency; hence, we have the use of picosecond lasers in some studies [15], [16]. Alternatively, one can employ a two-mode laser and a wave-mixing angle such that the grating constant Λ and therefore the sound frequency C_s/Λ matches the frequency separation of the two modes [17].

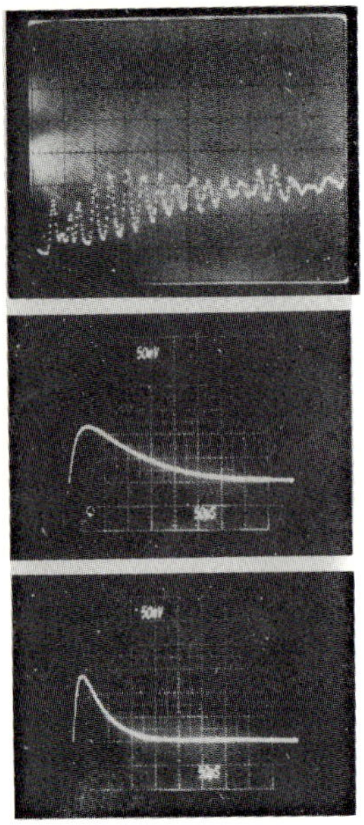

(a)

(b)

(c)

Fig. 3. (a) The temporal dependence of the diffraction of a CW He-Ne laser from a transient grating induced by two nanosecond Nd:YAG second harmonic laser pulses in a nematic film (PCB; 40 μm thick; homeotropically aligned) (after [17]). Time scale is 20 ns/div. Similar ultrasonic wave generation was also observed in a smectic liquid crystal film [17b]. (b) The dynamics of the thermal grating formation and decay in a nematic film as monitored by a CW He-Ne probe laser diffraction. The grating wave vector $K_1 - K_2$ is perpendicular to the director axis. (c) Same as in (b), but for $K_1 - K_2$ along the director axis.

Using the pump probe experiment, and from modulation data similar to Fig. 3(a), both the velocity of sound and the acoustic attenuation constant in the bulk nematic can be measured simultaneously. The sound velocities in the nematic and smectic phase obtained using these transient wave-mixing effects are in excellent agreement with those obtained by other techniques [18], [19]. Obviously, if the incident plane of the laser and the wave-mixing angle are chosen appropriately, and using a planar sample, one could also study the acoustic velocity and attenuation anisotropies. These transient wave-mixing interference effects, therefore, could be developed into very versatile and powerful techniques (cf. work by Nelson *et al.* [16] and Eichler and Stahl [15]) for studying bulk acoustic properties in nematics as well as smectics.

In these experiments, self diffractions from the two incident Nd:YAG lasers are also observed. This shows that the rise time of the thermal grating buildup is therefore on the order of the laser pulse duration, as mentioned earlier. To investigate further the rise and fall dynamics, we again monitor the diffraction from a CW He-Ne beam. Fig. 3(b) and (c) show the typical dynamics of the thermal buildup and decay for the grating wave vector $\vec{q}$ perpendicular and parallel to the nematic axis, respectively. The rise time is on the order of the incident laser pulses, while the decay time depends on the grating wavevector direction with respect to the director axis, as well as on the grating constant. Obviously, for a much smaller grating constant, the decay time can be considerably shorter than the 50 μs or so observed here.

In some nematics, e.g., PCB (4-cyano-4′-pentylbiphenyl), the natural absorption is quite small at the second harmonics of the Nd:YAG laser. To enhance the absorption constant and thus the nonlinearity, it is possible to dissolve some dyes that absorb at around 0.53 μm, e.g., rhodamine 6G. Experiments have been conducted [13], [17] in nematic films with traces of dissolved dyes which showed that the required laser energies for several nonlinear processes based on the thermal effect were drastically reduced. A detailed study of the rise portion of the He-Ne diffraction also shows that the intramolecular relaxation processes that transfer the excitation from the dye to the liquid crystal take place probably at much shorter time scales than the laser pulse length, as the diffraction from a dyed sample has exactly the *same* rise time as the undyed sample.

IV. Optical Wavefront Conjugation and Multiwave Mixings

Fig. 4 depicts a typical optical configuration for wavefront conjugation studies. The two incident waves E_1 and E_2 generate a grating (via thermal or reorientational effect), E_3 is the retroreflection of E_2, while E_4 is the generated fourth (conjugated or reflected) wave. In liquid crystalline systems, there have been several studies of wavefront conjugation based on different nonlinearity mechanisms. Fekete *et al.* [20] employed nanosecond laser-induced molecular reorientation in the *isotropic* phase, the mechanism and dynamics of which were first studied quantitatively by Wong and Shen [20]. These individual molecular reorientational effects exhibit some enhancement, owing to greater molecular correlation as the liquid is cooled towards the isotropic → nematic transition temperature T_C. On the other hand, Garibyan *et al.* [21] employed a liquid crystal light valve for optical wave-mixing processes. The reference and object beam set up an interference intensity grating on the light valve which, acting on the photoconductor material coating, changes the impedance of the (nematic) liquid crystal film. In conjunction with an applied dc field, the intensity grating leads to an orientational (and therefore a phase) grating that optical wavefront conjugation rising purely optical field-induced effect in the nematic phase was first demonstrated by our group where the phase aberration correction capability was also confirmed [22]. In a later study, Leith *et al.* [23] employed thermal nonlinearity to demonstrate the feasibility of speckle-noise removal in wavefront conjugation using a spatially partially coherent laser beam.

For highly nonlinear materials, it is possible to obtain amplified phase conjugation reflections [24]. Moreover, noise originating from the system could also interfere coherently with the pump beams and, with proper feedback

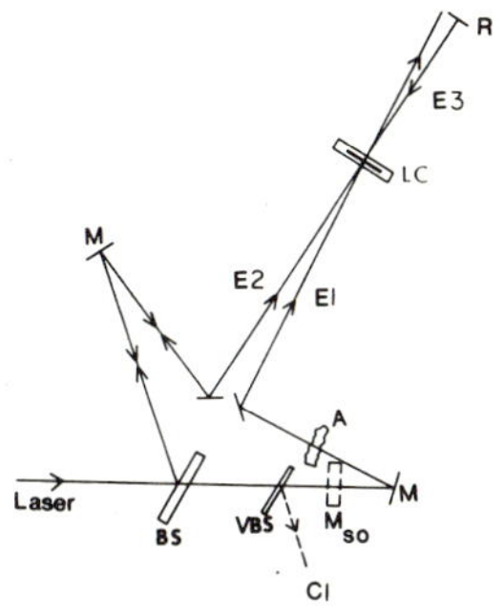

Fig. 4. Schematic of the setup for optical wavefront conjugation using a liquid-crystal (LC) film. E_1: object beam, E_2: reference beam, A: aberrator, M: mirror, M_{so}: mirror to be used for self-oscillation effect, VBS: variable beam splitter to control the intensity of E_1, BS: beam splitter, R: total reflector. For an example of wavefront conjunction result, see Fekete *et al.* [20] (isotropic phase of liquid crystal) or [20] (nematic phase).

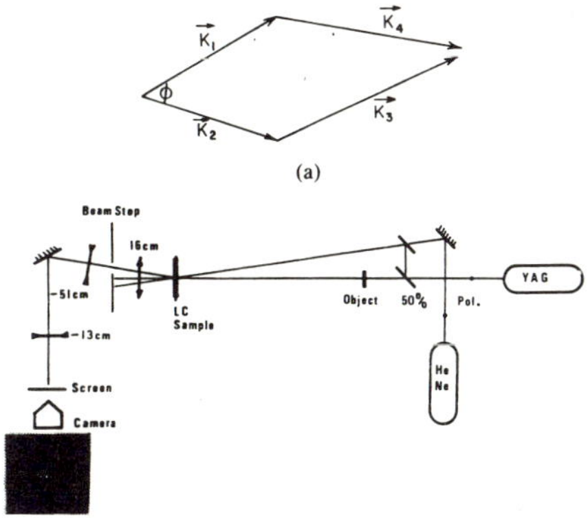

(a)

(b)

Fig. 5. (a) Phase matching in optical wavelength conversion using K_1 and K_2 to induce the grating and K_3 to generate K_4 via four-wave mixing process. (b) Experimental setup for infrared-to-visible image conversion using the transient grating induced by infrared (1.06 μm) Nd:YAG laser pulses in a dyed nematic liquid crystal film. The reconstructing beam is a CW He-Ne (0.63 μm) laser. The photo insert is a typical observed reconstructed image of the wire mesh object.

from a mirror, lead to self oscillations [25]. A study of amplified reflection based on thermal nonlinearity of nematic has been reported [26]. Another consequence of very strong two-wave mixing is that the probe beam will be amplified by the pump [26]. Herman and Serinko [5] first calculated the efficiency of this process in a nematic film just above a dc magnetic field-induced Freedericks transition. They predicted that amplification of the probe could arise with a very low power laser (with power on the order of a few watts). This was indeed verified (Khoo and Zhuang [2]). There are obviously several possible further studies based on these results, such as image amplification via phase conjugations and ring-laser oscillators. Extension of these studies involving visible lasers to lasers in the infrared regime are also envisioned.

It is possible to use beams at one wavelength to induce a refractive index grating (or hologram), which subsequently diffracts an incoming beam at another wavelength [27]. Fig. 5(a) shows the basic geometry required for phase matching used in our study using a nematic liquid crystal [28]. For a variation of the Bragg diffraction angle ϕ by $\delta\phi$ for which the phase-matched condition is still satisfied, i.e., $\Delta kl < \pi$, one gets

$$\delta\phi = 2\Pi\phi^{-1}(k_1 d(1 + k_1/k_2))^{-1}$$

where k_1 and k_2 are the wavevectors. The diffraction solid angle of the object beam which sends an area A on the sample is given by $\phi_D = 4\Pi^2/k_1^2A$. It is therefore possible, in principle, to have the number of resolution elements N, defined as $N \sim \delta\phi/\phi_D$, as

$$N \sim \frac{Ak_1}{l(1 + k_1/k_2)}.$$

For example, if $A \sim 1$ mm^2, $l \sim 100$ μm, and $k_1 = 2\pi/\lambda \sim 2\pi/(1.06$ μm), we have $N \sim 10^4$. Resolution of this order was reported by Martin and Hellwarth [27] using a slightly different (but basically similar) geometry where both K_3 and K_4 are exactly opposite in directions to those depicted in Fig. 5 here. The point about nematic liquid crystal film is that using liquid crystal "doped" with some infrared absorbing dyes, one could perform these real-time holographic processes with less laser energy (per unit area) because of the unusually high thermal indexes. Fig. 5(b) is a setup used in a recent experiment [28]. The liquid crystal used is a PCB sample with traces of dissolved infrared absorbing dyes. The object used for this real-time holographic imaging process is a wire mesh. The pump and probe beams are from a 20 ns, 1.06 μm Nd:YAG laser pulse, while the probe beam is a CW He-Ne laser (0.6328 μm). Visible images (at 0.6328 μm) are observed at input energies on the order of 50 and 15 mJ on a spot size of 0.5 cm^2. The energy per unit area is considerably less than that used in previous studies [27]. If the sample temperature is raised to near T_c, a decrease in the energy requirement by an order of magnitude is observed. This process of infrared-to-visible image conversion is by no means the only possible or the most desirable one, but it points to the possibility of other applications based on this nonlinear interaction between laser beams of different wavelengths in the transient or steady-state regime. It is also conceivable that one can utilize the orientational effect to achieve infrared-to-visible conversion, as briefly demonstrated also in [28].

V. Conclusion

We have presented a discussion of the theories and experiments on laser-induced optical nonlinearities and some recently observed wave-mixing processes in nematic liquid crystals based on the phase grating induced by two laser pulses. These studies have demonstrated again the unique and interesting physical characteristics of liquid crystals that have attracted the attention of fundamental and applied researchers alike. It is also clear that some practically useful nonlinear optical devices could be constructed. The nematic phase is but one of the several mesophases of liquid crystal that possess these interesting nonlinearities. Cholesterics and smectics [4] and other hybrid forms of nematics [6] also possess large nonlinearities. We anticipate that many more effects will be observed in the near future.

Acknowledgment

The author is grateful to all his co-workers and to Prof. Y. R. Shen of the University of California, Berkeley, for several constructive discussions over the years, and for the use of Fig. 1(c).

References

[1] I. C. Khoo and Y. R. Shen, "Nonlinear optical properties and processes," *Opt. Eng.*, vol. 24, pp. 579-585, 1985.

[2] E. C. Khoo, "Optically induced molecular reorientation and third order nonlinear process in nematic liquid crystals," *Phys. Rev.*, vol. A23, pp. 2077-2081, 1981; see also I. C. Khoo and S. L. Zhuang, "Nonlinear light amplification in a nematic liquid crystal above the Freedericks transition," *Appl. Phys. Lett.*, vol. 37, pp. 3-5, 1980; I. C. Khoo, "Theory of optically induced molecular reorientation and quantitative experiments on wave-mixings and self-focusing of light," *Phys. Rev.*, vol. A25, pp. 1636-1644, 1982; I. C. Khoo, "Reexamination of the theory and experimental results of optically induced molecular reorientation and nonlinear diffraction in nematic liquid crystals: spatial frequency and temperature dependence," *Phys. Rev.*, vol. A27, p. 2747, 1983.

[3] S. D. Durbin, S. M. Arakelian, and Y. R. Shen, "Optically induced birefringance and Freedericks transition in nematic liquid crystal," *Phys. Rev. Lett.*, vol. 47, pp. 1411-1415, 1981.

[4] B. Ya Zel'dovich, N. V. Tabiryan, and Yu S. Chilingaryan, *Zh. Eksp. Teor. Fiz.*, vol. 81, p. 72, 1981 (*Sov. Phys. JETP*, vol. 54, pp. 32-36, 1981; N. V. Tabiryan and B. Ya Zel'dovich, "The orientational optical nonlinearity of liquid crystals," *Molec. Cryst. Liq. Cryst.*, vol. 62, pp. 237-250, 1980; N. F. Pilipetski, A. V. Sukhov, N. V. Tabiryan, and B. Ya Zel'dovich, "The orientational mechanism of nonlinearity and the self-focusing of He-Ne laser radiation in nematic liquid crystal mesophase (theory and experiment)," *Opt. Commun.*, vol. 37, pp. 280-284, 1981; S. R. Galstyan, O. V. Garibyan, N. V. Tabiryan, and Yu S. Chilingaryan, "Light induced Freedericks transition in a liquid crystal," *JETP Lett.*, vol. 33, pp. 437-441, 1981.

[5] R. M. Herman and R. J. Serinko, "Nonlinear optical processes in nematic liquid crystals near a Freedericks transition," *Phys. Rev.*, vol. A19, pp. 1757-1769, 1969.

[6] G. Barbero, F. Simoni, and P. Aiello, "Nonlinear optical reorientation in hybrid aligned nematics," *J. Appl. Phys.*, vol. 55, pp. 304-309, 1984; see also G. Barbero and F. Simoni, *Appl. Phys. Lett.*, vol. 41, p. 504, 1982.

[7] H. L. Ong, "Optically induced Freedericks transition and bistability in a nematic liquid crystal," *Phys. Rev.*, vol. A28, pp. 2393-2407, 1983.

[8] A. S. Zolot'ko, V. F. Kitaeva, N. Kroo, N. N. Sobolev, and L. Chillag, "The effect of an optical field on the nematic phase of the liquid crystal OCBP," *JETP Lett.*, vol. 32, pp. 158-162, 1980; L. Csillag, J. Janossy, V. F. Kitaeva, N. Kroo, and N. N. Sobolev, "The influence of the finite size of the light spot size on the laser induced reorientation of liquid crystals," *Molec. Cryst. Liq. Cryst.*, vol. 84, pp. 125-135, 1982.

[9] Y. G. Fuh, R. F. Code, and G. X. Xu, "Time dependence and diffraction efficiency of optically-induced phase gratings in nematic liquid crystal films," *J. Appl. Phys.*, vol. 54, pp. 6388-6393, 1983.

[10] S. D. Durbin, S. M. Arakelian, and Y. R. Shen, "Strong optical diffraction in a nematic liquid crystal with high nonlinearity," *Opt. Lett.*, vol. 1, pp. 145-149, 1982.

[11] See, for example, E. B. Priestley, P. J. Wojtowicz, and P. Sheng, *Introduction to Liquid Crystals*. New York: Plenum, 1975; see also W. H. de Jeu, *Physical Properties of Liquid Crystalline Materials*. New York: Gordon and Breach, 1980, ch. 4 and references therein.

[12] H. Hsiung, L. P. Shi, and Y. R. Shen, "Transient laser-induced molecular reorientation and laser heating in a nematic liquid crystal," *Phys. Rev.*, vol. A30, pp. 1453-1462, 1984.

[13] I. C. Khoo and R. Normandin, "The mechanism and dynamics of transient thermal grating diffraction in nematic liquid crystal films," *IEEE J. Quantum Electron.*, vol. QE-21, pp. 329-335, 1985; I. C. Khoo and S. Shepard, "Submillisecond grating diffractions in nematic liquid crystal films," *J. Appl. Phys.*, vol. 54, pp. 5491-5494, 1983.

[14] R. G. Horn, "Refractive indices and order parameters of two liquid crystals," *J. Phys.*, vol. 39, pp. 105-109, 1978; W. Urbach, H. Hervet, and F. Rondelez, "Thermal diffusivity measurement in nematic and smectic phases by forced Rayleigh light scattering," *Molec. Cryst. Liq. Cryst.*, vol. 46, pp. 209-221, 1978.

[15] See, for example, H. Eichler and H. Stahl, "Time and frequency behavior of sound waves thermally induced by modulated laser pulses," *J. Appl. Phys.*, vol. 44, pp. 3429-3435, 1973.

[16] K. A. Nelson, R. J. D. Miller, D. R. Lutz, and M. D. Fayer, "Optical generation of tunable ultrasonic waves," *J. Appl. Phys.*, vol. 53, pp. 1144–1149, 1982.
[17] I. C. Khoo and R. Normandin, "Nanosecond laser induced optical wave mixing and ultrasonic wave generation in the nematic phase of liquid crystals," *Opt. Lett.*, vol. 9, pp. 285–287, 1984; see also I. C. Khoo and R. Normandin, "Nanosecond laser induced ultrasonic waves and erasable permanent gratings in smectic liquid crystal," *J. Appl. Phys.*, vol. 55, pp. 1416–1418, 1984.
[18] K. Miyano and J. B. Ketterson, "Ultrasonic study of liquid crystals," *Phys. Rev.*, vol. 12, pp. 615–635, 1975; M. E. Mullen, B. Luthi, and M. J. Stephen, "Sound velocity in a nematic liquid crystal," *Phys. Rev. Lett.*, vol. 28, pp. 799–801, 1972.
[19] A. E. Lord, Jr. and M. M. Labes, "Anisotropic ultrasonic properties of a nematic liquid crystal," *Phys. Rev. Lett.*, vol. 25, pp. 570–572, 1970; A. E. Lord, Jr. "Anisotropic ultrasonic properties of a smectic liquid crystal," *Phys. Rev. Lett.*, vol. 29, pp. 1366–1369, 1972.
[20] D. Fekete, J. AuYeung, and A. Yariv, "Phase conjugate reflection by degenerate four-wave mixing in a nematic liquid crystal in the isotropic phase," *Opt. Lett.*, vol. 5, pp. 51–53, 1980; P. Ye, G. Chu, Z. Chang, P. Fu, G. Ji, and X. Lin, "Four-wave mixing and its relaxation effect in liquid crystals," in *Tech. Dig., 11th Int. Quantum Electron. Conf.*, 1980, pp. 638–639; G. K. L. Wong and Y. R. Shen, "Optical-field-induced ordering in the isotropic phase of a nematic liquid crystal," *Phys. Rev. Lett.*, vol. 30, pp. 895–897; see also C. Flytzanis and Y. R. Shen, "Molecular theory of orientational fluctuations and optical Kerr effect in the isotropic phase of a liquid crystal," *Phys. Rev. Lett.*, vol. 33, pp. 14–17, 1974.
[21] O. V. Garikyan, I. N. Kompanets, A. V. Parfyonov, N. F. Pilipetsky, V. V. Sukunov, A. N. Sudarkin, A. V. Sukhov, N. V. Tabiryan, A. A. Vasiliev, and B. Ya Zel'dovich, "Optical phase conjunction by microwatt power of reference waves via liquid crystal light valve," *Opt. Commun.*, vol. 38, pp. 67–70, 1981.
[22] I. C. Khoo and S. L. Zhuang, "Wave front conjugation in nematic liquid crystal films," *IEEE J. Quantum Electron.*, vol. QE-18, pp. 246–248, 1981.
[23] E. N. Leith, H. Chen, Y. Cheng, G. Swanson, and I. C. Khoo, "Phase conjugation with light of reduced spatial coherence," in *Proc. 5th Rochester Conf. Coherence and Quantum Opt.*, Rochester, NY, June 1983.
[24] See, for example, B. Fischer, M. Cronin-Golomb, J. O. White, and A. Yariv, "Amplified reflection, transmission and self-oscillation in real-time holography," *Opt. Lett.*, vol. 6, pp. 519–521, and references therein.
[25] H. Rajbenbach and J. P. Huignard, "Self induced coherent oscillations with photorefractive $Bi_{12}SiO_{20}$ amplifier," *Opt. Lett.*, vol. 10, pp. 137–139, 1985; J. Feinberg and R. W. Hellwarth, "Phase-conjugating mirror with continuous-wave gain," *Opt. Lett.*, vol. 5, pp. 519–521, 1980.
[26] I. C. Khoo, "Wavefront conjugation with gain and self-oscillation with a nematic liquid crystal film," *Appl. Phys. Lett.*, vol. 47, pp. 908–910, 1985; I. C. Khoo and S. L. Zhuang, "Nonlinear light amplification in a nematic liquid crystal above the Freedericks transition," *Appl. Phys. Lett.*, vol. 37, pp. 3–5, 1980.
[27] G. Martin and R. W. Hellwarth, "Infra-red to optical image conversion by Bragg reflection from thermally induced index gratings," *Appl. Phys. Lett.*, vol. 34, pp. 371–373, 1979.
[28] I. C. Khoo and R. Normandin, "Infra-red to visible image conversion capability of nematic liquid crystal film," *Appl. Phys. Lett.*, vol. 47, pp. 350–352, 1985

Iam-Choon Khoo (M'85) was born in Penang, Malaysia, in 1949. He received the B.S. degree in physics with first class honors from the University of Malaya, Malaysia, in 1971, and the M.A. and Ph.D. degrees in physics from the University of Rochester, Rochester, NY, in 1973 and 1976, respectively.

He has held postdoctoral and research associate positions at Ames Laboratory, Iowa State University of Science and Technology, Ames, IA, the University of Southern California, Los Angeles, CA, and the University of Toronto, Ont., Canada. He then joined the Physics Department at Wayne State University, Detroit, MI, in 1979 as an Assistant Professor, becoming an Associate Professor in 1983. In 1984, he joined the faculty at The Pennsylvania State University, University Park, PA, as an Associate Professor with the Department of Electrical Engineering. His current research interests are in theoretical and experimental nonlinear optical processes, optical wave mixing, wavefront conjugation, optical bistability, and switching in liquid crystalline materials. He has published about 100 journal articles and conference proceedings.

Dr. Khoo is a member of the Optical Society of America, the American Physical Society, and the Society for Photo-Optical Instrumentation Engineers.

All-optical bistability in nematic liquid crystals at 20 μW power levels

A. D. Lloyd and B. S. Wherrett
Department of Physics, Heriot–Watt University, Edinburgh EH14 4AS, United Kingdom

(Received 12 April 1988; accepted for publication 3 June 1988)

Optical switches at power levels as low as 14 μW, allowing bistable operation with compact disk laser systems, are reported in optimized cavities containing planar-aligned nematic liquid crystals.

From optimization studies of nonlinear narrow passband interference filters[1,2] it has been concluded that the independent tailoring of finesse and cavity length, made possible by removal of absorption from within the cavity, can lead to a significant reduction in operating power levels.

A metallic (e.g., gold) partial mirror at the rear of the cavity is thus used as both a reflector and an absorber, with thermal conduction into the spacer leading to optothermal bistability. This design allowed the construction of rigid cavities which have been used to demonstrate optical bistability with a number of liquid phase thermo-optic materials.[3,4]

For a fixed cavity design of this type, the lowest power at which bistability is possible can be written:[5]

$$P_c = \frac{\lambda_v}{D} \frac{1}{|\partial n/\partial T| \partial T/\partial P_a} g[R_f, R_b, \exp(-\alpha D)], \qquad (1)$$

where λ_v is the wavelength of operation, D the spacer thickness, $\partial n/\partial T$ the thermo-optic coefficient, $\partial T/\partial P_a$ the temperature rise per unit power absorbed, and g a cavity factor discussed in Ref. 5.

That the critical switch power is inversely proportional to the magnitude of thermo-optic coefficient ($\partial n/\partial T$) was confirmed by experiment,[3] and led to the choice of nematic phase liquid crystals as suitable, low absorption, spacer materials. Their high thermo-optic coefficient ($\partial n_e/\partial T = -2\times10^{-3}$ K^{-1} at room temperature) produced all optical bistability at critical power levels of 140 μW, allowing the demonstration of submilliwatt laser diode operation of these devices.

For the application of these thermal devices to optical two-dimensional information processing, it will be necessary, however, to operate within the 1–100 μW regime in order to realize the parallelism that optics offers in principle, at moderate total power levels.

Study of the refractive indices of the liquid crystal 4-cyano-4′-pentylbiphenyl (K 15, 5CB or PCB) over its nematic range,[7] has shown an increase in magnitude of the thermo-optic coefficient as the nematic-isotropic phase transition is approached. From (1) it is clear that if $\partial T/\partial P_a$ is constant over this range, then a commensurate reduction in operating power should result.

Experimental confirmation of this thermo-optic enhancement was obtained by enclosing the entire cell within a specifically designed, controlled temperature environment. This allowed choice of ambient temperature to 1/100 of a degree and stability approaching 1/1000 of a degree.[6] Figure 1 shows the observed reduction in critical power, for both the ordinary and extraordinary refractive indices, as the phase transition is approached. This is compared with the theoretical reduction in operating power level due to the increasing thermo-optic coefficient, inferred from a set of high precision index measurements of K 15 at 633 nm.[7] Over an ambient range of just 8 °C the critical power reduces by approximately one order of magnitude to 13.3 μW at 34.72 °C, corresponding to a $\partial n_e/\partial T$ of approximately -2×10^{-2} K^{-1}. This low-power requirement allows bistable operation well below 100 μW and, when operated in reflection, these devices display an adequate switch contrast of ~2:1 (Fig. 2).

Operation with laser diodes simply requires construction of a cell of the same configuration as above, but with the dielectric stack reflectivity centered to match that of the gold at the wavelength used. Thus we were able to demonstrate laser diode driven optical bistability using the AlGaAs laser output from a demounted compact disk player read arm (Fig. 3). The higher critical switch power of 62.5 μW is due to the larger spot size of 117 μm ($1/e^2$ diameter) used.

Though the switching powers in Fig. 1 show a reduction by an order of magnitude, the temporal response of the device remains virtually unchanged over the same temperature range (Fig. 4). Elevating the temperature from 26.00 to 34.70 °C, for the same detuning in each case, reduces the 100% overdrive power from 963 to 87 μW with no penalty in response time. The characteristic switching time in fact decreased from 2.4 to 1.8 ms.

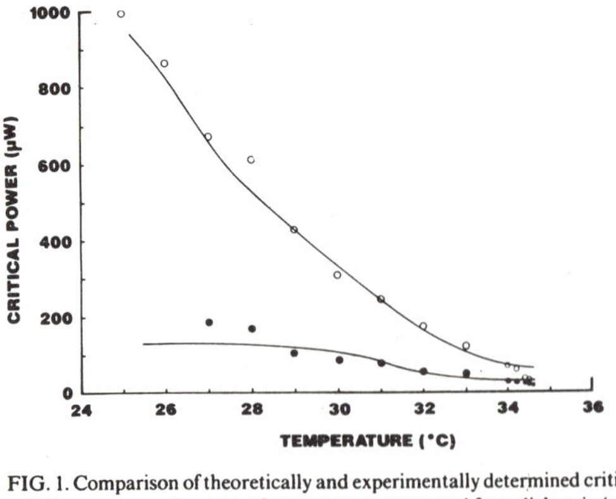

FIG. 1. Comparison of theoretically and experimentally determined critical switch powers as a function of temperature, measured for a dielectric/gold 26.5 μm cell containing K 15 ($R_f = 0.85$, $R_b = 0.90$). (○) $n_\perp$ (o-ray), (●) $n_\parallel$ (e-ray), (♦) n_i (isotropic phase).

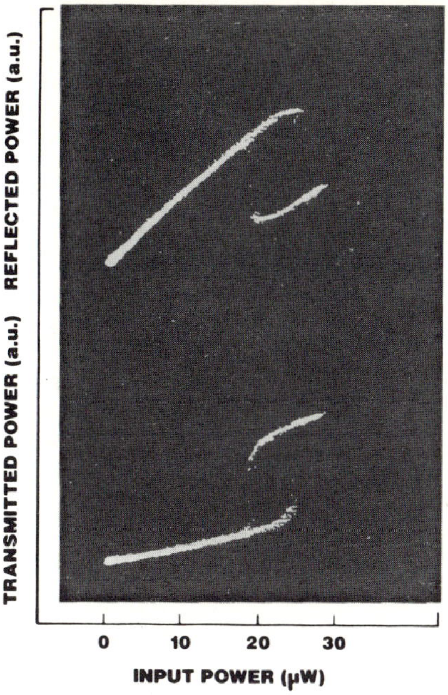

FIG. 2. Bistable operation of a dielectric/gold *K* 15 cell observed in transmission and reflection. Spot size is 25 μm.

Extended operation of these devices has shown a characteristic drift in detuning similar to that experienced in ZnSe interference filters,[8] but on a far longer time scale. For example, switching a cell at 1 Hz over a period of 24 h effected a slow reduction in the width of the bistable loop, though, at the end of this period, it was still possible to distinguish the minimum upper branch output from the maximum output in the original lower branch by a contrast of 1.7:1.

To take advantage of the inherent parallelism of optics with conventional cw laser sources requires devices with high sensitivity. Techniques such as pixellation and cavity/substrate optimization will produce significant increases in the sensitivity of all opto-thermal dispersive bistable systems. The limits are set by the sophistication of construction techniques and the magnitude of refractive index change with absorbed power. Liquid crystals have presently demonstrated the greatest exploitable thermo-optic nonlinearity and have produced switching levels nearly two orders of magnitude lower than that of typical ZnSe interference filter devices.[9]

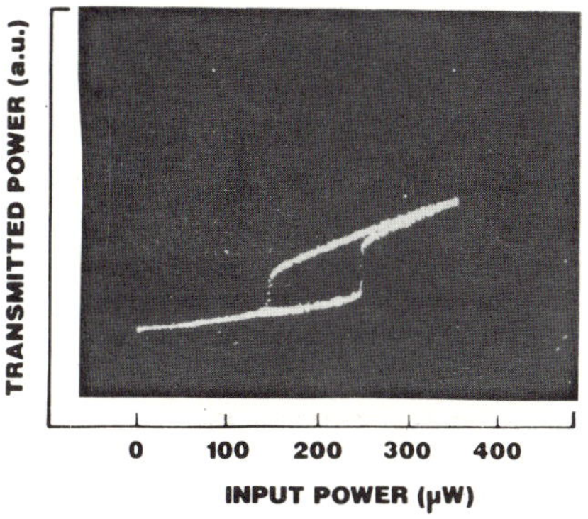

FIG. 3. Bistable operation of a dielectric/gold *K* 15 cell using a CD laser diode package. Spot size is 117 μm.

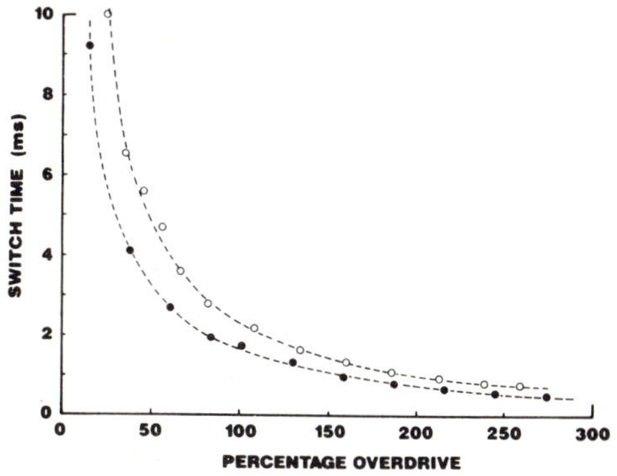

FIG. 4. Temporal response of a dielectric/gold *K* 15 cell at different ambient temperatures with constant detuning. (○) 26.00 °C, (●) 34.70 °C.

Planar-aligned nematic liquid crystals also display a nonlinear birefringence which has been used to produce switches in both transmitted power and polarization state.[10] This, allied with their dielectric anisotropy and thus sensitivity to applied electric fields, permits their application to a wide range of optical/optoelectronic devices.

A. D. Lloyd is grateful for an SERC CASE award in cooperation with STL Ltd., and partial funding through the SERC/DTI JOERS program is acknowledged. We would like to thank Linkam Scientific Instruments for assistance in the design of the temperature stage, and M. Brett (Mullard Ltd.) for the supply of the CD laser diode assembly. The *K* 15 liquid-crystal sample was obtained from BDH Chemicals Ltd., Poole, Dorset.

[1] B. S. Wherrett, D. Hutchings, and D. Russell, J. Opt. Soc. Am. B **3**, 351 (1986).
[2] A. C. Walker, Opt. Commun. **59**, 145 (1986).
[3] A. D. Lloyd, I. Janossy, H. A. MacKenzie, and B. S. Wherrett, Opt. Commun. **61**, 339 (1987).
[4] C. Somerton and D. L. Tunnicliffe, Opt. Commun. **65**, 143 (1988).
[5] D. C. Hutchings, A. D. Lloyd, I. Janossy, and B. S. Wherrett, Opt. Commun. **61**, 345 (1987).
[6] Model THW 200, Linkam Scientific Instruments Ltd., Tadworth, Surrey KT20 5HT, U.K.
[7] P. P. Karat and N. V. Madhusudana, Mol. Cryst. Liq. Cryst. **36**, 51 (1976).
[8] R. J. Campbell, J. G. H. Mathew, S. D. Smith, and A. C. Walker (unpublished).
[9] Y. T. Chow, B. S. Wherrett, E. W. Van Stryland, B. T. McGuckin, D. Hutchings, J. G. H. Mathew, A. Miller, and K. Lewis, J. Opt. Soc. Am. B **11**, 1535 (1986).
[10] A. D. Lloyd, Opt. Commun. **64** 302 (1987).

Photostimulated change of phase-transition temperature and "giant" optical nonlinearity of liquid crystals

S. G. Odulov, Yu. A. Reznikov, M. S. Soskin, and A. I. Khizhnyak

Physics Institute, Ukrainian Academy of Sciences
(Submitted 28 April 1983)
Zh. Eksp. Teor. Fiz. **85**, 1988–1996 (December 1983)

It is shown that the optical "giant" conformation nonlinearity of nematic liquid crystals is due to the change of the macroscopic order parameter of the crystal and to the high-frequency polarizability of the molecules that enter it during the conformation transition. The change of the order parameter leads to the experimentally observed shift of the phase-transition temperature of a nematic MBBA liquid crystal when acted upon by laser radiation of wavelength near the crystal absorption band. The magnitude and sign of the shift determine the experimentally obtained values and signs of the nonlinearity parameter, while the temperature dependence of the refractive index determines the observed growth of these parameters as the clearing point is approached from the mesophase side.

PACS numbers: 61.30.Eb, 64.70.Ew, 78.20.Dj, 42.65. — k

"Giant" nonlinearity of liquid crystals is being intensively investigated of late.[1–5] Among the mechanisms that cause the anomalously large values of ε_2 ($\varepsilon_2 = 2\Delta n\bar{n}/|\mathbf{E}|^2$, where Δn is the optically induced change of the refractive index of the nematic liquid crystal (NLC), $\bar{n}^2 = (1/3)(2n_o^2) + n_e^2)$ is the average refractive index, n_0 and n_e are the refractive indices of the ordinary and extraordinary waves, and $\mathbf{E}$ is the optical-wave field intensity) following action near the absorption edge of such media are optically induced conformation transformations of the molecules[2,6] of NLC.

It was shown earlier[2] that the change of the refractive indices of these media is due to the change of the polarizability of the liquid-crystal molecules upon phototransformation. We establish in the present paper that the appearance of a conformation transition in the optical nonlinearity is due to specific properties of the liquid-crystal state. An investigation of the temperature characteristics of the conformation nonlinearity of the NLC methoxybenzylidenebutylaniline (MBBA) has shown that besides the change of the high-frequency polarizabilities γ of the molecules, phototransformations of the molecules lead to a change of the macroscopic order parameter S. Each of these mechanisms contributes to the optical nonlinearity, and the contribution of the change of the order parameter to the nonlinearity increases as the "nematic-isotropic liquid" phase-transition point T_c is approached. A change in the value of S shifts the temperature T_c; this is interpreted as a manifestation of the dependence of the order parameter on the conformation state of the molecules.[7]

The model proposed explained the previously obtained[2,8] relations between the values of the components of the linear cubic susceptibility tensor $\chi^{(3)}$, as well as the signs of these components.

EXPERIMENTAL RESULTS

We investigated the characteristics of the conformation nonlinearity of the NLC MBBA as a function of temperature. Nominally pure samples ($T_c \approx 40$ °C) were placed in glass cuvettes 40–50 μm thick with covers that ensured planar orientation of the liquid-crystal director. The cuvette was placed in a thermostat whose temperature was monitored accurate to 0.05 °C.

The nonlinearity characteristics were investigated by a standard dynamic holography technique.[2] Sinusoidal thin phase gratings with period $\Lambda = 10$–$20\ \mu$m were recorded with polarized radiation of a helium-cadmium laser ($\lambda = 0.44\ \mu$m) having a power $P \lesssim 50$ mW. The lattice vector $|\mathbf{q}| = 2\pi/\Lambda$ was perpendicular to the NLC director. The cuvette was perpendicular to the bisector of the light-beam convergence angle, and the radiation polarization was either extraordinary or ordinary. This experimental geometry prevented the onset of orientational "giant" nonlinearity.[1]

The radiation intensity I_{d1} in first-order self-diffraction is uniquely connected with the optically induced change $\Delta n_{o,e}$ of the refractive index, viz.,

$$I_{d1} = (\pi z \Delta n_{o,e}/\lambda)^2 I_0, \tag{1}$$

where I_0 is the intensity of the incident radiation, z is the sample thickness, and the subscripts o and e label the ordinary and extraordinary light polarizations.

In the entire investigated temperature region below $T - T_c \approx -1$ °C the measured basic nonlinearity characteristics at fixed temperature did not differ qualitatively from those previously investigated for conformation nonlinearity.[2,8]

The experimental dependence of the self-diffraction intensity of the extraordinary polarized radiation on the normalized temperature $\tau = (T - T_c)/T_c$ is shown in Fig. 1. The same figure shows the temperature dependence of the characteristic hologram-erasure time t_H, which is equal to the writing time as well as to half the characteristic relaxation time of the nonlinearity.[2] The value of I_{d1}, which is proportional to $(\Delta n_{o,e})^2$, increases as the phase-transition point is approached, even though in this case the hologram-erasure time, which determines the stationary value of $\Delta n_{o,e}$ (Ref. 2), decreases. The dependence of the refraction intensity of the ordinary polarized radiation is similar in form.

In the phase transition region, which amounted to about 0.5 °C, as well as at temperatures close to it

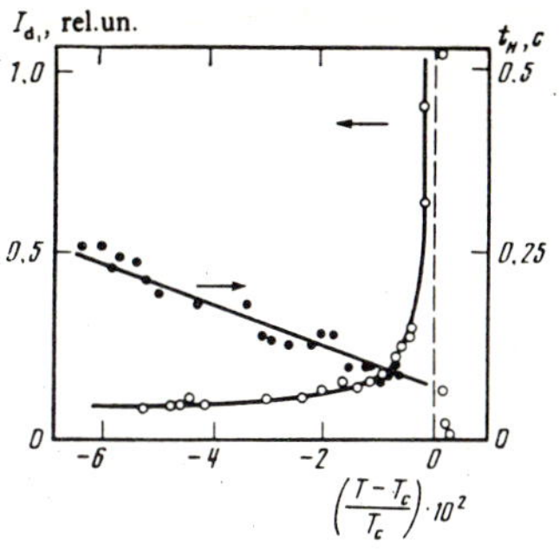

FIG. 1. Dependences of the self-diffraction intensity I_{d1} and of the effective time t_H of erasure of the holographic grating on the normalized temperature.

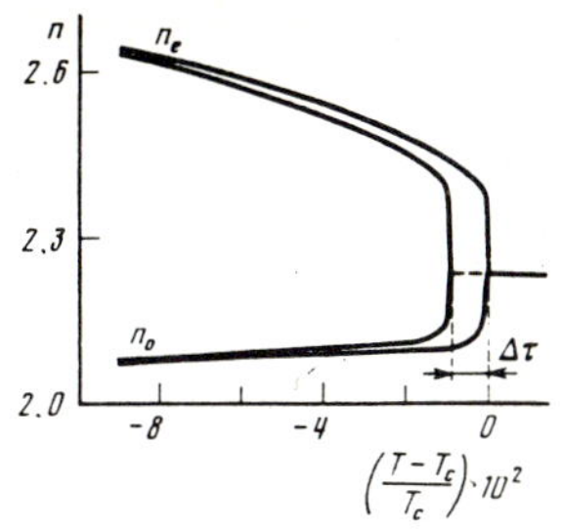

FIG. 2. Shift of temperature dependence of the refractive index of an NLC with changing temperature of the phase transition temperature T_c. The induced refractive index at a given temperature is determined by the shape of the $n(\tau)$ curve and the value of the shift of T_c, while the sign of $\Delta n_{o,e}$ is determined by the sign of the change of T_c.

$(-0.003 \lesssim \tau \lesssim 0.003)$, nonlinear effects with various characteristics appeared. Strong self-action effects with prolonged buildup and relaxation times ($t_H \gtrsim 10$ sec) were observed near the transition into the nematic state. At higher temperatures there appeared sometimes directly in the phase-transition region additional self-diffraction orders corresponding to writing holographic gratings with vector $\mathbf{q}_\perp$ prendicular to the basic vector $\mathbf{q}$, and equal to it in absolute value. We did not succeed in establihsing the conditions necessary for the onset of such gratings, which indicate the appearance of two-dimensional spatial structures in the NLC. Above the clearing temperature T_c, the self-diffraction intensity decreased to zero, while near $T_c (\tau \lesssim 0.003)$ (see Fig. 1.) an unusually strong nonlinearity was observed (the nonlinearity parameter was comparable with the parameter ε_2 in the mesophase). The nonlinearity lifetimes were longer than the characteristic times of the conformation nonlinearity.[1)]

DISCUSSION

1. We have shown earlier by the self-action procedure that extraordinary polarized radiation induces a diverging lens in an MBBA liquid crystal, on account of conformation nonlinearity, while ordinary polarized light induces a converging lens. This means that the charge of the MBBA refractive index is negative for the e-wave and positive for the o-wave.[8] In this case the increase of the nonlinearity on approaching the clearing point can be explained by assuming that the phase-transition temperature is lowered when photoconverted molecules appear in the NLC (see Fig. 2).

The shift of the clearing point can be qualitatively understood by recognizing that the photoconverted molecules act as impurities in the NLC volume. It is known that impurities can alter substantially the temperature T_c and the character of the phase transition, bringing the latter closer to a second-order transition.[10–14] The change of the character of the phase transition leads to a change of the form of the temperature dependence of the NLC refractive index, while the change of T_c shifts the $n(\tau)$ curve as a whole along the τ axis by an amount $\Delta\tau$. If the density of the photoconverted molecules is small or if the molecules themselves are mesogenic, the decisive factor in the change of the refractive index will be the shift of its temperature dependence as a result of the change of T_c. Let us obtain in this approximation an expression for the increment Δn_c of the refractive index, due to the lowering of T_c. The refractive indices of an NLC, for the o- and e-waves, can be written in the form[15]

$$\frac{n_{o,e}^2-1}{\bar{n}^2+2}=\frac{4\pi}{3}\frac{N_A\rho}{M}[\bar{\gamma}+\beta(\gamma_e-\gamma_o)S], \quad (2)$$

where N_A is the Avogadro number, ρ is the density, M is the molecular weight, $\gamma_{e,o}$ the polarizabilities of the mesogenic molecule, $\bar{\gamma}=1/3\gamma_e+2/3\gamma_o$, $\bar{n}^2=1/3n_e^2+2/3n_o^2$, $\beta=2/3$ for the o-polarization, and $\beta=1/3$ for the e-polarization.

The main contribution to the $\Delta n_{o,e}(\tau)$ dependence is made by the temperature dependences of the order parameter and of the polarizability of the individual molecules. The refractive-index growth increment is therefore

$$\Delta n_{o,e}=\frac{2\pi}{3}\frac{N_A\rho}{M}\frac{\bar{n}^2+2}{n_{o,e}}\beta\left[S\frac{\partial(\gamma_e-\gamma_o)}{\partial\tau}+(\gamma_e-\gamma_o)\frac{\partial S}{\partial\tau}\right]\Delta\tau. \quad (3)$$

The change of the temperature T_c can be treated as the result of an interaction of the macroscopic order parameter S with the microscopic order parameters Q that characterize the NLC intramolecular degrees of freedom.[7,11] Such a degree of freedom can be, in particular, the angle between the planes of the benzene rings in the liquid-crystal molecule[7] or the time-averaged deviation of end fragment of the mesogenic molecule.[16] A change of the parameter Q alters the local electric fields, while the macroscopic order parameter S, meaning also T_c, also changes because of the short-range character of the forces that bind the NLC molecules. The interaction of Q and S makes the quantities S and $\gamma_e-\gamma_o$ in expression (3) for the induced refractive index functions of the intramolecular paramter Q.

If it is assumed that the phase-transition temperature shift is proportional to the density of the photoconverted molecules, then

$$\Delta\tau=\frac{N}{N_0}\frac{\partial\tau^*}{\partial Q}\Delta Q,$$

where $(\partial\tau^*/\partial Q)\Delta Q$ is the shift of the phase-transition temperature when all the liquid-crystal molecules are converted into the new conformation state $\Delta\tau^*$.

Recognizing that in the stationary case $N = I_0\alpha_{o,e} t/h\nu$, where $\alpha_{o,e}$ is the absorption coefficient and $N_0 = N_A/\rho M$, we find ultimately that the induced change of the refractive index is proportional to the light intensity, i.e., is described by the cubic nonlinearity

$$\Delta n_{o,e} = \frac{2\pi\beta}{3}\frac{I_0\alpha_{o,e}t}{n_{o,e}h\nu}\left[S\frac{\partial(\gamma_e-\gamma_o)}{\partial\tau}+(\gamma_a-\gamma_o)\frac{\partial S}{\partial\tau}\right]\Delta\tau^*. \quad (4)$$

This coincides with the experimental results obtained for the conformational nonlinearity.[2] The dependence of the absorption coefficient $\alpha_{o,e}$ on the order parameter also leads to a change of the NLC refractive index. This change, however, is described by a sixth-order nonlinearity, and can be neglected at the intensities employed.

The temperature dependence of the square of the refractive index $(\Delta n_e)^2$ that determines the self-diffraction intensity in our experiments [see (1)] can be calculated by using for the employed wavelength ($\lambda = 0.44\ \mu$m) a relation approximated from the experimental data:

$$n_e = 2.3+0.38|\tau|^{0.5}+0.33|\tau|. \quad (5)$$

If it is assumed that the shift $\Delta\tau$ of the phase transition is small compared with the values of τ, the stationary value of the self-diffraction intensity will be proportional to the square of the first term of the expansion (5) in τ:

$$I_{d_1} \propto [(0.19|\tau|^{-0.5}+0.33)\Delta\tau]^2, \quad (6)$$

where, in accord with (4), the shift of the phase-transition temperature is $\Delta\tau = 2\Delta\tau^* t_H$, while $t_H(\tau)$ is the experimentally obtained erasure time of the holographic gratings.

The calculation corresponding to (6) is represented by the solid line in Fig. 1. The dark points are the experimental $t_H(\tau)$ and their variation was approximated by a straight line sloped to yield best agreement between calculation and experiment. It can be seen that the calculated curve fits the experimental data within the error limit.

We can thus state that the refractive-index change due to the conformational nonlinearity is described, within the experimental accuracy, by a shift of the $n(\tau)$ dependence along the τ axis, without changing the form of the function $n(\tau)$ itself; this can be treated as a lowering of the phase-transition temperature T.

2. We have performed an independent experiment to observe the phase-transition temperature shift following illumination of an NLC by a helium-cadmium laser. The change of the temperature T_c was recorded in terms of the charge of the birefringence of the liquid crystal by the action of this radiation.

The measurements were made with a nonlinear polarization interferometer (see Fig. 3). A thermostat with the liquid crystal was placed between two crossed polarizers p_1 and p_2. The beam from a helium-neon laser ($\lambda = 0.638\ \mu$m), whose ε_2 is vanishingly small, was broadened by a telescope T and passed through polarizers into a cell with the NLC. Polarized and intensity-modulated emission from a helium-cadmium laser was also guided into the cell with the crystal whose birefringence it altered on account of conformation

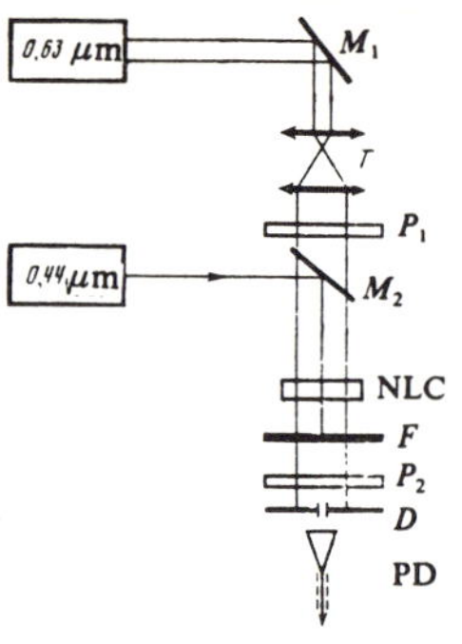

FIG. 3. Diagram of nonlinear polarization interferometer: M—mirrors, T—telescope, P—polarizers, F—filter absorbing the $\lambda = 0.44\ \mu$m radiation, D—diaphragm, PD—photodiode.

nonlinearity. The He–Cd laser emission was modulated in such a way that the induced birefringence reached alternately either its stationary value or a zero value. The shutter operating time was in this case substantially shorter than the characteristic nonlinearity-establishment times.

The diameter of the helium-cadmium laser beam was substantially less than the diameter of the helium-neon laser beam. The interferometer output was the interference pattern of the polarized beams of the helium-neon laser, against the background of which was observed a periodically time-varying intensity region corresponding to the action of the helium-cadmium laser.

With the shutter open, the emission intensity in the He–Cd laser illumination region varies as

$$I_1 \propto \cos^2\left\{\frac{\pi z}{\lambda}[\Delta n^0(\tau)+(\Delta n_e+\Delta n_o)(\tau)(1-e^{-\bar{t}/2\tau_H})]\right\}, \quad (7)$$

where Δn^0 is the MBBA intrinsic birefringence, $\Delta n_e + \Delta n_o$ is the stationary induced birefringence, and t is the time reckoned from the instant of shutter operation. At $\bar{t} \gg 2t_H$ this expression yields the stationary value of the birefringence of the mixture of the ground state and of the modified temperature-dependent conformational state of the liquid crystal.

When the shutter is closed, the intensity change is given by the expression

$$I_2 \propto \cos^2\left\{\frac{\pi z}{\lambda}[\Delta n^0(\tau)+(\Delta n_e+\Delta n_o)(\tau)e^{-\bar{t}/2t_H}]\right\}, \quad (8)$$

which determines in the stationary case the birefringence of the basic MBBA crystal at a given temperature.

The temperature dependences of I_1 and I_2 corresponding to the stationary values of (7) and (8) are shown in Fig. 4. The relative shift of these curves points to a change of the phase-transition temperature under the action of the helium-cadmium laser radiation. This change can be determined, starting from the equality of I_1 and I_2, from the condition

$$\frac{\pi z}{\lambda}[\Delta n^0(\tau-\Delta\tau)-\Delta n^0(\tau)] = \Delta(\Delta n). \quad (9)$$

Here $\Delta(\Delta n)$ is the birefringence change determined from the

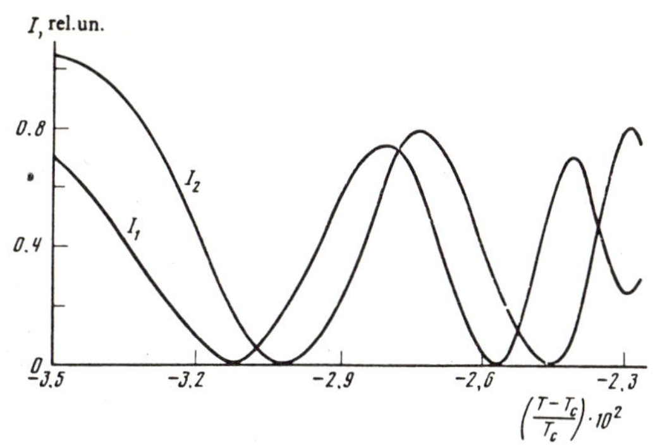

FIG. 4. Temperature dependence of the intensity of the light passing through the polarization interferometer: I_1—illumination at $\lambda = 0.44$ μm, I_2—with illumination.

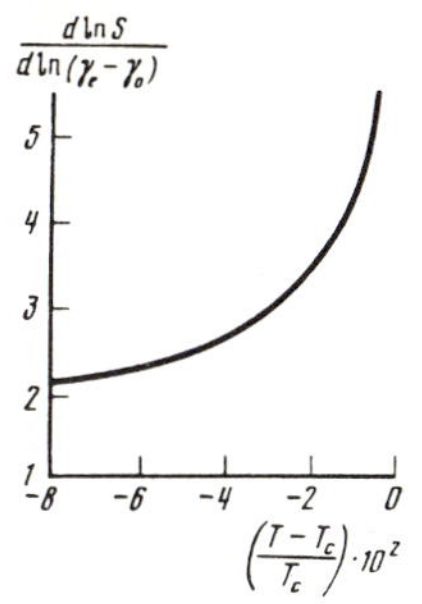

FIG. 5. Temperature dependence of the relative contribution of the change of the order parameter S and of the molecular polarizability γ to the conformational nonlinearity.

shift of the experimental curves and leading to a phase advance equal to that in the absence of external illumination, while $\Delta n^0(\tau)$ is described, on the basis of an approximation of the published data,[17,18] by the expression

$$\Delta n^0=0.11+0.43|\tau|^{0.5}+0.1|\tau|. \tag{10}$$

Solution of (9), assuming conservation of the function (10) under phototransformation, yields $\Delta\tau\approx 3\cdot 10^{-4}(\Delta T_c\approx 10^{-1}$ K). Such a shift of the phase-transition temperature leads, according to (6), to a refractive-index change $\Delta n_e \approx 5\cdot 10^{-4}(T_c = 22\,^\circ\mathrm{C})$. This value of Δn_e corresponds to a nonlinearity parameter $\varepsilon_2\approx 0.1$ cm^3/erg, of the same order as that previously obtained.[2,8]

We note that the obtained considerable changes of the birefringence under saturation conditions cannot be due to heating of the sample by light absorption. Inded, even in the absence of heat exchange, at a power density $I_0\approx 1$ W/cm^3, at an absorption coefficient $\alpha\approx 25$ cm^{-1}, and at a cell thickness 50 μm the sample heating during the time of action of the laser pulse did not exceed 0.02 K, lower by approximately an order of magnitude than the ΔT_c obtained by us. That the thermal contribution is small is indicated also by the fact that the nonlinearity parameter ε_2^T for the thermal mechanism is several orders lower than that measured in experiment.[2] The possible heating of the sample by the continuous emission of the testing helium-neon laser influences only the rate of sample cooling and introduces no measurement error.

Starting from the linear dependence, we can estimate the shift ΔT_c^* of the phase-transition temperature of photoconverted molecules relative to the temperature T_c of the unperturbed sample. The experimentally recorded shift $\Delta T_c\approx 0.1\,^\circ$C corresponds to a relative density $N/N_0\approx 0.1$, therefore $\Delta T_c^*\approx 1\,^\circ$C.

Measurement of the temperature dependences of the birefringence, as well as holography experiments, points thus to a shift of the temperature T_c upon phototransformation, and the value of shift leads to observable values of the nonlinearity parameter.

3. Expression (4) shows that the change of the refractive index is connected with the change of the NLC-molecule polarizability as well as with the change of the order parameter S. Using the temperature dependences of the refractive indices and the known $S(\tau)$ dependence[19] we can estimate the contribution of the change of the order parameter and of the molecular polarizabilities to the nonlinearity. The temperature dependence of the anisotropy of the molecular polarizability can be calculated starting from the Vuks formula[15]

$$(\gamma_e-\gamma_o)=S^{-1}(\bar{\alpha}_e-\bar{\alpha}_o),\quad \bar{\alpha}_e-\bar{\alpha}_o=\frac{3}{4\pi}\frac{M}{N_A\rho}\frac{n_e^2-n_o^2}{\overline{n^2}+2},$$

which contains the difference of the NLC polarizabilities. Figure 5 shows $d\ln S/d\ln(\gamma_e-\gamma_o)$ vs temperature. It can be seen that the contributions of both mechanisms are comparable in size, and when the phase transition temperature is approached the contribution to the nonlinearity of the change of the order parameter increases.

4. We obtain now an expression for the components of the NLC nonlinear-susceptibility tensor due to a shift of the phase-transition temperature. In our case the third-order nonlinear polarizability $P^{(3)}$ can be written in the form

$$P_l^{(3)}\sim\alpha_{ij}E_iE_j\frac{dn_{kl}}{d\tau}\Delta\tau E_k. \tag{11}$$

Starting from this relation we can formally introduce the cubic nonlinear susceptibility tensor $\chi^{(3)}$ whose components are given by

$$\chi_{ijkl}\sim\alpha_{ij}\frac{dn_{kl}}{d\tau}\Delta\tau.$$

The symmetry of the tensor $\chi^{(3)}$ for the conformation-phototransformation mechanism admits of only four nonzero components of the tensor χ_{ijkl}.[2] The relations between the components χ_{ijkl} and their signs can be qualitatively understood on the basis of Fig. 2. Quantitative estimates can be obtained by using the relations for the refractive indices n_o and n_e, as well as the fact that for the wavelength $\lambda = 0.44$ μm we have $\alpha_{11}/\alpha_{33}\approx 0.4$. This yields

$$\chi_{1111}=-0.075\chi_{3333},\quad \chi_{3311}=0.22\chi_{3333},$$

$$\chi_{1133}=-0.5\chi_{3333},\quad \chi_{1111}>0,\quad \chi_{3333}<0.$$

Earlier experimental data[2,8] yield the close values

$$\chi^{e}_{1111} = -0.1\chi^{e}_{3333}, \quad \chi^{e}_{3311} = 0.44\chi^{e}_{3333},$$

$$\chi^{e}_{1133} = -0.44\chi^{e}_{3333}, \quad \chi^{e}_{1111} > 0, \quad \chi^{e}_{3333} < 0.$$

The value of χ^{e}_{3311} does not agree with our estimate. The discrepancy is possibly due to the insufficient accuracy of the approximation of $n_o(\tau)$ as well as to the fact that the change of the refractive index for the ordinary wave is connected not only with the shift but also with the change of the form of the function $n_o(\tau)$ during phototransformation.

CONCLUSION

It can thus be regarded as established that the main cause of the "giant" nonorientational nonlinearity of nematic liquid crystals is the conformation change of the mesogenic molecule upon phototransformation, which leads to a change of the intramolecular order parameter. The interaction of this parameter, which characterizes the intramolecular degrees of freedom of the molecule, with the macroscopic order parameter of the NLC leads to a lowering of the phase transition temperature T_c, meaning also to a change of the refractive index of the liquid crystal. The change of the refractive index is due both to a change of the order parameter S and to a change of the anisotropy of the polarizability of the mesogenic molecule. The contributions of both mechanisms to the nonlinearity are comparable far from the phase transition, and as the latter is approached the contribution of the change of the order parameter S increases.

From the proposed model it follows naturally that the change of the refractive index is positive for the ordinary wave and negative for the wave with the extraordinary polarization. Also agreeing in the main with the experimental data are the relations between the values of the individual components of the nonlinear cubic susceptibility.

The observed nonlinearity mechanism can be used hereafter to investigate the effect of impurities on the character of phase transitions in liquid crystals, inasmuch as the experiments afford a unique possibility of controlling the impurity density. To realize such investigations it is necessary to explain the form of the photoconformation transformations of mesogenic molecules. Once the intermolecular order parameter responsible for the change of the refractive index is determined, dynamic holography methods can be used to study the interaction of this order parameter with the macroscopic parameter S. This problem is now exceedingly vital for the understanding of the nature of phase transitions in liquid crystals.

The authors are grateful to E. M. Aver'yanov for the possibility of becoming acquainted with Ref. 7 and S. V. Shiyanovskiĭ for helpful discussions.

[1)]Unfortunately, the insufficient accuracies of the temperature stabilization and of the material purity, as well as the uncontrollable presence of oxygen and moisture in MBBA, which lead to gradual lowering of the temperature T_c (Ref. 9), have made detailed investigation of these effects impossible so far.

[1]B. Ya. Zel'dovich, I. F. Pilipetskiĭ, A. V. Suhkov, and N. V. Tabiryan, Pis'ma Zh. Eksp. Teor. Fiz. **31**, 287 (1980) [JETP Lett. **31**, 263 (1980)].

[2]S. G. Odulov, Yu. A. Reznikov, M. S. Soskin,a nd A. I. Khizhnyak, Zh. Eksp. Teor. Fiz. **82**, 1475 (1982) [Sov. Phys. JETP **55**, 854 (1982)].

[3]A. S. Zolot'ko, V. F. Kitaeva, N. Kroo, P. P. Sobolev, and L. Chillag, Pis'ma Zh. Eksp. Teor. Fiz. **32**, 170 (1980) [JETP Lett. **33**, 162 (1980)].

[4]J. C. Khoo, S. L. Zhyang, and S. Shepard, Appl. Phys. Lett. **39**, 937 (1981).

[5]G. Barbero and F. Simoni, Appl. Phys. Lett. **41**, 504 (1982).

[6]H. Hervet, W. Urbach, and F. Rondelez, J. Chem. Phys. **68**, 2725 (1978).

[7]E. M. Aver'yanov, Fazovye perekhody i konformatsiya molekul v odnoosnykh zhidkikh kristallakh (Phase Transition and Molecule Conformation in Uniaxial) Krasnoyarsk, Preprint IFSO-186, 1982, p. 32.

[8]S. G. Odulov, Yu. A. Reznikov, M. S. Soskin, and A. I.Khizhnyak, Izvestiya AN SSSR, seriya fiz. **47**, 2274 (1983).

[9]V. de Jeu, Physical Properties of Liquid-Crystal Matter (Russ. transl.), Mir, 1982, Chap. II, §1.

[10]S. A. Pikin, Strukturnye prevrashcheniya v zhidkikh kristallakh (Structural Transformations in Liquid Crystals) Nauka, Vol. III, p. 3, (1981).

[11]M. A. Anisimov, Usp. Fiz. Nauk **114**, 249 (1974) [Sov. Phys. Usp. **17**, 722 (1975)].

[12]A. A. Kovalev, G. L. Nekrasov, Yu. V. Razvin, V. A. Grozhik, and S. V. Serak, Opticheskie metody obrabotki informatsii Minsk: Nauka i tekhnika, 1978, p. 21.

[13]W. E. Haas, K. F. Nelson, J. E. Adams, and G. A. Dir, J. Electrochem. Soc. **121**, 1667 (1974).

[14]P. E. Cladis, Phys. Let. **48A**, 179 (1974).

[15]L. M. Blinov, Elektro- i magnitooptika zhidkikh kristallov, Nauka, 1978, Ch. II, §2.4.

[16]E. M. Aver'yanov, V. A. Shutkov, A. Ya. Korets, V. F. Shabanov, and P. V. Adomenas, Pis'ma Zh. Eksp. Teor. Fiz. **31**, 511 (1980) [JETP Lett. **31**, 480 (1980)].

[17]J. Haller, M. A. Huggins, and M. J. Freiser, Mol. Cr. Liq. Cr. **16**, 53 (1972).

[18]L. M. Blinov, V. A. Kizel', V. G. Rumyantsev, and V. V. Titov, Kristallografiya **20**, 1245 (1975) [Sov. Phys. Crystallogr. **20**, 750 (1975)].

[19]R. Chang, Mol. Cr. Liq. Cr. **30**, 155 (1975).

Translated by J. G. Adashko

SUBJECT INDEX

AUTHOR INDEX

PERSPECTIVES IN CONDENSED MATTER PHYSICS

Published Volumes

1. G. Margaritondo, *Electronic Structure of Semiconductor Heterojunctions*
2. A.H. MacDonald, *Quantum Hall Effect: a Perspective*
3. M. Ronchetti and G. Jacucci, *Simulation Approach to Solids*
4. W. Mönch, *Electronic Structure of Metal-Semiconductor Contacts*
5. I. Jánossy, *Optical Effects in Liquid Crystals*

Forthcoming Volumes

H. Neddermeyer, *Scanning Tunneling Microscopy*
H. R. Ott., *The Last Ten Years of Superconductivity*

Executive Editor: L. Miglio

Dipartimento di Fisica dell'Università di Milano
Via Celoria, 16 I-20133 MILANO
Fax + 39/2/2366583; **Telex** 334687 INFNMI; **Tel.** + 39/2/2392.408